Advances in

ECOLOGICAL RESEARCH

VOLUME 22

Advances in

ECOLOGICAL RESEARCH

series edited by

M. BEGON
Department of Zoology
University of Liverpool
Liverpool L69 3BX
UK

A. H. FITTER
Department of Biology
University of York
York, YO1 5DD
UK

A. MACFADYEN
23 Mountsandel Road
Coleraine
Northern Ireland

VOLUME 22

The Ecological Consequences
of Global Climate Change

edited by

F. I. WOODWARD
Department of Animal and Plant Sciences
University of Sheffield, P.O. Box 601
Sheffield, S10 2UQ
UK

ACADEMIC PRESS
Harcourt Brace Jovanovich, Publishers
London   San Diego   New York   Boston
Sydney   Tokyo   Toronto

ACADEMIC PRESS LTD.
24/28 Oval Road
London NW1

United States Edition published by
ACADEMIC PRESS INC.
San Diego, CA 92101

British Library Cataloguing in Publication Data
Advances in ecological research.
Vol. 22
1. Ecology
I. Begon, Michael
574.5

ISBN 0–12–013922–7

This book is printed on acid-free paper

Typeset by Latimer Trend & Company Ltd, Plymouth
Printed in Great Britain by
T.J. Press (Padstow) Ltd, Padstow, Cornwall.

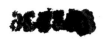

Contributors to Volume 22

J. M. ADAMS, *The Department of Botany, University of Cambridge, Downing Street, Cambridge CB2 3EA, UK.*

J. M. ANDERSON, *The Department of Biological Sciences, University of Exeter, Exeter EX4 4PS, UK.*

G. B. BONAN, *The National Center for Atmospheric Research, P.O. Box 3000, Boulder, Colorado 80307, USA.*

M. E. CAMMELL, *Silwood Centre for Pest Management, The Department of Biology, Imperial College at Silwood Park, Ascot SL5 7PY, UK.*

C. M. GOODESS, *Climatic Research Unit, University of East Anglia, Norwich NR4 7TJ, UK.*

P. M. HOLLIGAN, *Plymouth Marine Laboratory, Prospect Place, West Hoe, Plymouth PL1 3DH, UK.*

J. D. KNIGHT, *Silwood Centre for Pest Management, Department of Biology, Imperial College at Silwood Park, Ascot SL5 7PY, UK.*

J. P. PALUTIKOF, *Climatic Research Unit, University of East Anglia, Norwich NR4 7TJ, UK.*

M. L. PARRY, *Atmospheric Impacts Research Group, Environmental Change Unit, University of Oxford, Oxford OX1 2JP, UK.*

W. A. REINERS, *Department of Botany, University of Wyoming, P.O. Box 3156, Aven Nelson Building, Laramie, Wyoming 82071–3165, USA.*

S. H. SCHNEIDER, *The National Center for Atmospheric Research, P.O. Box 3000, Boulder, Colorado 80307, USA.*

J. B. SMITH, *Department of Environmental Sciences, University of Virginia, Charlottesville, Virginia 22903, USA.*

T. M. SMITH, *Department of Environmental Sciences, University of Virginia, Charlottesville, Virginia 22903, USA.*

H. H. SHUGART, *Department of Environmental Sciences, University of Virginia, Charlottesville, Virginia 22903, USA.*

F. I. WOODWARD, *Department of Animal and Plant Sciences, University of Sheffield, P.O. Box 601, Sheffield, S10 2UQ, UK.*

Preface

The concepts and concerns regarding the global effects of a continued increase in the atmospheric concentrations of greenhouse gases have enjoyed a high visibility in newspapers and scientific journals. These concerns are now being translated into big-science projects. These international projects aim to understand better the processes of climate and ecosystem changes and impacts and are being designed under the aegis of the World Climate Research Programme and the International Geosphere Biosphere Programme. Biological and climatic systems are intertwined in processes leading to impacts and feedbacks and so it has emerged that climatologists, atmospheric scientists, terrestrial and marine ecologists must collaborate in research programmes, else the bases of their future projections are incomplete. This special volume of *Advances in Ecological Research* brings together eight chapters which propose and demonstrate the two major components of current climate change research, future prediction and the interdisciplinary approach.

The first chapter by S. H. Schneider describes the approaches and problems in making future world-scale climatic projections by general circulation models (GCMs). One major problem of current GCMs is that for them to operate in realistic amounts of computer time and over the world scale, the spatial resolution is limited. Therefore the information is too coarse a scale for predicting ecosystem impacts. C. M. Goodess and J. P. Palutikof show approaches that will need to be used to attempt some scaling down from the large scale of the GCM to the regional scale required by ecologists, catchment-scale hydrologists and agriculturalists.

This scaling down is as yet incomplete but it is important for biologists to work in parallel, developing realistic models suitable for predicting regional-scale spatial changes in vegetation and agricultural crops, when GCMs are more appropriately scaled. M. L. Parry demonstrates the current capacity of models to predict changes in the distribution of agricultural crops, whilst current approaches to the more complex natural and semi-natural ecosystems are described by T. M. Smith, H. H. Shugart, G. B. Bonan and J. B. Smith.

The actual responses of agricultural crops to changes in geographical patterns of acceptance by farmers and policy makers will also be strongly dependent on another scale of ecosystem impact—the effects of pests and pathogens. M. E. Cammell and J. D. Knight describe our current understanding of the ways in which these pests will respond to changes in climate and crop type.

A current problem for both the agricultural and natural ecosystem models of climate response is a poor or limited handling of the dynamics of change. T. M. Smith *et al.* describe small-scale patch models suitable for investigating this dynamic. Another crucial feature of the dynamic responses of ecosystems is the response of the soils. J. M. Anderson describes the dynamic role of soils, both in terms of their capacity to sequester carbon and in their capacity to feedback to climate through the release of greenhouse gases, and to ecosystems through changes in nutrient cycling.

The modelling approach for predicting the wide range of effects caused by increases in greenhouse gases in the atmosphere is a crucial approach for changing global energy policy and providing forewarning of serious impacts to humanity. The coastal zone represents such a crucial zone, suffering from changes in land-use practice on on side and sea-level rise on the other. P. M. Holligan and W. A. Reiners show the importance and complexity of the zone, drawing on a wide and disparate area of evidence and information. Such a disparate source of information is also discussed in the final chapter by J. M. Adams and F. I. Woodward, which discusses the use of historical information as a source of ecological understanding and testing for models of future change.

Contents

The Climatic Response to Greenhouse Gases

STEPHEN H. SCHNEIDER

The Development of Regional Climate Scenarios and the Ecological Impact of Greenhouse Gas Warming

C. M. GOODESS and J. P. PALUTIKOF

The Potential Effect of Climate Changes on Agriculture and Land Use

MARTIN PARRY

Modeling the Potential Response of Vegetation to Global Climate Change

T. M. SMITH, H. H. SHUGART, G. B. BONAN and J. B. SMITH

CONTENTS

Effects of Climatic Change on the Population Dynamics of Crop Pests

M. E. CAMMELL and J. D. KNIGHT

Responses of Soils to Climate Change

J. M. ANDERSON

Predicting the Responses of the Coastal Zone to Global Change

P. M. HOLLIGAN and W. A. REINERS

The Past as a Key to the Future: The Use of Palaeoenvironmental Understanding to Predict the Effects of Man on the Biosphere

J. M. ADAMS and F. I. WOODWARD

The Climatic Response to Greenhouse Gases

STEPHEN H. SCHNEIDER

I. WHY BUILD A MODEL?

To understand or predict the result of some event in nature it is common to build and perform an experiment. But what if the issues are very complex or the scale of the experiment unmanageably large? To forecast the effect of human pollution on climate poses just such a dilemma, for this uncontrolled experiment is now being performed on "laboratory earth". How then can we be anticipatory, if no meaningful physical experiment can be performed? While nothing can provide certain answers, we can turn to a surrogate lab, not a room with test tubes and Bunsen burners, but a small box with transistors and microchips. We can build mathematical models of the earth and perform our "experiments" in computers.

Mathematical models translate conceptual ideas into quantitative statements. Models usually are not faithful simulations of the full complexity of reality, of course, but they can tell us the logical consequences of explicit sets of plausible assumptions. To me, that certainly is a big step beyond pure conception—or to put it more crudely, modeling is a major advance over "hand-waving".

ADVANCES IN ECOLOGICAL RESEARCH VOL. 22
ISBN 0–12–013922–7

The kinds of problems that can be studied by climate models include the downstream atmospheric effects of unusual ocean surface temperature patterns (e.g. so-called "El Nino" events), climatic effects of volcanic explosions, ice age–interglacial sequences, ancient climates, the climatic effects of human land uses or pollutants such as carbon dioxide (CO_2) and even the climatic after effects of nuclear war (e.g. see Schneider, 1987 from which parts of this chapter are adapted). We will consider several of these examples later.

II. BASIC ELEMENTS OF MODELS

To simulate the climate, a modeler needs to decide which components of the climatic system to include and which variables to involve. For example, if we choose to simulate the long-term sequence of glacials and interglacials, our model needs to include as explicitly as possible the effects of all the important interacting components of the climatic system operating over the past million years or so. Besides the atmosphere, these include the ice masses, upper and deep oceans, and the up and down motions of the earth's crust. Also, life influences the climate and thus must be included too; plants, for example, can affect the chemical composition of the air and seas as well as the reflectivity (i.e. albedo) and water-cycling character of the land. These mutually interacting subsystems form part of the internal components of the model. On the other hand, if we are only interested in modeling very short-term weather events—say, over a single week—then our model can ignore any *changes* in the glaciers, deep oceans, land shapes, and forests, since these variables obviously change little over one week's time. For short-term weather, only the atmosphere itself needs to be part of the model's internal climatic system.

The slowly varying factors such as land forms or glaciers are said to be external to the internal part of the climatic system being modeled. Modelers also refer to external factors as boundary conditions, since they form boundaries for the internal model components. These boundaries are not always physical ones, such as the oceans, which are at the bottom of the atmosphere, but can also be mass or momentum or energy fluxes across physical boundaries. An example is the solar radiation impinging on the earth. Solar radiation is often referred to by climatic modelers as a *boundary forcing function* of the model for two reasons: the energy output from the sun is not an interactive, internal component of the climatic system of the model; and the energy from the sun forces the climate toward a certain temperature distribution.

We could restrict a model to predict only a globally averaged temperature that never changes its value over time (i.e. is in equilibrium). This very simple model would consist of an internal part that, when averaged over all of the atmosphere, oceans, biosphere, and glaciers, would describe two characteristics: the average reflectivity of the earth and its average greenhouse properties. The boundary condition for such a model would be merely the incoming solar energy. Such a model is called *zero dimensional*, since it collapses the east–west, north–south, and up–down space dimensions of the actual world into one point that represents some global average of all earth–atmosphere system temperatures in all places. If our zero-dimensional model were expanded to resolve temperature at different latitudes and longitudes and heights, then it would be three-dimensional. The *resolution* of a model refers to the number of dimensions included and to the amount of spatial detail with which each dimension is explicitly treated.

Modelers speak of a hierarchy of models that ranges from simple earth-averaged, time-independent, temperature models up to high-resolution, three-dimensional, time-dependent models known as general circulation models (GCMs). While three-dimensional, time-dependent models are usually more physically, chemically, and (somewhat) biologically comprehensive, they are also much more complicated and thus more expensive to construct and use (e.g. Washington and Parkinson, 1986). Choosing the optimum combination of factors is an intuitive art that trades off completeness and (the modelers hope) accuracy for tractability and economy (e.g. Land and Schneider, 1987). Moreover, the theoretical feasibility of long-range simulation must also be evaluated; in other words, some problems are inherently unpredictable. For those problems where predictability is not ruled out in principle, such a trade-off between accuracy and economy is not "scientific" *per se*, but rather is a value judgement, based on the weighing of many factors. Making this judgement depends strongly on the problem the model is being designed to address.

III. THE PROBLEMS OF PARAMETERIZATIONS AND CLIMATIC FEEDBACK MECHANISMS

Clouds, being very bright, reflect a large fraction of sunlight back to space, thereby helping control the earth's temperature. Thus, predicting the changing amount of cloudiness over time is essential to reliable climate simulation. But most individual clouds are smaller than even the smallest area represented by the smallest resolved element (i.e. "grid box") of a global climate or weather prediction model. A single thunderstorm is typically a few kilometers in size, not a few hundred—the size of many "high-resolution" global model grids. Therefore, no global climate model available now (or likely to be available in the next few decades) can explicitly resolve every individual cloud. These important climatic elements are therefore called *sub-*

grid-scale phenomena. Yet, even though we cannot explicitly treat all individual clouds, we can deal with their collective effects on the grid-scale climate. The method for doing so is known as *parameterization*, a contraction for "parametric representation". Instead of solving for sub-grid-scale details, which is impractical, we search for a relationship between climatic variables we do resolve (e.g. those whose variations occur over larger areas than the grid size) and those we do not resolve. For instance, climatic modelers have examined years of data on the humidity of the atmosphere averaged over large areas and have related these values to cloudiness averaged over that area. It is typical to choose an area the size of a numerical model's grid—a few hundred kilometers on a side. The vertical temperature and humidity structure of the atmosphere also affect its stability and thus may be included in cloud parameterization schemes. While it is not possible to find a perfect correspondence between these averaged variables, reasonable relationships have been found in a wide variety of circumstances. These relationships typically require a few factors, or parameters, some of which are derived empirically from observed data, not computed from first principles. The parameterization method applies to almost all simulation models, whether dealing with physical, biological or even social systems. Figure 1, for example, is a schematic of the kinds of biophysical parameters that must be specified or parameterized at grid points in order to calculate fluxes of energy and materials between and through the atmosphere soils and biota (e.g. Dickinson *et al.*, 1986). The most important parameterizations affect processes called *feedback mechanisms*. This concept is well known outside of computer-modeling circles. The word feedback is vernacular. As the term implies, information can be "fed back" to you that will possibly alter your behavior.

So it is in the climate system. Processes interact to modify the overall climatic state. Suppose, for example, a cold snap brings on a high albedo snow cover which tends to reduce the amount of solar heat absorbed, subsequently intensifying the cold. This interactive process is known to climatologists as the *snow-and-ice/albedo/temperature feedback mechanism*. Its destabilizing, *positive-feedback* effect is becoming well understood and has been incorporated into the parameterizations of most climatic models. Unfortunately, other potentially important feedback mechanisms are not usually as well understood. The most difficult one is so-called *cloud feedback*, which could be either a positive or negative feedback process, depending on circumstances.

Biogeochemical cycle processes could also serve as climatic feedbacks. For example, increasing atmospheric concentration of CO_2 from fossil fuel burning could be mitigated somewhat by CO_2 fertilization of photosynthesizing organisms. This added uptake of CO_2 is a negative feedback that could slow the rate of global warming from CO_2 emissions. On the other hand,

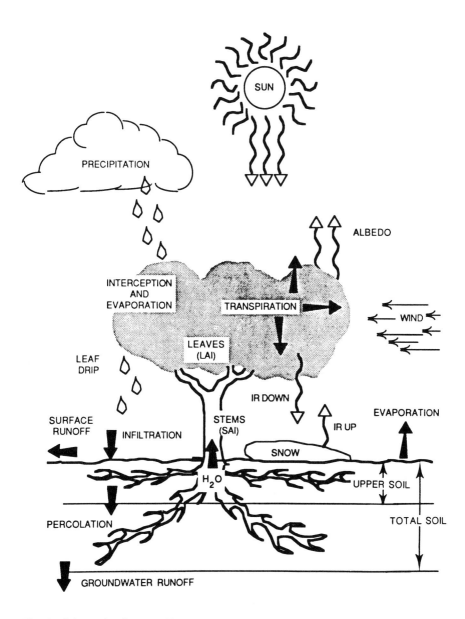

Fig. 1. Schematic diagram illustrating the features included in the land-surface parameterization scheme used here. LAI = leaf area index, SAI = stem area index, IR = infrared index. (Source: Dickinson *et al.*, 1986.)

increased surface temperature from greenhouse gas increases could enhance
the metabolic activity of soil microbes, thus speeding up the decomposition
of dead organic matter into CO_2 (in aerobic soils) or CH_4 (in anaerobic
places). This positive biological feedback on climate change could enhance
global warming rates (e.g. Lashof, 1989).

IV. SENSITIVITY AND SCENARIO ANALYSIS

Although the ultimate goal of any forecasting simulation may be to produce
a single, accurate time-series projection of some evolving variable, a lesser
goal may still be quite useful and certainly is more realizable: to specify
plausible scenarios of various uncertain or unpredictable variables and then
to evaluate the sensitivity of some predicted variable to either different
scenarios or different model assumptions.

For example, in order to predict the societal impact of climatic changes
from increasing concentrations of certain trace gases like CO_2 it is first
necessary to invoke behavioral assumptions about future population, economic and technological trends. (Such factors are external to the climatic
forecast model, of course, but must be forecasted none the less.) Although
these may be impossible to forecast with confidence, a set of plausible
scenarios can be derived. The differential consequences for the climatic
forecast of each of these scenarios can then be evaluated.

Another sort of sensitivity analysis involves building a model which
incorporates important, but highly uncertain, variables into its internal
structure. Key internal factors such as the cloud feedback or vertical oceanic
mixing parameters, can be varied over a plausible range of values in order to
help determine which internal processes have the most importance for the
sensitivity of the climate to, say, CO_2 build-up. Even though one cannot be
certain which of the simulations is most realistic, sensitivity analyses can (a)
help set up a priority list for further work on uncertain internal model
elements and (b) help to estimate the plausible range of climatic futures to
which society may have to adapt over the next several decades. Given these
plausible futures, some of us might choose to avoid a low-to-moderate
probability, high consequence outcome associated with some specific scenario. Indeed public policy or private purchase of insurance is often motivated
to avoid plausible, high-cost scenarios. On the other hand, more risk-prone
people might prefer extra scientific certainty before asking society to invest
present resources to hedge against uncertain, even if plausible, climatic
futures. At a minimum, cross-sensitivity analysis, in which the response of
some forecast variables to multiple variations in uncertain internal and/or
external parameters, allows us to examine quantitatively the differential
consequences of explicit sets of plausible assumptions.

In any case, even if we cannot produce a reliable single forecast of some future variables, we might be able to provide much more credible sensitivity analyses, which can have practical applications in helping us to investigate a range of probabilities and consequences of plausible scenarios. Such predictions may simply be the best "forecasts" that honest natural or social scientists can provide to inform society on a plausible range of alternative futures of complex systems. How to react to such information, of course, is in the realm of values and politics.

Let us proceed then to several examples of this process.

V. MODEL VALIDATION

The principal reason that advocates of concern over the prospects of global warming—and I am unabashedly one of them (e.g. see Schneider, 1990a,b)—stand before groups such as Congressional or Parliamentary committees and take their time with our concerns is not based solely on speculative theory. Rather, such concern is based on the fact that the models which we use to foreshadow the future have already been validated to a considerable degree, although not to the full satisfaction of any responsible scientist. At least several methods can be used, and none by itself is sufficient. First, we must check overall model-simulation skill against the real climate for today's conditions to see if the control experiment is reliable. The seasonal cycle is one good test. Figure 2, from Manabe and Stouffer (1980), shows how remarkably well a three-dimensional global circulation model can simulate the regional distribution of the seasonal cycle of surface air temperature—a well-understood climate variation that is, when averaged over the northern hemisphere, several times larger than global averaged ice age–interglacial changes. The seasonal-cycle simulation is a necessary test of some processes that can be called "fast physics", but it does not tell us how well the model simulates slow changes in forest or ice cover or deep ocean temperatures, since these variables do not change much over a seasonal cycle, though they do influence long-term trends.

A second method of verification is to test in isolation individual physical subcomponents of the model (such as its parameterizations) directly against real data and/or more highly resolved process models. This still is not a guarantee that the net effect of all interacting physical subcomponents has been properly treated, but it is an important test. For example, the upward infrared radiation emitted from the earth to space can be measured from satellites or calculated in a climate model. If this quantity is subtracted from the emitted upward infrared radiation at the earth's surface, then the difference between these quantities, G, can be identified as the "greenhouse effect", as on Fig. 3 from Raval and Ramanathan (1989). Although radiative

Fig. 2. A three-dimensional climate model has been used to compute the winter to summer temperature differences all over the globe. The model's performance can be verified against the observed data shown below. This verification exercise shows that the model quite impressively reproduces many of the features of the seasonal cycle of surface air temperature. These seasonal temperature extremes are mostly larger than those occurring between ice ages and interglacials or for any plausible future CO_2 change. (Source: Manabe and Stouffer, 1980.)

processes may not be the only ones operative in nature or in the general circulation model (GCM) results for G, the close agreement among the satellite results (line labeled ERBE on Fig. 3), a GCM (labeled CCM) and line-by-line radiative transfer calculations on Fig. 3 give strong evidence that this physical subcomponent is well modeled at grid scale. Another just-emerging set of validation tests of internal model processes is to compare model-generated and observed statistics of grid-point daily variability (e.g.

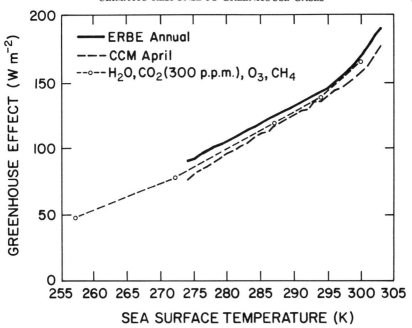

Fig. 3. Comparison of greenhouse effect, heat-trapping parameter, G, and surface temperature, obtained from three sources: ($-$) ERBE annual values, obtained by averaging April, July and October 1985 and January 1986 satellite measurements, (- - -) three-dimensional climate-model simulations for a perpetual April simulation (National Center for Atmospheric Research Community Climate Model (CCM), (- o -) line-by-line radiation-model calculations by Dr A. Arking using CO_2, O_3 and CH_4. The line-by-line model results come close to the CCM and the ERBE values. (Source: Raval and Ramanathan, 1989.)

Rind *et al.*, 1989; Mearns *et al.*, 1990). These have provided some examples of excellent agreements between model and observed variability (e.g. surface temperatures) as well as examples of poorer agreement (e.g. relative humidity).

As a third validation method, some researchers express more confidence *a priori* in a model whose internal makeup includes more spatial resolution or physical detail, believing that "more is better". In some cases and for some problems this is true, but by no means for all. The "optimal" level of complexity depends upon the problem we are trying to solve and the resources available to the task (e.g. Land and Schneider, 1987). All three methods of validation must constantly be used and reused as models evolve if we are to improve the credibility of their predictions. And to these we can add a fourth method: the model's ability to simulate the very different climates of the ancient earth or even those of other planets.

As an example of the fourth method of validation, we know from observations of nature, a point neglected by some critics of global warming research, that the last ice age, which globally was about 5°C colder than the past 10 000 year interglacial era, also had CO_2 levels about 25% less than the pre-industrial values. Methane, another important greenhouse gas, also was reduced by nearly a factor of 2 at glacial maxima relative to pre-industrial interglacial values. Ice cores in Antarctica have shown us, since these cores contain gas bubbles that are records of the atmospheric composition going back over 150 000 years, that the previous interglacial age some 125 000–130 000 years ago had CO_2 and methane levels comparable with those in the present interglacial (Barnola et al., 1987). The close correlation of change in these greenhouse gases and in planetary temperature over geological epochs (see Fig. 4a) is for a temperature change of a magnitude that one would project based on the CO_2-temperature sensitivity of today's generation of computer models (e.g. Lorius et al., 1990). However, we still cannot assert that this greenhouse gas–temperature change coincidence is proof that our model's sensitivity is quantitatively correct, since other factors were operating during the ice age–interglacial cycles. The best that can be said is that the evidence is strong, but circumstantial.

Because of all these lines of validation, I believe that we are entitled to assume a rough factor of 2–3 confidence that the magnitude of global surface temperature change projected at most official assessments (i.e. 1·5–4·5°C equilibrium surface warming from CO_2 doubling—e.g. see IPCC, 1990) has a better than even chance of being correct—indeed, that is why in a popular context (e.g. Schneider, 1990a,b) I often refer to the prospect for global warming in the twenty-first century as "coin-flipping odds of unprecedented change", since a sustained global change of more than 2°C would be unprecedented during the past 10 000-year era of human civilization's development.

One final comment about the recent scientific debate over the plausibility of global warming projections. Raval and Ramanathan's (1989) satellite observations of the water vapor–greenhouse feedback mechanism, a process that is central to most models' estimates of some 3°C plus or minus 1·5°C equilibrium warming from doubling CO_2, led them to conclude that:

> The greenhouse effect is found to increase significantly with sea surface temperature. The rate of increase gives compelling evidence for the positive feedback between surface temperature, water vapour and the greenhouse effect; the magnitude of the feedback is consistent with that predicted by climate models.

In other words, the heat-trapping capacity of the atmosphere is fairly well modeled and measured on earth, and much of the sometimes polemical (e.g.

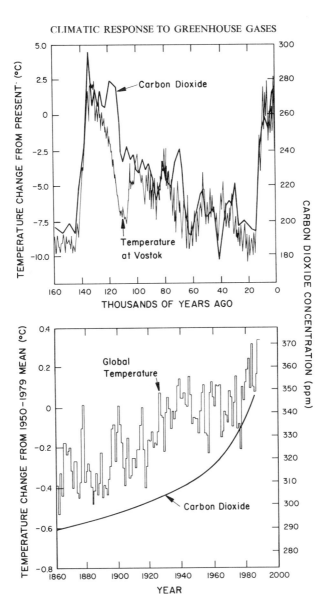

Fig. 4. CO_2 and temperature are closely correlated over the past 160 000 years (top) and, to a lesser extent, over the past 100 years (bottom). The long-term record, based on evidence from a Soviet core at Vostok, Antarctica and interpreted by the glaciology group at Grenoble, France (Barnola *et al.*, 1987) shows how the local temperature and atmospheric CO_2 rose nearly in step as an ice age ended about 130 000 years ago, fell almost in synchrony at the onset of a new glacial period and rose again as the ice retreated about 10 000 years ago. The recent temperature record shows a global warming trend ($\sim 0.5°C$), as traced by workers at the Climatic Research Unit at the University of East Anglia. Whether the accompanying build up of CO_2 and other trace greenhouse gases in the atmosphere caused the 0·5°C warming is hotly debated. (Source: Schneider, 1990a,b.)

12   S. H. SCHNEIDER

Brookes, 1989 or *Detroit News*, 1989) debate in the media over the greenhouse effect has little reality. This empirical confirmation of the natural
greenhouse effect, which is consistent with the greenhouse effect of climate
models, stands in contrast to the theoretical and physical deductions of
Ellsaesser (1984) or Lindzen (1990) that negative temperature–water vapor
feedback processes in parts of the tropics will reduce present model estimates
of global warming by a factor of "one-half to one-fifth"; in this context
Lindzen (Lindzen, 1990, p. 296) asserted his belief that water is "nature's
thermostat". However, the satellite and model results suggest otherwise.

But recently Ramanathan and Collins (1991) showed that tropical cumulus clouds could provide strong negative feedback over the 10–20% of the
world covered by such clouds if the surface temperatures warmed to more
than 30°C. While this potential negative feedback will not substantially alter
the likelihood of global warming of the typically calculated range, it will
complicate projecting regional patterns of climate change.

VI. CLIMATIC MODEL RESULTS

To investigate future climatic changes, let us first turn not to the present but
to the ancient times when the dinosaurs ruled the planet, when climatic
conditions were perhaps as warm as they have been since life started on
earth. The time we are about to focus on had abundant plant life (and much
of the fossil fuels available today were laid down); the continents were in
substantially different positions; sea levels were hundreds of meters higher;
and only 20% or so of the world was not covered with seas. This is the mid-
Cretaceous period about 100 million years ago. At that time it is not believed
that any permanent ice existed in significant quantities on earth, and even
Antarctica was unglaciated. Broad-leafed, tropical plants were found in the
middle of mid- to high-latitude continents, and famous fossil finds have
included alligators in Ellesmere Island, which is now above the Arctic Circle
and hardly able to sustain even a limited summer appearance of leafy
vegetation, let alone these subtropical reptiles. What could possibly explain
such a warm climatic era? If the same basic models that we use to estimate
greenhouse effect on earth can be applied to the Cretaceous and a reasonable
simulation is obtained, then both scientific explanation of the ancient
paleoclimate and some verification of the model's ability to reproduce
radically different future climatic periods can be obtained.

A. Paleoclimatic Modeling: Is the Climatic Past the Key to the Future?

1. Mid-Cretaceous Warming Period

Three-dimensional atmospheric circulation model studies which explicitly resolve land and sea and also explicitly calculate atmospheric motions have been applied to the continental configurations of the mid-Cretaceous by Barron and Washington (1982). Although exhibiting greater sensitivity to paleogeography than earlier energy balance models (e.g. Barron *et al.*, 1981), such three-dimensional models do confirm the basic inferences of the energy balance climate model experiments. Barron and Washington (1982), for example, describe experiments with an assumed minimum sea surface temperature of 10°C (an implied large poleward ocean heat transport) and found that mid-latitude continental interior regions, such as Asia, still became cold in winter. This result brings into question whether greater oceanic heat transport alone is a solution to the problem of keeping mid-Cretaceous, high latitude continental interiors warm, even in winter.

The Cretaceous paleocontinental positions are shown in Fig. 5. The Barron and Washington Cretaceous case surface temperature was 4·8°C warmer than the control; and the poles warmed up by several-fold more. Surface temperatures over land and oceans in the higher northern latitudes were typically between 270 and 280 K, and well below freezing in the Antarctic. However, in this "swamp ocean" case, no heat is stored or explicitly transported poleward by oceans. Thus, it is natural to inquire whether increased poleward transport of heat due to oceanic currents might not alter this result.

Since this climatic model from the National Center for Atmospheric Research (NCAR) did not include a dynamical oceanic circulation model coupled to the atmospheric circulation model, it was necessary to assume how oceanic dynamics could alter sea surface temperatures. In the spirit of sensitivity analysis, Schneider *et al.* (1985) ran a new set of GCM calculations in which we made the extreme assumption that mid-Cretaceous oceanic motions were perfectly efficient in transporting heat within the oceans; that is, despite a large latitudinal gradient in solar heating, we assumed a globally uniform oceanic surface temperature fixed at 293 K (20°C), a reasonable warm world *average* surface temperature.

Our simulations produced a warm world compared to today, but continental interiors in middle to high latitudes and (especially) in higher topographic regions, still had surface temperatures significantly below freezing (see Fig. 5a). Despite the very high implicit oceanic heat transport implied by the all-293 K oceans, the northern hemisphere mid-continental temperatures in Schneider *et al.* (1985) had actually *decreased* relative to the colder

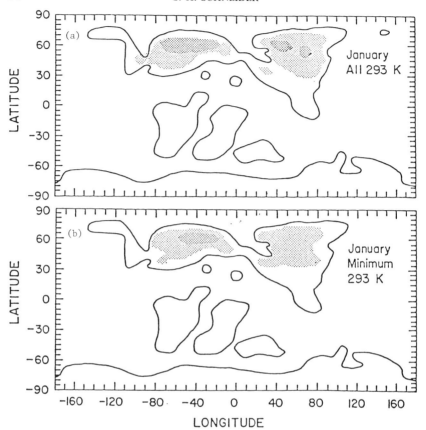

Fig. 5. Surface temperature for January solar forcing computed by averaging the model-generated weather statistics over a 50-day period with continents where they were 100 million years ago. For all cases, CO_2 levels and the latitudinal distribution of solar insolation are based on today's values, and for Fig. 5a all oceanic surface temperatures are held fixed at 293 K (20°C), equivalent to a super-efficient heat-transporting ocean. Dark shading implies temperatures at some time in the simulation were less than 260 K and lighter shading implies less than 275 K. Note that substantially subfreezing surface temperatures are computed in areas of mid- to high-latitude zones. In Fig. 5b, ocean temperatures are first calculated by a "swamp" ocean model, but are then never allowed to drop below 293 K.

swamp ocean case of Barron and Washington. The reason our warmer ocean case has colder land surface temperatures is associated with atmospheric circulation changes. The simulated sluggishness of atmospheric motions in the all-293 K case is insufficient to export the extremely warm mid- to high-latitude oceanic surface temperatures into mid-continental areas.

We performed an additional set of simulations in which oceanic surface temperatures were explicitly calculated, but never allowed to drop below

293 K, even in polar night waters. With tropical oceanic temperatures now greater than 300 K, a substantial surface temperature difference from equator to pole is set up on a very warm planet. The resulting atmospheric circulation is much more vigorous in this case, as might be expected. But despite the much enhanced atmospheric circulation strength and the much warmer average planetary surface temperatures, subfreezing continental interiors in midwinter are still widespread, even though there has been a considerable warming everywhere (see Fig. 5b). Clearly, even these exaggerated, "high-transport" warm oceans scenarios are unable to route enough heat via atmospheric motions to offset the midwinter, continental interior infrared radiative cooling to space which creates some areas of subfreezing land surface temperatures.

For midwinter solar input, it appears from all our cases run so far that additional forcing mechanisms are required to produce near- or above-freezing temperatures on land, as the paleoclimatic evidence strongly suggests. One favorite hypothesized forcing is a much enhanced atmospheric "greenhouse effect" from greatly increased CO_2 concentration (Budyko and Ronov, 1980). Geochemical models have independently suggested that the mid-Cretaceous was a time when perhaps five to ten times more CO_2 was in the atmosphere than at present (Berner *et al.*, 1983). Thus, the NCAR Cretaceous experiments contribute some small measure of confidence about the overall question of modeling large greenhouse effects. Unfortunately, the present model is not comprehensive enough to produce fully credible surface temperature simulations. Therefore, to further check these CO_2 experiments against the background of nature's large climatic changes, modelers have turned to yet another paleoclimatic case, but one of a very different character from the very warm Cretaceous case.

2. Climatic Optimum

One of the most successful paleoclimatic simulations to date was performed by John Kutzbach and several colleagues at the University of Wisconsin in Madison (e.g. Kutzbach and Guetter, 1986; Prell and Kutzbach, 1987). Kutzbach and colleagues attempted to explain the warmest period in recent climatic history, the so-called "climatic optimum" that occurred between about 5000 and 9000 years ago. It was a time when summertime northern continental temperatures were probably several degrees warmer than at present and monsoon rainfall was more intense throughout Africa and Asia. Kutzbach found that the optimum could be explained simply by the fact that the tilt of the earth's axis (its obliquity) was slightly greater then than now. Also, the orbit was such that the earth was closer to the sun (i.e. perihelion) in June rather than in January as it now is. These variations, these slight perturbations in the earth's orbit, do not make substantial changes in the annual amount of solar radiation received on the earth, but do change substantially the difference between winter and summer heating periods. About 5% more solar heat over much of the northern hemisphere summer and a comparable amount less in winter occurred 9000 years ago compared to the present. This change was sufficient in Kutzbach's simulations to alter substantially mid-continental warming in the summer months, which led to enhanced monsoonal rainfall and river runoff in the models. His results, as seen in Fig. 6, matched quite well with considerable amounts of paleoclimatic evidence gathered by an international team of scientists (COHMAP, 1988), and are helping to explain one of the important mysteries in the paleoclimatic record.

3. Younger Dryas: A Mini-Ice Age Fluctuation

When the earth emerged from the grip of the last ice age about 12 000 years ago, conditions around the globe were rapidly turning toward the warmer interglacial values that we have been experiencing for the last 10 000 years. There were some exceptions, such as the great glaciers in northeast Canada which took several thousand more years finally to melt, but by and large by 11 000 years ago much of the warm weather flora and fauna had begun to return to northern latitudes, especially in western Europe. Then suddenly, around 11 000 years ago, a dramatic cooling of near ice-age intensity struck rapidly and lasted for almost 1000 years before the uninterrupted present interglacial finally won out. This large climatic fluctuation is known to paleoclimatologists as the Younger Dryas. It was most intense in Europe, especially northwest Europe and England, and to a lesser extent the northeastern part of North America. Although a Younger Dryas climatic "signal" is observed in some geologic records throughout the world, the most severe effects seem to have been in the North Atlantic sector and especially on the west coast of Europe. This is certainly suggestive of an oceanic cause. Indeed, a number of paleoclimatologists, such as Ruddiman and McIntyre (1981) had suggested that rapid break up of the Laurentide and Fenno-Scandinavian ice sheets around 10 000–12 000 years ago dumped a high volume of fresh melt-water into the North Atlantic. Since fresh water leads to conditions in which freezing can more easily occur than for the normally saline North Atlantic ocean, the obvious interpretation is that such a "melt-water spike" caused an extensive cover of sea ice over the North Atlantic (e.g. Broecker *et al.*, 1985), blocking the normal pattern whereby Gulf Stream heating of the northeastern part of the North Atlantic keeps the Iceland,

---

Fig. 6. Surface temperature departures (experiment-minus-control K) (a) and precipitation minus evaporation (b) for July for 9 kyr BP. The shading on the map indicates the departures are statistically significant, based upon the model's own natural variability. Heavy cross hatching shows remnant of North America ice sheet.

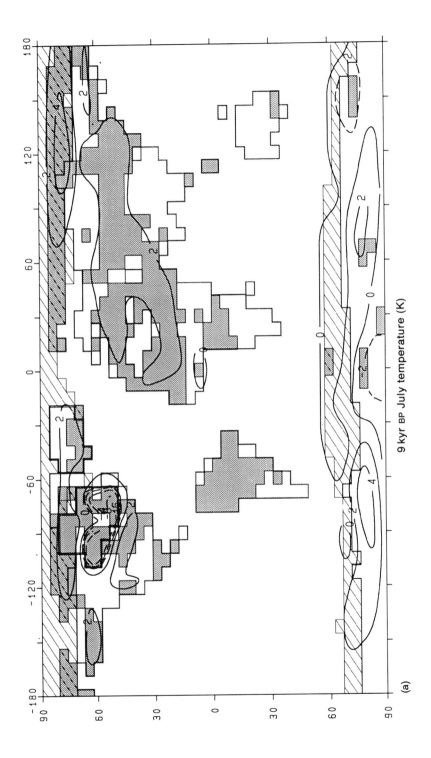

(a)

9 kyr BP July temperature (K)

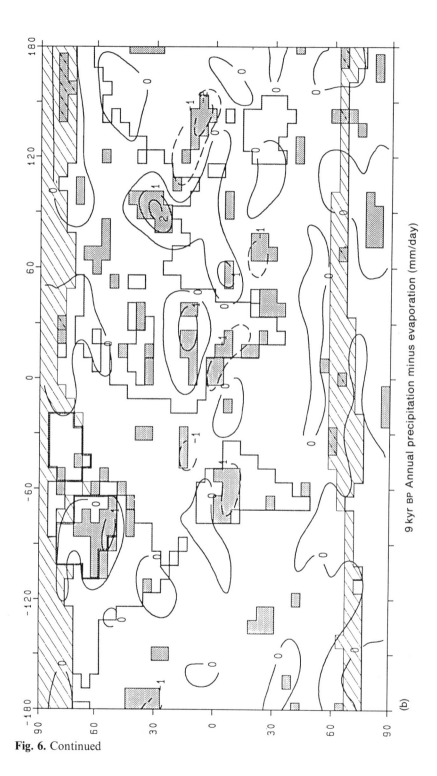

9 kyr BP Annual precipitation minus evaporation (mm/day)

(b)

Fig. 6. Continued

England and northwest Europe climate much more equable than the climate of, say, Hudson Bay, which is at the same latitude but feels no benefit of the warming current.

If sea ice were suddenly imposed throughout the north half—or at least the northeast quadrant—of the North Atlantic, then substantial drops in air temperature over the ice and immediately downwind would be expected. To test this hypothesis quantitatively, Starley Thompson and I imposed just such an assumption on the NCAR three-dimensional climate model (e.g. see Schneider *et al.*, 1987). Independently, Rind and Peteet (1986) at the Goddard Institute for Space Studies, NASA, New York City used a three-dimensional model developed by James Hansen and colleagues to simulate the Younger Dryas. Figure 7 shows the NCAR result for both winter and summer. In this model sea ice was imposed down to 45° N latitude over the whole North Atlantic—not necessarily because we believe that to have been precisely what had occurred during the Younger Dryas, but simply as a climate sensitivity experiment to show how a regional change in the character of the North Atlantic could cause a climatic response that could be studied differentially from place to place and from winter to summer. Note in these results that during the summer (Fig. 7a) the climatic effects of this incredibly cooled North Atlantic are primarily felt right along the coastline, hardly spreading inland. In this summer case, the solar heating of the land is very intense, and even though the assumed ice-covered oceans are extremely cold, the primary mechanism controlling the mid-continental temperatures is summertime solar radiation. As Fig. 7 shows, the oceans have a relatively smaller influence outside of coastal areas. On the other hand, in the wintertime when the solar input is very small, the normally moderate winter oceans maintain a more equable climate, particularly in the western parts of the continents. Thus, if the Younger Dryas really was characterized by a substantial increase of North Atlantic sea ice, then one would expect the European winter temperatures to be very severely depressed, as indeed they are in the simulations of Fig. 7b for much of Europe and even northeastern United States and Canada. Indeed, much paleoclimatic evidence suggests a return to near ice-age conditions near the European coast and in England at this time, which would tend to confirm the model's basic patterns. Similar results were obtained simultaneously at the Goddard Institute for Space Studies (GISS) and elsewhere.

Although differing in detail from the NCAR model, the GISS results show similar differential sensitivity between winter and summer. Moreover, both sets of results show that there is substantial seasonal and regional differences in the climatic response for these particular forcings. To lend even further support, L.D.D. Harvey from the University of Toronto and Gerald North, then at Goddard Space Flight Center, independently used two-dimensional energy balance models and found similar patterns of regional and seasonal

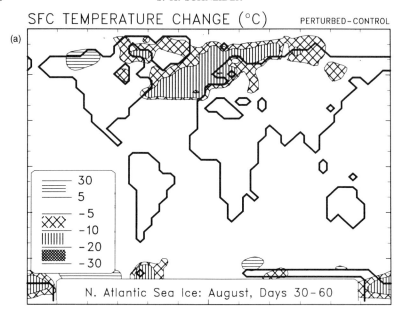

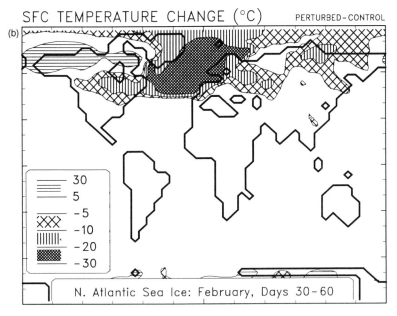

Fig. 7.(a) Surface temperature change in the NCAR Community Climate Model for an imposed sea ice cover in the North Atlantic Ocean down to about 45° N for August. (b) is the same as (a) except for February.

changes (see Schneider *et al.*, 1987). This is another example of the basic modeling methodology employed by most climate modelers today: sensitivity analysis across a hierarchy of climatic models. Because the climatic signals are so different from season to season and from place to place, these models' results help to suggest locations where paleoclimatologists should dig through the geologic record in order to uncover remains of the seasonal and regional distribution of past climates. As that evidence becomes available, then not only will we have an opportunity for an explanation of a very intriguing, rapid and dramatic past climatic change, but we will have another opportunity to verify independently the ability of our models to reproduce large climatic changes.

B. Modeling the Future Greenhouse Effect

The most important question surrounding the greenhouse gas controversy is simply: What will be the regional distribution of climatic changes associated with significant increases in CO_2 and other trace greenhouse gases? (Other trace gases such as chlorofluorocarbons, nitrogen oxides, or ozone can, all taken together, have a comparable greenhouse effect to CO_2 over the next century; Dickinson and Cicerone, 1986.) To investigate such possibilities one needs a model with regional resolution. It needs to include processes such as the hydrologic cycle, the transfer of heat and moisture through biota, and the storage of moisture in the soils—since these factors are so critical to atmospheric and ecological systems. To investigate plausible climatic scenarios, modelers have typically run "equilibrium simulations" in which instantaneously increased values of carbon dioxide are imposed at an initial time and held fixed while the model is allowed to approach a new equilibrium. Manabe and Wetherald (1986), for example, in one of the most widely quoted results (shown in Fig. 8), find a summer "dry zone" in the middle of North America, as well as increased moistness in some of the monsoon belts. All of this was from a quadrupling of CO_2 held fixed over time. The model was allowed to run several decades of simulated time in order to reach equilibrium. But Manabe's team used an artificially constrained ocean that consisted of a uniform mixed layer of depth of 70 m with no heat flow to or from the deep ocean below. While this shallow "sea" allows the seasonal cycle to be satisfactorily simulated (see Fig. 2), a purely mixed layer ocean does not include the important processes whereby water is transported horizontally from the tropics toward the poles or vertically between the mixed layer and the abyssal depths. The latter processes slow the approach toward thermal equilibrium of the surface waters, and certainly would affect the transient evolution of the surface temperature changes in the ocean due to the actual time evolving increase of trace greenhouse gases.

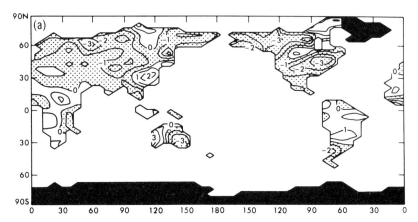

Fig. 8. Geographical distribution of the difference of soil moisture (cm) between the 4 × CO$_2$ and the standard experiments obtained for the (June–July–August) period for Geophysical Fluid Dynamics Laboratory (GFDL) GCM. Dark shading represents decreases in soil moisture.

Thus, during the transient phase of warming the surface temperature increases from latitude to latitude and land to sea could have a different pattern than at equilibrium. This, in turn, could cause significantly different climatic anomalies during the transient phase than would be inferred from equilibrium sensitivity tests to fixed increases in trace gases.

To answer this transient response question more reliably, it has been proposed that three-dimensional atmospheric models be coupled to fairly realistic three-dimensional oceanic models. To date, only a handful of such model experiments have been run (e.g. Washington and Meehl, 1989; Stouffer et al., 1989), but none over the century or two time scale needed to address this important issue adequately. Even so, these preliminary transient simulations do find very different regional patterns of climatic anomalies over time than in equilibrium. One reason for detailing the CO$_2$ transient/regional climate anomaly example here is to exemplify the philosophical remarks made earlier about the need for various sensitivity experiments across a hierarchy of models. Such methods are especially essential during the development phase of modeling. For example, Schneider and Thompson (1981) ran quasi-one-dimensional models to illustrate the potential importance of this transient issue. However, these models could not provide reliable simulations because of their physical simplicity. But the one-dimensional models were economically efficient enough to be able to be run over the century time span needed to explore the CO$_2$ transient issue. On the other hand, the more complex three-dimensional general circulation models are not at an adequate phase of development to have trustworthy coupled atmosphere/ocean models that are both well verified and economical enough to be run repeatedly over the 100 years needed for greenhouse-gas transient issues. Thus, the simple model helps to identify potentially important problems, provides some quantitative sensitivity studies, and helps set priorities for three-dimensional coupled model research over the next few decades. Since the ecological and other environmental impacts of increasing greenhouse gases depend on the specific regional and seasonal distribution of climatic change (e.g. see Root and Schneider, submitted paper), resolution of the transient debate, among others, is critical for climatic impact assessment and ultimate policy responses to the advent or prospect of increasing greenhouse gases.

VII. CONCLUDING REMARKS

It is well known that the 25% increase in CO_2 that is documented since the industrial revolution, the 100% increase in methane since the industrial revolution and the introduction of man-made chemicals such as chlorofluoro-carbons (also clearly implicated for stratospheric ozone depletion and the Antarctic ozone hole) since the 1950s should all together have trapped about two extra watts of infrared radiant energy over every square meter of the earth–troposphere system. However, what is less well accepted is how to translate that 2 W or so of heating into "X" degrees of temperature change, since this involves assumptions about how that heating will be distributed among surface temperature rises, evaporation increases, cloudiness changes, ice changes and so forth. The factor of 2–3 uncertainty in global temperature rise projections as cited in typical National Academy of Sciences' reports or the United Nations-sponsored Intergovernmental Panel on Climate Change (IPCC, 1990) reflects a legitimate estimate of uncertainty held by most in the scientific community. Indeed, recent modifications of the British Meteoro-logical Office climate model to attempt to mimic the effects of cloud droplets and ice particles roughly halved their model's sensitivity to doubled CO_2— but is still well within the often-cited 1·5–4·5°C range. However, this model, like all current GCMs, does not yet include parameterizations for incremen-tal (i.e. small vertical displacements at each vertical level) cloud-top height feedback. This process has long been known to have the potential to alter substantially the heat-trapping efficiency of clouds (e.g. Schneider, 1972), although in what direction local cloud heights would change with increased greenhouse gases remains unclear.

Lindzen suggests on physical grounds that "warming is also associated with deeper cumulus convection", which would dry out the upper tropos-phere in the tropics and reduce the infrared heat-trapping capacity of the atmosphere, which, Lindzen continues, "are a negative rather than a positive feedback to CO_2 heating, and should diminish the effect of CO_2 warming rather than magnify it by a factor of approximately 3, as occurs in present models" (Lindzen, 1990, p. 297). What this argument neglects is the possible positive feedback effects of incremental cloud-top height increases should this "deeper" cloud postulate prove true. Increasing cloud-top height *per se* (e.g. Schneider, 1972) is likely to increase heat trapping, a positive feedback. Lindzen also neglects any possible radiative heat-trapping effects of cirrus cloud shields that might be generated as a result of the postulated enhanced cumulus activity (e.g. Manabe and Wetherald, 1967). Ramanathan and Collins (1991) suggest such shields might be important negative feedback processes in parts of the humid tropics. In this vein, Mitchell *et al.* (1989) wisely pointed out that although their "revised cloud scheme is more detailed, it is not necessarily more accurate than the less sophisticated scheme". I am yet to see this forthright and important caveat quoted by any of the global warming critics who cite the British work as a reason to lower our concern by a factor of 2 or so. Recently, Lindzen himself told a US senate hearing that he no longer has confidence in his cumulus mechanism for negative feedback.

Finally, as stated earlier, prediction of the detailed regional distribution of climatic anomalies, that is, where and when it will be wetter and drier, how many more floods might occur in the spring in California or forest fires in Siberia in August, is simply highly speculative, although some plausible scenarios can be given. Some such scenarios are given in Table 1, from the National Academy of Sciences, 1987, assessment.

In an effort to shed some light on these questions, Schneider *et al.* (1990) offered a set of "forecasts" on changes in some important meteorological variables, over a range of temporal, spatial, and statistical scales. I believe that carefully qualified, explicit scenarios of plausible future climatic changes are preferable to impact speculations based on implicit or casually formulated forecasts. Therefore, Table 2 was prepared to provide impact assessment specialists with ranges of climate changes that reflected their interpretation of state-of-the-art modeling results. These projections were based on an analysis of then available results and provide what were believed to be plausible estimates about the direction or magnitude of some important anthropogenic climatic changes over the next 50 years or so—a typical time estimate for an equivalent doubling of CO_2—together with a simple high, medium, or low level of confidence for each variable. (By "equivalent doubling" it is meant that CO_2 together with other trace greenhouse gases have a radiative effect equivalent to doubling the pre-industrial value of CO_2 from about 280 to 560 p.p.m.) As another measure of the nature of the uncertainties, a subjective estimate is included of the time that may be necessary to achieve a widespread scientific consensus on the direction and magnitude of the change. In some cases—such as the magnitude and

Table 1
Possible climate changes from doubling of CO_2 (Source: National Academy of Sciences, 1987)

Large Stratospheric Cooling (virtually certain)
Reduced ozone concentrations in the upper stratosphere will lead to reduced absorption of solar ultraviolet radiation and therefore less heating. Increases in the stratospheric concentration of CO_2 and other radiatively active trace gases will increase the radiation of heat from the stratosphere. The combination of decreased heating and increased cooling will lead to a major lowering of temperatures in the upper stratosphere.

Global-Mean Surface Warming (very probable)
For a doubling of atmospheric CO_2 (or its radiative equivalent from all the greenhouse gases), the *long-term* global-mean surface warming is expected to be in the range of 1·5 to 4·5°C. The most significant uncertainty arises from the effects of clouds. Of course, the *actual* rate of warming over the next century will be governed by the growth rate of greenhouse gases, natural fluctuations in the climate system, and the detailed response of the slowly responding parts of the climate system, i.e. oceans and glacial ice.

Global-Mean Precipitation Increase (very probable)
Increased heating of the surface will lead to increased evaporation and, therefore, to greater global mean precipitation. Despite this increase in global average precipitation, some individual regions might well experience decreases in rainfall.

Reduction of Sea Ice (very probable)
As the climate warms, total sea ice is expected to be reduced.

Polar Winter Surface Warming (very probable)
As the sea ice boundary is shifted poleward, the models predict a dramatically enhanced surface warming in winter polar regions. The greater fraction of open water and thinner sea ice will probably lead to warming of the polar surface air by as much as three times the global mean warming.

Summer Continental Dryness/Warming (likely in the long term)
Several studies have predicted a marked long-term drying of the soil moisture over some mid-latitude interior continental regions during summer. This dryness is mainly caused by an earlier termination of snowmelt and rainy periods, and an earlier onset of the spring-to-summer reduction of soil wetness. Of course, these simulations of long-term equilibrium conditions may not offer a reliable guide to trends over the next few decades of changing atmospheric composition and changing climate.

High-Latitude Precipitation Increase (probable)
As the climate warms, the increased poleward penetration of warm, moist air should increase the average annual precipitation in high latitudes.

Rise in Global Mean Sea Level (probable)
A rise in mean sea level is generally expected due to thermal expansion of sea water in the warmer future climate. Far less certain is the contribution due to melting or calving of land ice.

---

direction of changes in sea level and global annual-averaged temperature and precipitation—such a consensus has virtually been reached. In other cases, such as changes in the extent of cloud cover, time-evolving patterns of regional precipitation, or the daily, monthly, or interannual variance of many climatic variables, the large uncertainties surrounding present projections will only be reduced with decades of considerably more research. (IPCC, 1990, figs 11·1 and 11·2, suggest 10–20 years for major research progress as well.)

Let us consider, for example, the first row on Table 2: temperature change. The global average change of $+2$ to $+5°C$ is typical of that in most national and international assessments for an equivalent doubling of greenhouse gases, neglecting transient delays. While no probability bounds are usually given, we intuitively suggest this 2–5°C range as a ± 1 standard deviation estimate. The neglect of transients means that the range given is based on the assumption that trace gases have been increased over a long enough period for the climate to come into equilibrium with the increased concentration of greenhouse gases. In reality, as noted earlier, the large heat capacity of the oceans will delay realization of most of the equilibrium warming by at least many decades. This implies that at any specific time when we reach an equivalent CO_2 doubling (by say 2030), the actual global temperature increase may be considerably less than the $+2$ to $+5°C$ listed in Table 2. However, this "unrealized warming" (Hansen et al., 1985) will eventually be experienced when the climate system thermal response catches up with the greenhouse gas forcing.

Forecasts of regional or watershed-scale changes in temperature, evaporation, or precipitation are most germane to impact assessment. But, as Table 2 suggests, such regional forecasts are much more uncertain than global equilibrium projections. Regional temperature ranges given in Table 2 are much larger than global changes, and even allow for some regions of negative change. This regional uncertainty is reinforced by suggestions that SO_2 pollution from industrial activities (largely in industrial areas of the northern hemisphere) might increase cloud albedos near the pollution sources (Wigley, 1989; Charlson et al., 1991). Higher northern latitude surface temperature increases are up to several times larger than the global average response, at least in equilibrium. Because of the importance of regional or local impact information, techniques need to be developed to evaluate smaller-scale effects of large-scale climatic changes. (For example, Gleick (1987) employed a regional hydrology model driven by large-scale climate change scenarios from various GCM inputs.)

Even more uncertain than regional details, but perhaps most important, are estimates of climatic variability such as the frequency and magnitude of severe storms, enhanced heat waves, or reduced frost probabilities (Mearns et al., 1984; Parry and Carter, 1985). For example, some modeling evidence suggests that hurricane intensities will increase with climate changes (Emanuel, 1987).

Table 2

Range of climate changes

Phenomena	Projection of probable global annual average change[a]	Distribution of change			Significant transients	Confidence of projection		Estimated time for research that leads to consensus (years)
		Regional average	Change in seasonality	Interannual variability[e]		Global average	Regional average	
Temperature[f]	+2 to +5°C	−3 to +10°C	Yes	Down?	Yes	High	Medium	0 to 10
Sea level	0 to 80 cm[g]	[b]	No	d	Yes[g]	High	Medium	5 to 20
Precipitation	+7 to +15%	−20 to +20%	Yes	Up?	Yes	High	Low	10 to 40
Direct solar radiation	−10 to +10%	−30 to +30%	Yes	d	Possible	Low	Low	10 to 40
Evapo-transpiration	+5 to +10%	−10 to +10%	Yes	d	Possible	High	Low	10 to 40
Soil moisture	d	−50 to +50%	Yes	d	Yes	d	Medium	10 to 40
Runoff	Increase	−50 to +50%	Yes	d	Yes	Medium	Low	10 to 40
Severe storms	d	d	c	d	Yes	d	d	10 to 40

[a] For an "equivalent doubling" of atmospheric CO_2 from the pre-industrial level.
[b] Increases in sea level at approximately the global rate except where local geological activity prevails of if changes occur to ocean currents.
[c] Some suggestions of longer season and increased intensity of tropical cyclones as a result of warmer sea surface temperatures.
[d] No basis for quantitative or qualitative forecast.
[e] Inferences based on preliminary results for the United States from Rind *et al.* (1989).
[f] Based on three-dimensional model results. If only trace gas increases were responsible for twentieth century warming trend of about 0·5°C, then this range should be reduced by perhaps 1°C.
[g] Assumes only small changes in Greenland or West Antarctic ice sheets in twenty-first century. For equilibrium, hundreds of years would be needed and up to several meters of additional sea level rise could be accompanied by centuries of ice sheet melting from an equilibrium warming of 3°C or more.
Source: Modified (by Schneider) from Schneider, S. H., Gleick, P. and Mearns, L. O. in Waggoner, P. E. (1990). *Climate Change and U.S. Water Resources*, pp. 41–73. Wiley, New York.

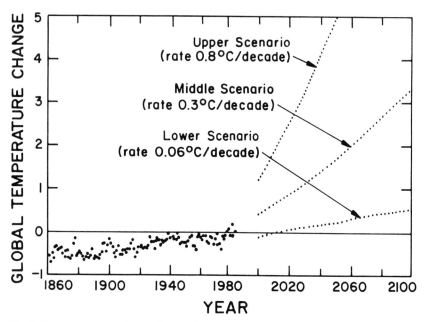

Fig. 9. Three scenarios for global surface air temperature change to the year 2100 derived from combining uncertainties in future trace greenhouse gas projections with uncertainties of modeling the climatic response to those projections. Sustained global temperature changes beyond 2°C would be unprecedented during the era of human civilization. The middle to upper range represents climatic change at a pace 10 to 100 times faster than typical long-term natural average rates of change. (Source: Jaeger, 1988.)

Such issues are just now beginning to be considered and evaluated from equilibrium-climate model results, and will, of course, have to be studied again for realistic transient cases to be of maximum value to ecologists or impact assessors.

Another uncertainty raised by the transient nature of the actual trace gas forcing is the emission and removal rates of CO_2, methane and other greenhouse gases. Figure 9 shows three plausible scenarios based on high, medium, and low emission rates. These uncertainties have been added to those associated with estimates of climate sensitivity and the delay associated with oceanic heat capacity. In any case, since the earth has apparently not experienced a global average temperature more than 1–2°C warmer than at present over the past glacial cycle (150 000 years), all but the slowest scenario in Fig. 9 represents a rapid, large climatic change to which the environment and society will have to adapt.

Although climatic models are far from fully verified for future simulations, the seasonal and paleoclimatic simulations are strong evidence that state-of-the-art climatic models already have considerable skills, particularly at larger scales and for longer averaging periods. An awareness of just what models are and what they can and cannot do is probably the best we can ask of the public and its representatives. Then, the tough policy problem is how to apply the society's values in choosing to face the future given the possible outcomes that climatic models foretell.

Modelers will continue to develop and refine new models by turning to larger computers to run them and more observations to improve and verify them. We must ask the indulgence of society to recognize that immediate, definitive answers are not likely, as coupling of higher resolution atmosphere, ocean, ecosystems, land surface and chemistry submodels will take a decade or more to develop fully—let alone validate. In essence, what climate models and their applications typify is a growing class of problems not unique to climate but also familiar in other disciplines: nuclear waste disposal, safety of food additives or drugs, efficacy of strategic defense technologies and so forth. These are problems for which objective "scientific" probabilities and outcomes cannot be obtained—except by performing the experiment on ourselves. To deal with these complex socio-technical problems requires a new understanding of the central role of (bounded) uncertainty and the willingness to deal with heuristic probalistic estimation; it also calls for more modeling—provided the context of that modeling is understood and parallel observational programs for validation are concurrently pursued. All of this research will have to be highly interdisciplinary, spanning physical, biological and social scientific disciplines (e.g. Schneider, 1988; NAS, 1988). Efficient functioning of society will depend upon our understanding of both the utility and limitations of models—both natural and social scientific disciplines need to create scenarios of global change to perform impact assessments (e.g. Clark, 1988). However, in some senses more than economic efficiency is at stake, as the very survival of some people—particularly in low lying coastal areas—or some species near their climatic tolerance limits will depend on present decisions we make to deal with plausible climatic futures.

With regard to ecological impacts, the traditional approach has been to estimate the implications of some climate scenario on a particular species (e.g. Davis and Zabiniski (1991) for tree species, or Root (1988a,b) for birds). However, as Root and Schneider (submitted paper) have argued, because of the differential rates at which different species will respond to climatic changes, the most important ecological issues are likely to be the tearing apart of communities. Therefore, major research emphasis is needed on the conservation implications of differential migration rates in response to climatic change.

If the public is totally ignorant of the nature, use, or validity of climatic (or many other kinds of) models, then public-policy debates based on model results will be haphazard at best. In this case, the decision-making process tends to be dominated by special interests, exaggerated media debates or a technically trained elite. Expanded interdisciplinary research to narrow uncertainties should not be used as an excuse for inaction in my opinion. However, nearly all would agree that improved understanding and predictive capabilities of earth systems can put decision-making of all kinds—adaptive or preventative—on a firmer factual basis.

REFERENCES

Barnola, J.M., Raynaud, D., Korotkevich, Y.S. and Lorius C. (1987). Vostok ice core provides 160 000-year record of atmospheric CO_2. *Nature* **329**, 408–414.

Barron, E.J. and Washington, W.M. (1982). The Cretaceous atmospheric circulation: Comparisons of model simulations with the geologic record. *Palaeogeogr. Palaeoclim. Palaeoecol.* **40**, 103–134.

Barron, E.J., Thompson, S.L. and Schneider, S.H. (1981). An ice-free Cretaceous, results from climate model simulations. *Science* **212**, 501–508.

Berner, R.A., Lasaga, A.C. and Garrels, R.M. (1983). The carbonate–silicate geochemical cycle and its effect on atmospheric carbon dioxide over the past 100 million years. *Amer. J. Sci.* **283**, 641–683.

Broecker, W.S., Peteet, D.M. and Rind, D. (1985). Does the ocean–atmosphere system have more than one stable mode of operation? *Nature* **315**, 21–25.

Brookes, W.T. (1989). The global warming panic. *Forbes*, 25 December 1989, 96–102.

Budyko, M.I. and Ronov, A.B. (1980). Chemical evolution of the atmosphere in the Phanerozoic. *Geochemistry International 1979* **16**, 1–9.

Charlson, R.J., Largner, J., Rodhe, H., Leovy, C.B. and Warren, S.G. (1991). Perturbation of the northern hemisphere radiative balance by backscattering from anthropogenic sulfate aerosols. *Tellus* **43a,b**, 152–163.

Clark, W. (1988). The human dimensions of global environmental change. In: *Toward an Understanding of Global Change: Initial Priorities for US Contributions to the International Geosphere–Biosphere Program*, pp. 134–200. National Academy Press, Washington, DC.

COHMAP Members, (P.M. Anderson *et al.*) (1988). Climatic changes of the last 18 000 years: Observations and model simulations. *Science* **241**, 1043–1052.

Davis, M.B and Zalinski, C. (1991). Changes in geographical ranges resulting from greenhouse warming effects on biodiversity in forests. In: *Consequences of Greenhouse Warming to Biodiversity* (Ed. by Peters and Lovejoy), Yale University Press, in press.

Detroit News (1989). Loads of media coverage. Editorial, 22 November 1989.

Dickinson, R.E. and Cicerone, R.J. (1986). Future global warming from atmospheric trace gases. *Nature* **319**, 109–115.

Dickinson, R.E., Henderson-Sellers, A., Kennedy, P.J. and Wilson, M.F. (1986). Biosphere–Atmosphere Transfer Scheme (BATS) for the NCAR Community Climate Model. NCAR Technical Note 275, December 1986, National Center for Atmospheric Research, Boulder, Colorado.

Ellsaesser, H.W. (1984). The climatic effect of CO_2: A different view. *Atmos. Environ.* **18**, 431–434.

Emanuel, K.A. (1987). The dependence of hurricane intensity on climate. *Nature* **326**, 483–485.

Gleick, P.H. (1987). Regional hydrologic consequences of increases in atmospheric CO_2 and other trace gases. *Climatic Change* **10**, 137–160.

Hansen, J., Russell, G. Lacis, A. Fung, I. and Rind, D. (1985). Climate response times: dependence on climate sensitivity and ocean mixing. *Science* **229**, 857–859.

Intergovernmental Panel on Climate Change (IPCC) (1990). *Scientific Assessment of Climate Change*. Report prepared by Working Group 1. World Meteorological Organization, Geneva.

Jaeger, J. (1988). Developing Policies for Responding to Climatic Change: A Summary of the Discussions and Recommendations of the Workshops Held in Villach, 28 September to 2 October 1987 (WCIP-1, WMO/TD-No. 225).

Kutzbach, J.E. and Guetter, P.J. (1986). The influence of changing orbital parameters and surface boundary conditions on climate simulations for the past 18 000 years. *J. Atmos. Sci.* **43**, 1726–1759.

Land, K.C. and Schneider, S.H. (1987). Forecasting in the social and natural sciences: An overview and analysis of isomorphisms. *Climatic Change* **11**, 7–31.

Lashof, D.A. (1989). The dynamic greenhouse: feedback processes that may influence future concentrations of atmospheric trace gases and climatic change. *Climatic Change* **14**, 213–242.

Lindzen, R.S. (1990). Some coolness concerning global warming. *Bull. Amer. Meteor. Soc.* **77**, 288–299.

Lorius, C., Jouzel, J., Raynaud, D., Hansen, J. and Le Treut, H. (1990). The ice-core record: climate sensitivity and future greenhouse warming. *Nature* **347**, 139–145.

Manabe, S. and Stouffer, R.J. (1980). Sensitivity of a global climate model to an increase in CO_2 concentration in the atmosphere. *J. Geophys. Res.* **85**, 5529–5554.

Manabe, S. and Wetherald, R.T. (1967). Thermal equilibrium of the atmosphere with a given distribution of relative humidity. *J. Atmos. Sci.* **24**, 241–259.

Manabe, S. and Wetherald, R.T. (1986). Reduction in summer soil wetness induced by an increase in atmospheric carbon dioxide. *Science* **232**, 626–628.

Mearns, L.O., Katz, R.W. and Schneider, S.H. (1984). Changes in the probabilities of extreme high temeperature events with changes in global mean temperature. *J. Clim. Appl. Meteorol.* **23**, 1601–1613.

Mearns, L.O., Schneider, S.H. Thompson, S.L. and McDaniel, L.R. (1990). Analysis of climate variability in general circulation models: Comparison with observations and changes in variability in 2 × CO_2 experiments. *J. Geophys. Res.* **95**, 20–469.

Mitchell, J.F.B., Senior, C.A. and Ingram, W.J. (1989). CO_2 and climate: A missing feedback? *Nature* **341**, 132–134.

National Academy of Sciences (NAS) (1987). *Current Issues in Atmospheric Change*. National Academy Press, Washington, DC.

National Academy of Sciences (NAS) (1988). *Toward an Understanding of Global Change: Initial Priorities for US Contributions to the International Geosphere–Biosphere Programme*. National Academy Press, Washington, DC.

Parry, M.L. and Carter, T.R. (1985). The effect of climatic variations on agricultural risk. *Climatic Change* **7**, 95–110.

Prell, W.L. and Kutzbach, J.E. (1987). Monsoon variability over the past 150 000 years. *J. Geophys. Res.* **92**, 8411–8425.

Ramanathan, V. and Collins, W. (1991). Thermodynamic regulation of ocean warming by cirrus clouds deduced from observations of the 1987 El Niño. *Nature* **351**, 27–32.

Raval, A. and Ramanathan, V. (1989). Observational determination of the greenhouse effect. *Nature* **342**, 758.

Rind, D., and Peteet, D. (1986). Comments on S.H. Schneider's Editorial, "Can Modeling of the Ancient Past Verify Prediciton of Future Climates?". *Climatic Change* **9**, 357–360.

Rind, D., Goldberg, R. and Ruedy, R. (1989). Change in climate variability in the 21st century. *Climatic Change* **14**, 5–37.

Root, T. (1988a). Environmental factors associated with Avian Distributional Boundaries. *J. Biogeograph.* **15**, 489–505.

Root, T. (1988b). Energy constraints on avian distributions and abundances. *Ecology* **169**, 330–339.

Root, T. and Schneider, S.H. (submitted). Global climatic change and ecosystems impacts: A critical review. *Cons. Biol.*

Ruddiman, W.F. and McIntyre, A. (1981). The north Atlantic Ocean during the last deglaciation. *Paleogeogr., Paleoclim., Paleocol.* **35**, 145–214.

Schneider, S.H. (1972). Cloudiness as a global climate feedback mechanism: The effects on the radiation balance and surface temperature of variations in cloudiness. *J. Atmos. Sci.* **29**, 1413–1422.

Schneider, S.H. (1987). Climate modeling. *Sci. Amer.* **256** (5), 72–80.

Schneider, S.H. (1988). The whole earth dialogue. *Issues in Science and Technology* **IV** (3), 93–99.

Schneider, S.H. (1989). The changing climate. *Sci. Amer.* **260** (3), 70–79.

Schneider, S.H. (1990a). *Global Warming: Are We Entering the Greenhouse Century?* Vintage Books, New York.

Schneider, S.H. (1990b). *Global Warming: Are We Entering the Greenhouse Century?* Lutterworth Press, London.

Schneider, S.H. and Thompson, S.L. (1981). Atmospheric CO_2 and climate: Importance of the transient response. *J. Geophys. Res.* **86**, 3135–3147.

Schneider, S.H., Thompson, S.L. and Barron, E.J. (1985). Mid-Cretaceous, continental surface temperatures: Are high CO_2 concentrations needed to simulate above-freezing winter conditions? In: *The Carbon Cycle and Atmospheric CO_2: Natural Variations Archaen to Present* (Ed. by E. Sundquist and W. Broecker), Geophysical Monograph Series, **32**, 554–559. American Geophysical Union, Washington, DC.

Schneider, S.H., Peteet, D.M. and North, G.R. (1987). A climate model intercomparison for the Younger Dryas and its implications for paleoclimatic data collection. In: *Abrupt Climatic Change* (Ed. by W.H. Berger and L.D. Labeyrie), pp. 399–417. Reidel, Dordrecht.

Schneider, S.H., Gleick P.H. and L.O. Mearns (1990). Prospects for climate change. In: *Climate Change and US Water Resources* (Ed. by P.E. Waggoner), pp. 41–73. John Wiley, New York.

Stouffer, R.J., Manabe, S. and Bryan, K. (1989). Interhemispheric asymmetry in climate response to a gradual increase of atmospheric CO_2. *Nature* **342**, 660–662.

Washington, W.M. and Meehl, G.A. (1989). Climate sensitivity due to increased CO_2: Experiments with a coupled atmosphere and ocean general circulation model. *Climate Dynamics* **4**, 1–38.

Washington, W.M. and Parkinson, C.L. (1986). *An Introduction to Three-Dimensional Climate Modeling*. University Science Books, Mill Valley, CA and Oxford University Press, New York, NY.

Wigley, T.M.L. (1989). Possible climate change due to SO_2-derived cloud condensation nuclei. *Nature* **339**, 365–367.

The Development of Regional Climate Scenarios and the Ecological Impact of Greenhouse Gas Warming

C.M. GOODESS and J.P. PALUTIKOF

I. SUMMARY

In this chapter we assess the ability of climatologists to provide detailed projections of anthropogenically induced greenhouse warming suitable for use in ecological impact studies. Alternative approaches to the development of regional scenarios are reviewed. Despite their many uncertainties, general circulation models (GCMs) are considered to offer the greatest potential. The use of the present generation of GCMs, equilibrium models, in scenario development is discussed. Major uncertainties exist, particularly with regard to the pattern of climate change at the regional level. One major reason for these uncertainties is the influence of feedback effects, many of which involve interactions between climate and ecological processes. Model results from transient GCMs are just becoming available and these suggest that the

ADVANCES IN ECOLOGICAL RESEARCH VOL. 22
ISBN 0–12–013922–7

regional pattern of change differs markedly if time-dependent effects such as the thermal inertia of the ocean are taken into account. We conclude that none of the currently available models provide scenarios of sufficient reliability for use in quantitative regional impact studies. They may, however, be used to assess the sensitivity of ecosystems to climate change and in the development of impact assessment methodologies, with full awareness of their limitations.

II. THE GREENHOUSE EFFECT

Over the last decade, both the scientific and political communities have come to recognize that human activities have the potential to cause climate change. There is now broad agreement amongst scientists that the anthropogenically induced greenhouse warming effect is real (Slade, 1990) although, at the present time, we cannot claim that the greenhouse effect has been detected (Karl *et al.*, 1989; Wigley and Raper, 1990).

Natural greenhouse gases in the atmosphere are necessary to sustain life: they maintain the earth at a temperature around 33°C higher than it would be if they were not present. This is because they are largely transparent to shortwave solar radiation, but absorb longwave radiation from the earth. Due to human activities, concentrations of the greenhouse gases in the atmosphere are increasing. Atmospheric CO_2 concentrations have risen from around 280 parts per million by volume (p.p.m.v.) in pre-industrial times (around AD 1750), to about 353 p.p.m.v. today. Concentrations of methane, nitrous oxide and the chlorofluorocarbons are also rising. We speak of the "equivalent CO_2 concentration", this being the concentration of all the radiatively active trace gases in the atmosphere in equivalent concentrations (allowing for their differing radiative properties) of CO_2. It is estimated that CO_2 emissions account for 80% of the contribution to global warming of current greenhouse gas emissions (Lashof and Ahuja, 1990).

Concern about the potential impacts of greenhouse warming has given the impetus to a wide range of environmental and economic impact studies. However, such impact studies can only be carried out if detailed *regional* scenarios of the expected changes in climate are available. In this chapter, we assess the ability of climatologists to provide ecologists with reliable projections of climate change at the regional level for use in impact studies.

The range of information required by ecologists is wide. Basic climate parameters such as annual and seasonal mean temperature and precipitation are likely to be of only limited value. Information may be required on a daily or even an hourly time-scale and for a wide range of climate and climate-related variables such as maximum and minimum temperature, wind-speed, soil moisture, evapotranspiration, runoff, snow-fall and frost occurrence.

The spatial resolution of studies may typically be of the order of tens of kilometres or less.

III. APPROPRIATE METHODS OF REGIONAL SCENARIO DEVELOPMENT

In the opening chapter of this volume, Schneider describes and assesses the currently available computer-based models of the climate system. He concludes that, despite their many problems, general circulation models (GCMs) are the most valuable modelling tool for the development of regional scenarios for impact assessment. GCMs have indeed been widely used in early assessments of the agricultural, ecological and hydrological impacts of greenhouse warming (Warrick *et al.*, 1986a,b; Parry *et al.*, 1988a,b; Gleick, 1989). There are, however, two alternative approaches to scenario development (Wigley *et al.*, 1986; Lamb, 1987).

The simplest approach is to impose an arbitrary change in certain climate variables. It might, for example, be reasonable to assume a temperature increase of 2°C and a precipitation increase of 10% for northern Scotland in a world in which the greenhouse gas concentrations are double the pre-industrial level. On this basis, the ecological impacts on the region could be assessed. It is, however, difficult to identify internally consistent changes for a range of climate parameters. Therefore, we consider this approach to be of limited use.

The second approach involves the use of past periods of relatively warm climate as analogues of a future high greenhouse gas world (Webb and Wigley, 1985). This approach rests on the fundamental assumption that, given constant boundary conditions, the climate system responds in a similar way to different forcing factors. That is to say, provided such characteristics as the distribution of land and sea ice, and the patterns of ocean currents are the same, it is irrelevant whether the warming is caused by changes in greenhouse gas concentrations or, for example, the solar irradiance: the regional climate change patterns will be the same. Potential analogues can be taken from the instrumental record or, more commonly, from the historical and geological records of palaeoclimate (Wigley *et al.*, 1986).

The Holocene thermal maximum at about 6000 BP has been suggested as a suitable high greenhouse gas world analogue (Budyko *et al.*, 1987). The mean global temperature was then about 1°C warmer than today. However, as Schneider (this volume) notes, the major cause of this relative warmth was the seasonal redistribution of incoming solar radiation: summer radiation increased at the expense of winter radiation. This pattern of forcing is very different from that expected because of global increases in the atmospheric concentration of greenhouse gases. In fact, as we describe in Section V, GCM results indicate that warming will be strongest in the winter half year. Furthermore, it is likely that boundary conditions at 6000 BP were substantially different from those of the present day. Modelling studies indicate that the residual North American ice sheet, for example, reduced the expected warming in northwest Europe by 1–2°C (Mitchell *et al.*, 1988). Similar problems are associated with the use of the previous interglacial (the Eemian, ∼125 000 BP) as a potential analogue. Even if these problems could be resolved, the lack of regional detail in the available data severely limits the use of palaeoclimate analogues in ecological studies. Climate reconstructions based, for example, on pollen analyses are most reliable for temperature, a climatic variable which is spatially coherent. Reconstructions of precipitation, which lacks spatial coherence, tend to be highly uncertain, whilst no reconstructions are available for many of the climate-related parameters of interest to ecologists.

These data problems should not, however, arise if the instrumental record rather than the palaeoclimate record is used (Wigley *et al.*, 1986). A typical approach is that of Lough and colleagues (Lough *et al.*, 1983; Palutikof *et al.*, 1984) who compared the warmest (1934–1953) and coldest (1901–1929) 20-year periods in the record of mean land-based northern hemisphere surface temperatures (Jones *et al.*, 1982). An advantage of this method is that many different climate-related variables can be investigated. Lough *et al.* (1983) looked at various energy-related meteorological parameters, such as changes in the number of heating degree days. The same 20-year analogue periods were used in a study of cloud cover changes (Henderson-Sellers, 1986). A major disadvantage, however, concerns the relatively small temperature difference between warm and cold periods over the instrumental record. In the case of the Lough *et al.* study, this difference is only 0·4°C. Instrumental scenarios can, therefore, only be used as a guide to conditions during the early stages of greenhouse warming. In a later section of this chapter we discuss evidence suggesting that the pattern of change during the early stages of warming may differ from that of the final stages.

Given the limitations of these alternative approaches to regional scenario construction, we conclude that GCMs offer the greatest potential for ecological impact studies.

IV. THE RELIABILITY OF GENERAL CIRCULATION MODELS

The basic characteristics and the performance of GCMs are outlined by Schneider (this volume) and have been extensively reviewed elsewhere (e.g. see MacCracken and Luther, 1985; Schlesinger and Mitchell, 1987; Cubasch and Cess, 1990; Mitchell *et al.*, 1990).

GCMs and numerical weather prediction models grew from similar origins in the 1950s. The aim of GCMs is to simulate the full three-dimensional character of the climate, making them the most sophisticated existing models of the atmosphere. They solve the primitive equations that describe the movement of energy and momentum, and the conservation of mass and water vapour. Physical processes, such as cloud formation and heat and moisture transports within the atmosphere and between the atmosphere and the surface, are also described. Atmospheric conditions are specified at a number of "grid points" on a regular grid over the earth's surface, and at several levels in the atmosphere. The primitive equations are then solved at each grid point, using numerical techniques. Although various methods are available, all models use a time step approach and an interpolation scheme between grid points (Henderson-Sellers and Robinson, 1986).

The current generation of GCMs model the equilibrium rather than the transient response of climate to greenhouse gas forcing. For the control run, the atmospheric concentrations of CO_2[1] are set at pre-industrial levels and the model is then run until it reaches equilibrium. This procedure is repeated using doubled CO_2 concentrations, the perturbed run. The differences in the results for the perturbed and control runs are taken to be the climate change due to the greenhouse effect. In reality, atmospheric concentrations of the greenhouse gases are steadily increasing with time, and the atmospheric response lags behind, due to the oceans' thermal inertia. As discussed later in this chapter, under this circumstance of transient response, the patterns of climate change may be very different from those predicted by GCMs for the equilibrium response. Results from transient-response GCM runs are becoming available.

Five principal groups have developed GCMs. These are the UK Meteorological Office (UKMO), Goddard Institute of Space Studies (GISS), National Center for Atmospheric Research (NCAR), Geophysical Fluid Dynamics Laboratory (GFDL) and Oregon State University (OSU). There are many different versions of each model. Improvements in computing speed and power in recent years have allowed all groups to implement substantial improvements to their models. The key characteristics of the most recent models are discussed by Schneider (this volume) and are summarized in table form by Cubasch and Cess (1990, Table 3·2).

Despite the great advances which have been made in climate modelling, a number of major shortcomings remain. These are widely recognized by climatologists and are summarized in Table 1.

In comparison to the requirements of impact assessment, all models suffer from poor horizontal resolution. Typical resolutions range from 8° latitude × 10° longitude (GISS GCM, Hansen *et al.*, 1984) to 4° latitude × 5° longitude (OSU GCM, Schlesinger and Zhao, 1989). The higher horizontal resolution of the latter model is only achieved at the expense of a reduced

Table 1
Outstanding problems with equilibrium GCMs

Low spatial resolution
Poor representation of sub-grid scale processes
Lack of relief and unrealistic geography
Poor representation of feedback processes, particularly cloud feedbacks
Failure to reproduce details of present-day climate
Regional discrepancies between perturbed climate, particularly for parameters such
as precipitation and soil moisture
Do not fully take into account time-dependent effects such as thermal inertia of
the oceans or transient nature of greenhouse-gas forcing

---

vertical resolution (two rather than nine vertical layers). A limited number of high-resolution models have been run. The UKMO model, for example, has been run with a horizontal resolution of 2·5° latitude by 3·75° longitude and a vertical resolution of 11 layers (Mitchell *et al.*, 1989). Thus the horizontal resolution of GCMs is in the order of 300–1000 km.

The sensitivity of model processes to horizontal resolution can be demonstrated. In a recent study, for example, Rind (1988) found that doubling model resolution (to 4° latitude × 5° longitude) caused less overall convection in the control climate, leading to differences in the temperature and wind fields. However, the author concluded that the higher resolution brought "mixed benefits" and did not necessarily produce more reliable results.

Schneider (this volume) discusses the implications of poor spatial resolution for the representation of climate processes such as convective precipitation and cloud formation, and describes methods for the parameterization of such processes. None of these parameterizations can, however, be considered as ideal physical representations. In Section V of this chapter we assess the potential of methods to estimate sub-grid scale values from mean grid-point values.

Climatologists describe the geography of the present generation of GCMs as "realistic". However, to an ecologist, model geography must appear highly simplified. The Mediterranean and Hudson Bay, for example, are generally treated as closed lakes. The Panama Isthmus, together with islands such as Japan and New Zealand, are missing from some models. Orography is also highly simplified, restricted to major features such as the Alps, the Tibetan Plateau and the Rockies. In Section V we demonstrate that direct comparisons between the simplified GCM geography and real-world geography cannot be made.

The concept of climate feedback mechanisms is discussed by Schneider (this volume). A number of different feedback mechanisms can be identified which are likely to affect the response of climate to greenhouse-gas forcing (Table 2).

Table 2
Climate feedback effects and the response to greenhouse warming

Feedback	Direction	Reference
Clouds (high level)	Positive	Platt, 1989
		Ghan et al., 1990
Clouds (low level)	Negative	Mitchell et al., 1989
		Slingo, 1990
Clouds (net effect)	Positive or negative	Cess et al., 1989
Water vapour	Positive	Cess, 1989
Ice-albedo	Positive	Dickinson et al., 1987
		Ingram et al., 1989
Biogeochemical		Lashof, 1989
Ocean CO_2 uptake/release:	Positive	
Terrestrial biota:		
(albedo, carbon storage, CO_2		
fertilization):	Positive	
CH_4 emissions from		
wetlands and rice:	Positive	
Tropospheric chemistry:	Negative	

One of the most important feedbacks is the water vapour feedback mechanism, in which warming leads to enhanced evaporation and higher concentrations of atmospheric water vapour which, because water vapour is itself a greenhouse gas, in turn leads to further warming. This mechanism is relatively simple and can readily be parameterized in climate models (Cess, 1989; Mitchell et al., 1990). It is the water vapour feedback, rather than the direct effects of increased greenhouse gas concentrations, which is responsible for much of the warming observed in the perturbed run of GCMs. Land surface, soil moisture and biogeochemical feedbacks involve interactions between climate and ecological processes and are discussed in Section V of this chapter.

Cloud feedbacks may be both negative and positive (Fig. 1) and remain controversial (Mitchell et al., 1990). At the present day, the net global effect of cloud feedbacks is considered to be negative (Fig. 1; Ramanathan et al., 1989). However, the relative balance of the two competing mechanisms is likely to change in a high greenhouse gas world. Furthermore, the relative strength of the different cloud feedback effects is likely to show regional variation. Mitchell et al. (1989), for example, suggest that negative cloud feedbacks associated with low-level cloud will dominate at mid-latitudes. The model-based study of Wetherald and Manabe (1986) indicates a decrease in low-level cloud (negative feedback) and an increase in high-level cloud (positive feedback), particularly at high latitudes. Such varying responses will affect both the magnitude and pattern of climate change. Cloud feedbacks

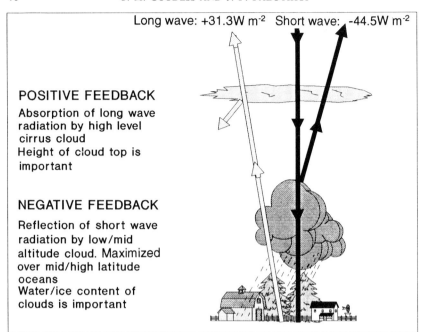

Fig. 1. Schematic representation of cloud feedback effects involving changes in the earth's radiation balance. Estimates of global net radiation from the Earth Radiation Budget Experiment (Ramanathan *et al.*, 1989) are given and indicate that, at the present time, the net effect of clouds is to cool the earth.

remain one of the greatest uncertainties in climate modelling and are considered to be responsible for much of the variation in climate sensitivity observed between different GCMs (Cess *et al.*, 1989).

One of the principal methods of model validation is to compare control run results with observational data sets. Because of the coarse resolution of GCMs, and because the output from the model runs is generated at each grid point only, it is not always appropriate to compare the results directly with station observations of climate variables. Gridded data sets have been produced from observations of a limited number of variables, for example, the combined land and sea temperature data set produced by Jones and colleagues (see Folland *et al.*, 1984; Jones *et al.*, 1985, 1986). The preparation of gridded data sets for parameters such as precipitation and wind-speed is constrained by the lack of comprehensive observations over the ocean. However, in some cases it may be possible to compare GCM output with data sets which are themselves generated by models. Palutikof *et al.* (1990), for example, compare near-surface wind speeds interpolated from GCM grid-point output with those estimated from a boundary-layer model using observed geostrophic winds.

Inter-model comparisons and comparisons with observations tend to be qualitative rather than quantitative in nature. Statistical methods are, however, being developed (Santer and Wigley, 1990; Wigley and Santer, 1990). The wider application of these methods should allow the *relative* reliability of individual models to be assessed.

Despite the technical problems of validation, as GCMs have improved so has their ability to reproduce the major features of the observed climate, such as the seasonal cycle. None the less, the models still persistently fail to reproduce many of the details of present-day climate. Many models, for example, perform particularly badly over Europe (Gates, 1985). Discrepancies in the predicted and observed strength and location of the Icelandic Low and Azores High are commonly observed (Santer, 1988). Northwest Europe, and the UK in particular, are very sensitive to atmospheric circulation changes related to shifts in the position and intensity of these systems.

Regional discrepancies between the perturbed runs of the different GCMs can be explained in part by unrealistic aspects in the control climates. This is a particular problem so far as variables such as precipitation and soil moisture are concerned (Mitchell and Warrilow, 1987). The GFDL model, for example, predicts a 30–50% reduction in summer soil moisture over western Europe, but this summer drying is not apparent in the NCAR model simulations (Manabe and Wetherald, 1987; Meehl and Washington, 1988). The differences in the perturbed runs of the two models can be directly linked to differences in the control runs. Model discrepancies also arise due to variations in the parameterization of climate and climate-related processes. Consensus between models tends to be particularly poor in sensitive regions such as northwest Europe where relatively minor circulation changes can have a major effect. In contrast, in some parts of the world such as the midwest of North America, GCMs may produce a consensus prediction.

The final outstanding problem with equilibrium-mode GCMs is quite fundamental. By their very nature, they do not fully take into account the thermal inertia of the oceans or their role as a carbon dioxide sink. In the real world, greenhouse gas emissions increase incrementally over time, they do not instantaneously double. As Schneider (this volume) notes, the actual global temperature increase at the expected time of CO_2 doubling will be less than the estimated equilibrium change. In a recent expert assessment of climate change, output from the perturbed $2 \times CO_2$ runs of a number of GCMs has been scaled down to correspond to a mean global warming of $1 \cdot 8°C$ (Mitchell *et al.*, 1990). Assuming a "business-as-usual" greenhouse gas emissions scenario, this is the actual warming estimated to occur by 2030, from a final $2 \times CO_2$ equilibrium warming of $2 \cdot 5°C$. However, as we discuss in Section VI, there is some evidence to suggest that both the magnitude and *pattern* of regional climate change simulated by transient models differ from those of equilibrium models.

Given all the uncertainties associated with equilibrium GCMs, can they possibly be of any use in climate change impact assessments? The answer is clearly yes, but only to a limited extent and with full awareness of their limitations. In the next section we assess the extent to which they can justifiably be used in the development of regional scenarios of greenhouse-gas induced warming.

V. EQUILIBRIUM CLIMATE CHANGE

A. Global and Large-scale Changes

It can be argued that the greatest confidence can be placed in those results which are consistently duplicated in all model studies. However, it is possible that all models may produce the same persistent errors. Duplication does not necessarily imply that a result is correct. Relative confidence in a particular result will, however, be increased if it can be both duplicated and explained by a plausible physical mechanism. Enhanced high-latitude warming, for example, can be explained by processes such as ice-albedo feedbacks. On this basis, the relative degree of confidence that can be placed in various aspects of GCM results can be assessed (Table 3). Similar assessments have been made by other groups (Mitchell et al., 1990; Schneider, this volume).

Schneider (this volume) discusses these consensus results and best estimates of various parameters in some detail. We will not repeat this discussion here but wish to emphasize the point about scale. We have reasonable confidence in global and large-scale estimates of temperature and, to a lesser extent, precipitation. However, the relative level of confidence decreases as one moves from the global to the regional scale and when parameters other than mean temperature and precipitation are considered.

B. Analysis of Grid-point Data

In order to demonstrate some of the problems which may arise in the use of grid-point data for the construction of regional scenarios we present seasonal values of temperature and precipitation for northwest Europe from the control and $2 \times CO_2$-perturbed runs of five GCMs (Table 4). As we have already stressed, there are many different versions of each model. The results we present here are not from the most recent versions and are used purely for illustrative purposes.

The location of the grid points for which data are presented are shown in Fig. 2. The area covered by these grid points is $50 \cdot 0$–$67 \cdot 5°N \times 11 \cdot 25°W$–$10 \cdot 0°E$. The GISS model has six grid points within this area, while the other

Table 3
GCM results from experiments for a doubling of atmospheric CO_2 concentration and an estimate of the relative confidence that can be placed in these results

Model results	Confidence
Global scale	
Warming of lower troposphere	High
Increased precipitation	High
Cooling of stratosphere	High
Zonal-regional scale	
Reduced sea ice	High
Enhanced NH polar warming (especially in winter half year)	High
Increased precipitation minus evaporation at high latitudes	High
More maximum temperature extremes	High
Increased continental summer dryness	Moderate
Stronger monsoon	Moderate
Regional pattern in detail	Unknown
Change in interannual variability	Unknown
Precipitation extremes	Unknown

NH = Northern hemisphere.

models all have nine points. The NCAR and GFDL models share the same network.

The $2 \times CO_2$-control run results for each model and for each season are shown in Figs 3 and 4. In these composite diagrams, the GFDL model, for example, has nine values plotted for each season, corresponding to the nine grid points in Fig. 2.

For temperature, it can be seen that the greatest increases are shown by the GFDL and NCAR models, particularly in the winter season (Fig. 3). For the whole year the increases are between 2·6°C and 7·9°C. The smallest temperature increases occur in the OSU GCM, with changes between 2·0°C and 4·0°C averaged over the whole year.

There does not appear to be any consistent relationship between predicted $2 \times CO_2$ temperature change over northwest Europe and either global $2 \times CO_2$ temperature change or mean global temperature in the control runs. It can be assumed that the global $2 \times CO_2$ temperature change reflects the sensitivity of a particular GCM to climate change (Cess and Potter, 1988). A model which predicts the greatest greenhouse warming is assumed to have the highest sensitivity and vice versa. The differences between the model results presented here for northwest Europe do not, however, appear to be directly attributable to differences in global model sensitivity.

Table 4

Key characteristics of the versions of the GCMs from which northwest Europe grid-point data are taken

	UKMO	GISS	NCAR	GFDL	OSU
Lat. × long.	5° × 7·5°	7·83° × 10°	4·5° × 7·5°	4·5° × 7·5°	4° × 5°
Vertical layers	11	9	9	9	2
Insolation	Annual and diurnal	Annual and diurnal	Annual	Annual	Annual
Ocean model	Prescribed heat exchange	Prescribed heat exchange, varying mixed layer	Mixed layer	Mixed layer	Six-layer ocean GCM
Global warming 2 × CO_2	5·2°C	4·2°C	3·5°C	4·0°C	2·8°C
Global precipitation change 2 × CO_2	+15%	+11%	+7·1%	+8·7%	+7·8%
For model description see:	Wilson and Mitchell, 1987	Hansen et al., 1984	Washington and Meehl, 1984	Wetherald and Manabe, 1986	Schlesinger and Zhao, 1989

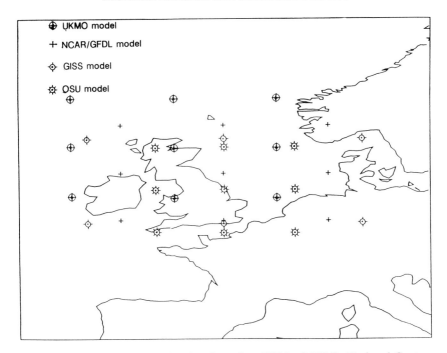

Fig. 2. Western European grid points from five GCMs. (NCAR: National Center for Atmospheric Research, GFDL: Geophysical Fluid Dynamics Laboratory, GISS: Goddard Institute for Space Studies, OSU: Oregon State University, UKMO: United Kingdom Meteorological Office.)

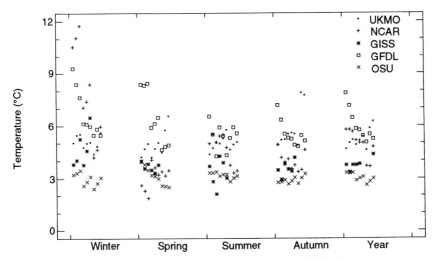

Fig. 3. 2 × CO_2 temperature increases estimated by five GCMs for the western European grid points shown in Fig. 2.

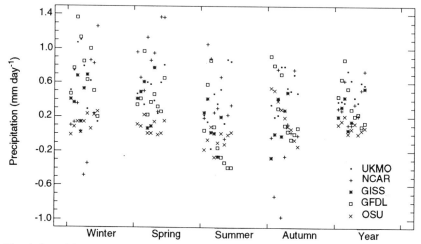

Fig. 4. 2 × CO_2 precipitation changes estimated by five GCMs for the western European grid points shown in Fig. 2.

The predictions for precipitation are much more variable (Fig. 4). All the models show lower rainfall in summer and autumn at some of the grid points, generally those located in the south or east of the region. The NCAR model, for example, shows a decrease in rainfall of as much as 1 mm day^{-1} in autumn at the grid point located close to the western Danish coast.

In Figs 5 and 6, we show longitudinal profiles taken at 55·55° N across three grid points of the GFDL and NCAR models (7·5°W, 0·0° and 7·5°E). The magnitude of the temperature changes predicted by the GFDL model is much greater than predicted by the NCAR model in all seasons except winter (Fig. 5). There are also some interesting differences in the seasonal distribution of change in the two models. The NCAR model predicts maximum change in the winter and summer, whereas in the GFDL model the warming in each season is similar.

In the case of precipitation, the changes predicted by the two models are entirely different (Fig. 6). The NCAR model predicts a decrease in precipitation in winter and autumn with a maximum increase in spring. The GFDL model predicts a small decrease in summer and an increase in the other seasons, peaking in the winter. The impacts of the climate changes predicted by each of these two models are likely to be very different.

Before the results from any GCM can be used in climate impact studies, the ability of each GCM to simulate present-day climate should be assessed. As an example of the problems which may be encountered, we consider the ability of one particular version of one GCM to simulate observed seasonal variations of temperature over the UK (Fig. 7). Here we compare GCM grid-point output with station data; a procedure not recommended in model

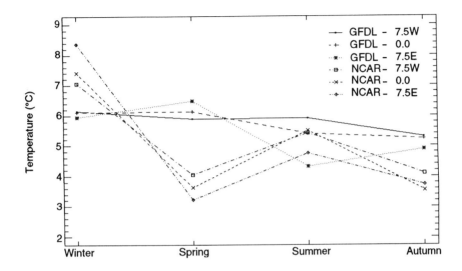

Fig. 5. $2 \times CO_2$ temperature increases estimated by the GFDL and NCAR GCMs for three grid points across the latitude band 55·5°N.

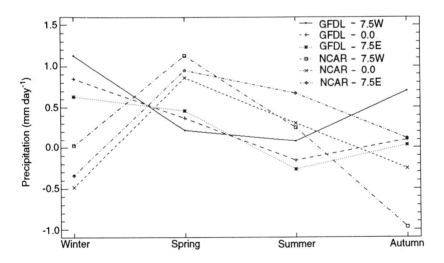

Fig. 6. $2 \times CO_2$ precipitation changes estimated by the GFDL and NCAR GCMs for three grid points across the latitude band 55·5°N.

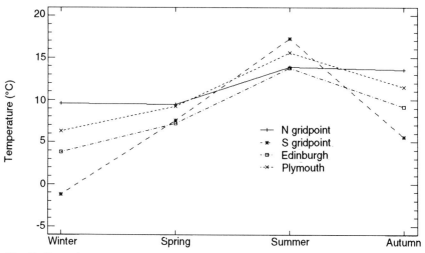

Fig. 7. Control run temperature output from the UKMO grid points shown in Fig. 8 compared with seasonal data for Edinburgh and Plymouth.

validation but legitimate when the suitability of GCMs for regional scenario construction is being assessed. The GCM control run data are from a version of the UKMO GCM and are compared with instrumental data from two stations representing the north–south/east–west range of UK climate and are shown in Fig. 8. The chosen grid points are located at 57·5°N 3·75°E and at 52·5°N 3·75°E. It should be stressed that this model version is typical of its generation. Its performance is considered to be no worse, and no better, than that of other contemporary models.

Neither grid point provides a reliable estimate of present-day UK conditions. This failure and the differences in the seasonal cycle at the two grid points can, in part, be attributed to the fact that the northernmost grid point represents a grid square which is treated by the model as an ocean box whilst the southernmost square is treated as a land box. The seasonal cycle of temperature is, therefore, too weak in the northern box and too strong in the southern box. Clearly the simplified model geography does not match real-world geography.

The examples we have presented in this section demonstrate that it would be unwise to produce a regional estimate of change simply by taking model output from the grid points located closest to the region of interest. A number of questions must be addressed. Which model, for example, is most reliable? Which grid points are most representative of the region of interest? Can sub-grid scale values be interpolated? There are no easy answers to any of these questions but in the next section we discuss appropriate methodologies for investigating such issues.

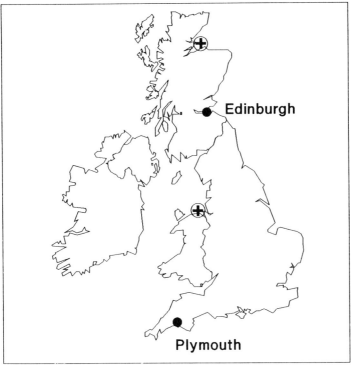

Fig. 8. Location of the two UKMO GCM grid points and the two observing stations for which seasonal data are plotted in Fig. 7.

C. Regional Scenario Development

At the present time, there are no a priori reasons to assume that any one GCM is better than any other. It is possible, however, that comparison of control run output and observational data may identify one model which performs best over a particular region. As numerical techniques for spatial validation become more widely available (Santer and Wigley, 1990; Wigley and Santer, 1990) this approach may become more practical.

If we consider the results from all available models, then at least we are aware of the range of variations in model estimates. The International Panel on Climate Change (IPCC) recently produced best estimates of changes in mean surface temperature, precipitation and soil moisture over five selected regions (Mitchell *et al.*, 1990) from three high-resolution equilibrium GCMs. As explained in Section IV of this chapter, these results, which are reproduced in Table 5, were adjusted to represent the expected global warming at 2030 assuming a "business-as-usual" scenario for greenhouse gas emissions.

Table 5

Estimates of changes in areal means of surface temperature, precipitation and soil moisture over selected regions, from pre-industrial times to 2030, assuming the IPCC "business-as-usual" scenario. Estimates are from three high-resolution GCMs—CCC: Canadian Climate Center; GFDL: Geophysical Fluid Dynamics Laboratory; UKMO: UK Meteorological Office. See text for discussion. (Source: Mitchell *et al.*, 1990.)

Region	Model	Temperature (°C)		Precipitation (%)		Soil moisture (%)	
		DJF	JJA	DJF	JJA	DJF	JJA
Central N.	CCC	4	2	0	−5	−10	−15
America	GFDL	2	2	15	−5	15	−15
(35–50N,	UKMO	4	3	10	−10	−10	−20
80–105W)							
SE Asia	CCC	1	1	−5	5	0	5
(5–30N,	GFDL	2	1	0	10	−5	10
70–105E)	UKMO	2	2	−15	15	0	5
Sahel	CCC	2	3	−10	5	0	−5
(10–20N,	GFDL	2	1	−5	5	5	0
20W–40E)	UKMO	1	2	0	0	10	−10
Southern	CCC	2	2	5	−15	0	−15
Europe	GFDL	2	2	10	−5	5	−15
(35–50N,	UKMO	2	3	0	−15	−5	−25
10W–45E)							
Australia	CCC	1	2	15	0	45	5
(12–45S,	GFDL	2	2	5	0	−5	−10
110–155E)	UKMO	2	2	10	0	5	0

We can see that over southern Europe in summer, for example, temperature is estimated to warm by 2–3°C, precipitation to decrease by 5–15% and soil moisture to decrease by 15–25%. It should be noted that these results apply only to each region as a whole and not to subregions.

In order to gauge the reliability of these regional estimates the IPCC expert group assumes that the range of best estimates of global mean warming for a CO_2 doubling (1·5–4·5°C) also represents the range of uncertainty associated with regional estimates of all climate variables (Mitchell *et al.*, 1990). The 1·5–4·5°C range implies that the regional estimates may actually lie within 70–145% of the given values. The group concludes that "confidence in these estimates is low" (Mitchell *et al.*, 1990).

More sophisticated methods for synthesizing grid-point results from GCMs into regional climate change scenarios are being developed. Wigley *et al.* (1992), for example, have developed a statistical method for averaging standardized patterns of temperature change from five different models to produce estimates which are not biased by the individual model sensitivities. This approach alone is not appropriate for precipitation because of the poor agreement between the different models, particularly in the tropics (30°S–30°N). For precipitation, therefore, both the probability of a change in precipitation in a specific direction and the percentage change are estimated (Wigley et al., 1992).

The methods of scenario development described above are all constrained by the spatial resolution and the particular grid-point mesh of GCMs. The control run data from the two grid points presented in Fig. 7, for example, are unrepresentative of UK geography and climate. In this particular case, however, it would be possible to extrapolate between the two grid points (one land-based and one ocean-based) to obtain a more realistic estimate of UK conditions. Given the location of the UK on the west coast of a continent such an extrapolation has some physical justification. This general concept has been employed in the development of numerical techniques for the interpolation of sub-grid scale values from GCM grid-point data (Kim et al., 1984; Wilks, 1989; Wigley et al., 1990).

The need to interpolate sub-grid scale values from coarse resolution grids (or point values from area means) is often referred to as the "climate inversion" problem (Gates, 1985) and has been tackled by two groups of scientists using present-day climate data. The first statistical techniques to be proposed were developed by Kim et al. (1984) using climate data from Oregon in the US. The method is based on principal component and regression analysis of seasonal temperature and precipitation observations.

Using an observational data set for a range of climate variables, regression equations are constructed in which station temperature and precipitation variations are explained by regionally averaged variations in, for example, temperature, precipitation and surface pressure. Then, the regionally averaged variations are replaced by GCM $2 \times CO_2 - 1 \times CO_2$ differences, averaged from a number of grid points. The predicted station values can be used to produce a contoured map showing sub-grid scale climate perturbations.

The initial Kim et al. analysis only utilized spatial variations in the amplitude of seasonal cycles in the regression equations and is, furthermore, considered inadequate for the production of physically reasonable results for precipitation (Wilks, 1989; Wigley et al., 1990).

Wilks (1989) has extended the method by the use of daily maximum and minimum temperature and precipitation data, by the introduction of a stochastic precipitation model and by the treatment of data intercorrelations. In this study, data are taken from three regions of Kansas, Iowa and Dakota, equivalent to three grid squares of the NCAR GCM. The improved method is based on rotated principal component analysis of the correlation matrix of observed station data with area mean data. The stochastic model allows the observed locations and intensities of convective precipitation (a sub-grid scale process) to be simulated as random phenomena. This method does, however, require a dense network of precipitation data to match the spatial scale of convective precipitation. Wilks considers it unwise to interpolate to points not used in developing the correlation matrix.

A somewhat different approach has been adopted by Wigley et al. (1990). This particular method was also developed using data from Oregon and relies on interannual climate variations. Area averaged variations in temperature, precipitation, mean sea level pressure, 700 mb height pressure and the zonal and meridional pressure gradients across the region are used to predict sub-grid scale changes in monthly temperature and precipitation in a high greenhouse gas world.

Whilst both techniques (Wilks, 1989; Wigley et al., 1990) produce useful results, a question-mark concerns their transferability. These techniques have been developed for regions of the US where the local climate pattern is determined by topographic/thermal forcing factors which are themselves relatively simple in form. In Oregon, for example, the coastline and the principal mountain ranges run north–south. The same forcing factors in other regions, such as the UK, may be much more complex. In particular, meteorological data from a very dense network of stations may be required in order to capture the spatial complexity of both the forcing factors and the climate itself.

The studies described above indicate that, in some parts of the world at least, it is possible to establish meaningful relationships between present-day area mean values and local station values of climate variables. In order to use these techniques in studies of climate change it is necessary to assume that these relationships will be unchanged. This is not an easy assumption to test. Wigley et al. (1990) do, however, put forward two arguments in support of this assumption. First, they say, the estimated mean values of climate variables such as temperature and precipitation in a high greenhouse gas world lie within the range of present-day extremes. Therefore, the statistical relationships developed on the basis of the present-day data can be applied to the investigation of greenhouse-gas induced changes without extrapolating beyond the bounds of the data used in the construction of the regression model. Second, they argue that the relationships must reflect the influences of orography, geography and land surface characteristics and that these are unlikely to change. It is certainly reasonable to assume that no major orographic or geographical changes will occur over the time-scale of greenhouse warming. Some minor changes in geography may occur due to coastal flooding as sea level rises. It is, however, less reasonable to assume that land surface characteristics will not change. We return to this issue in a later section of this chapter.

So far we have restricted our discussion to the parameters most commonly recorded during GCM experiments: temperature, precipitation and soil moisture. Estimates of other parameters can sometimes be obtained from modelling groups. If results are unobtainable, or are considered unreliable, then it may be possible to estimate particular parameters from those parameters which are available. A numerical boundary layer model, for example, has been used to estimate regional-scale surface wind speed variations in the UK from grid-point vector mean wind speeds predicted by a GCM (Palutikof *et al.*, 1990).

This brief discussion of regional scenario construction demonstrates the recent great advances in the development of methodologies for using GCM data in regional impact assessments. At the present time, however, all of these methodologies are constrained by the reliability of the current generation of equilibrium GCMs. Without major improvements in modelling, these methodologies can only be used to explore the sensitivity of the ecological system to climate change. Reliable, quantitative regional predictions are still beyond our grasp.

D. Interactions of Climate and Ecological Processes

One of the greatest uncertainties of the present generation of GCMs concerns the representation of feedback effects. The problems associated with cloud feedbacks were outlined in Section IV of this chapter. It has recently been suggested that a negative biological feedback mechanism involving planktonic algae and cloud albedo could help to offset greenhouse warming (Charlson *et al.*, 1987). The proposed mechanism involves variations in plankton productivity which, it is suggested, affect the production of dimethylsulphide (DMS), believed to be a major source of cloud condensation nuclei (CCN) over the oceans. DMS production is highest over the warmest and most saline oceans. So, it is argued, greenhouse warming will increase planktonic productivity and DMS production. Increased DMS production over the oceans means that more CCN will be produced. The increased CCN concentrations will in turn lead to an increase in cloud albedo, thus more radiation will be reflected back to space so offsetting the greenhouse effect (negative feedback). This theory is controversial (Schwartz, 1988) and more observational data are needed before it can be proved or disproved.

There are, however, many more certain examples of the interaction between climate and ecological processes which have potential implications for climate change. We know that changes in both flora and fauna can affect the rate of emission of greenhouse gases and their uptake. It has, for example, been suggested that methane emission rates from wetlands and rice paddy fields will increase due to greenhouse warming (Lashof, 1989). The role of the oceans in the global carbon cycle is currently the focus of intensive research. Emerging modelling results suggest that the oceans' biological carbon pump may play an important role in the response to climate change (Heinze and Maier-Reimer, 1989).

Recent modelling work also demonstrates the importance of land surface parametrizations in GCMs (Lean and Warrilow, 1989; Warrilow and Buckley, 1989). Variations in surface albedo and roughness, for example, influence the energy and moisture exchanges between the surface and the atmosphere (Warrilow and Buckley, 1989). The interactions between ecological and climate processes are complex. Despite recent advances, many of the crucial processes are still poorly parameterized in GCMs. The typical soil moisture scheme, for example, consists of a 15-mm "bucket" which fills or empties according to a highly idealized relationship between precipitation, evaporation and runoff (Warrilow and Buckley, 1989).

In the early GCMs all the land surface parameterizations were relatively simple and most surface parameters lacked geographical variation. More recently, geographic variations of important parameters such as albedo, roughness length and root depth have been introduced, together with more sophisticated parameterization schemes (Dickinson, 1984; Sellers *et al.*, 1986; Rowntree, 1988).

Albedo is now specified for individual vegetation types (Wilson and Henderson-Sellers, 1985; Dickinson *et al.*, 1986). Models also allow for the effects of snow on the albedo of vegetation, although the sophistication of schemes used varies. The availability of such vegetation data sets also allows the specification of appropriate roughness lengths for each vegetation type. In the earliest models, the interception of rainfall by vegetation was not included, whereas all models now include some form of interception capacity. Evaporation processes within GCMs are now explicitly controlled by soil and plant processes, including the role of foliage and vegetation canopy (Rowntree, 1988). The biosphere-atmosphere transfer scheme or BATS (Dickinson, 1984; Dickinson *et al.*, 1986) and simple biosphere model or SiB (Sellers *et al.*, 1986) are typical of the schemes currently employed in GCMs.

In the BATS, evaporation occurs in the wet parts of the canopy and from the soil, while transpiration occurs in the dry parts of the canopy. Stomatal resistance in the BATS is dependent on a prescribed minimum stomatal resistance factor for each vegetation type, solar radiation, temperature and the area of the transpiring surface. The SiB scheme is based on similar principles but is more complex. Transpiration, for example, is also dependent on vapour pressure deficit.

Despite the recent advances, land surface parameterizations are still highly simplified. Although, for example, some parameters now vary with vegetation type, in effect a single value for each parameter is assumed for each grid square (Rowntree, 1988). Further improvements to parameterizations which incorporate more realistic physical mechanisms and sub-grid scale variability are, however, being developed (Entekhaki and Eagleson, 1989;

Warrilow and Buckley, 1989). Improved observational/empirical data-sets of parameters such as snow albedo, surface roughness, canopy moisture capacity and soil moisture capacity and their variability are also becoming available (Warrilow and Buckley, 1989).

We also await the development of GCMs in which the vegetation cover and related parameters change in response to climate change. The importance of these interactions is demonstrated by a GCM simulation of the regional climatic impact of Amazon deforestation (Lean and Warrilow, 1989). A high resolution version of the UKMO GCM (2·5° latitude by 3·75° longitude) was used to explore the effect of replacing all Amazon tropical forests with savannah and pasture (Lean and Warrilow, 1989). Improvements were first made to the model's land surface hydrological scheme. A four-layer soil temperature scheme was used together with a parameterization of precipitation interception by the vegetation canopy. The latter scheme, however, was found to overestimate canopy evaporation in the control run. The spatial variability of present-day land surface and soil types was used in the control run. Areas of tropical forest were replaced by savannah and pasture in the perturbed run.

The local climate response to deforestation was dominated by a weakened hydrological cycle and by an increase in surface temperature. Evaporation was found to decrease, largely due to the reduction in surface roughness. The increase in surface albedo was considered to be the principal cause of reduced moisture flux and of a 20% decrease in precipitation. These results are qualitatively similar to those of other studies (Dickinson and Henderson-Sellers, 1988; Mitchell et al., 1990; Shukla et al., 1990). The consensus view is that the direct effects of tropical deforestation on regional climate may be large but that the impact on global climate will be relatively small. In a recent expert assessment, for example, it is argued that removal of all the tropical forests could only warm the global climate by about 0·3°C (Mitchell et al., 1990).

The above discussion illustrates the importance of interactions between climate and ecological processes. This is an area in which the performance of GCMs requires improvement, although this can only be achieved through interdisciplinary research.

So far we have only considered model simulations of the equilibrium response of climate to increased atmospheric CO_2. We now consider the transient response.

VI. TRANSIENT CLIMATE CHANGE

Equilibrium models are unrealistic in terms of their representation of greenhouse gas forcing. Atmospheric concentrations of greenhouse gases will not instantaneously double but are increasing at a steady rate, currently estimated to be about 0·4% per annum (Bretherton et al., 1990). The climate response to this transient forcing will be delayed by several decades because of ocean heat storage. Therefore, in order to model the transient response, ocean heat transport and dynamics must be incorporated into models.

Relatively simple energy balance and box-diffusion type models can be used to model the global and zonal transient response to greenhouse forcing (Bretherton et al., 1990). However, in order to investigate the transient regional response, coupled atmosphere/ocean (A/O) GCMs must be developed. Coupled A/O GCMs have very large computing requirements and to date only a few fully coupled models have been run (Harvey, 1989; Bretherton et al., 1990).

There are a number of technical problems associated with the use of transient GCMs. Many of these relate to differences in the response time between the atmosphere (rapid) and the oceans (very slow). When attempting to couple atmosphere and ocean GCMs there is a tendency for the models to "drift" out of radiative balance. Validation of coupled models is complicated by the lack of observations over the oceans. Where data are available, comparison reveals substantial control run errors. Washington and Meehl (1989) found, for example, that many ocean models display errors in the present-day sea surface temperature simulations. Coupled A/O GCMs are at an early stage of development and confidence in their results remains low. It is, however, interesting to compare the response of these models to greenhouse gas forcing with that of equilibrium-mode models.

If we were to compare $2 \times CO_2$ results from an equilibrium GCM with results from a transient model for the year in which the model estimates CO_2 will double, the magnitude of change should always be lower in the transient model. The direction of change should, however, be the same in both transient and equilibrium models: temperature and precipitation both increasing with greenhouse warming. Although the large-scale pattern of change in both types of model should be broadly similar, major differences may occur. In Section V we concluded that, since there is a plausible explanation, we can be relatively confident in the GCM result that shows enhanced greenhouse warming at higher latitudes. This feature is not, however, seen in recent A/O GCM experiments (Stouffer et al. 1989; Washington and Meehl, 1989). Stouffer et al. (1989) investigated the response of the GFDL atmospheric GCM coupled with a dynamic ocean model to a CO_2 increase of 1% year^{-1} over a 100-year period. This forcing is equivalent to a doubling of CO_2 by 2030. A marked asymmetry was found in the response of the two hemispheres. Warming reduced towards high latitudes in the southern hemisphere and was very slow in the circumpolar oceans of the southern hemisphere. These results reflect the large thermal inertia and the vast extent of the southern hemisphere oceans. In the northern hemisphere warming increased with latitude, with the notable exception of the northern North Atlantic. In this region warming was slow and, by the seventh decade of warming, a 25% reduction in the strength of the ocean thermohaline circulation was observed. The North Atlantic changes are considered to be related to feedback mechanisms involving evaporation, salinity and thermal advection (Stouffer *et al.*, 1989).

Similar results for the southern hemisphere were found by Washington and Meehl (1989) using the NCAR GCM. In this study, however, minimum warming over high latitude oceans occurred in both hemispheres. Again this effect was associated with a reduction in the strength of the thermohaline circulation. With transient CO_2 forcing (increasing at 1% year^{-1}) a net cooling was observed in the zone 60–80° N (Washington and Meehl, 1989).

Although confidence in the regional pattern of change predicted by these transient models is low, these results are disquieting. It seems possible that the equilibrium GCM results in which we have greatest relative confidence, such as enhanced high-latitude warming, may be misleading.

The first results from transient models are also interesting in what they can tell us about potential rates of climate change and the patterns of the time-dependent changes. It seems possible, for example, that the regional pattern of change in the early stages of greenhouse forcing may differ markedly from that during the later stages of warming (Washington and Meehl, 1989). There is also evidence to suggest that the rate of response to a doubling of CO_2 may be fastest over the earlier years (Washington and Meehl, 1989).

The expected rate of climate change is important when considering the ecological response. The IPCC recently assessed estimated rates of warming assuming a "business-as-usual" emissions scenario (Bretherton *et al.*, 1990). The estimates come from a global energy-balance upwelling diffusion model, not from a GCM. Over the period 1990–2030 the estimated warming is 0·7–1·5° C (best estimate 1·1° C) and over the period 1990–2070 it is 1·6–3·5° C (best estimate 2·4° C). Depending on the actual rate of warming, vegetation changes may lag (Davis, 1989) or keep pace with (Overpeck *et al.*, 1990) climate change. The rate of vegetation change itself potentially has implications for feedbacks and processes involving land surface parameterizations in GCMs (see Section V).

The results emerging from this new generation of models raise many complex and intriguing questions about the nature of climate change which are of direct interest to ecological impact studies.

VII. CONCLUSIONS

Despite their many uncertainties and known errors, GCMs are considered to offer the greatest potential for developing regional scenarios of climate change for use in ecological impact studies. The present generation of

GCMs, run in equilibrium mode, has been widely used in the construction of high-resolution regional scenarios. Increasingly sophisticated methodologies are being developed to overcome problems associated with the low spatial resolution of the GCM grid-point output.

More fundamental problems remain related to the validity of the model results themselves. Comparisons of control run results with present-day data suggest that GCMs are not always able to reproduce the present-day features of the global climate. Whereas considerable improvement has been achieved by increasing model resolution, doubts still remain concerning the representation of feedback effects, many of which involve complex interactions between climate and ecological processes. Until substantial advances in modelling occur, none of the available GCMs can provide reliable high resolution scenarios suitable for use in quantitative regional impact studies. They may, however, be used to assess the sensitivity of ecosystems to climate change and in the development of impact assessment methodologies, with full awareness of their limitations.

Fully coupled atmospheric/ocean GCMs which incorporate the transient nature of greenhouse gas forcing and the associated warming are just becoming available. The results suggest that the regional pattern of change differs markedly if time-dependent effects such as the thermal inertia of the ocean and the transient nature of the forcing are taken into account. The technical problems associated with these models are, however, great and they are not yet considered sufficiently reliable for use in the construction of regional climate change scenarios.

This chapter has dealt with the standard climatological parameters: temperature, precipitation, radiation, soil moisture and so on. We have not considered other aspects of a warming climate, such as rising sea levels. These may also have a major influence on the ecological response to greenhouse warming. It is also possible that other forms of pollution, such as acid rain and stratospheric ozone depletion by chlorofluorocarbons, may affect the way in which ecosystems respond.

It is unlikely that climatologists will ever be able to provide highly reliable and detailed predictions of future climate change over spatial scales of hundreds to tens of kilometres or less. We are, however, confident that climatologists can provide valuable information for assessing the sensitivity of ecological processes and systems to climate change.

NOTES

[1]In the discussion of climate models in this chapter, references to $2 \times CO_2$ and $1 \times CO_2$ can generally be taken to refer to equivalent CO_2 concentrations.

REFERENCES

Bretherton, F., Bryan, K. and Woods, J. (1990). Time-dependent greenhouse-gas-induced climate change. In: *Climate Change, the IPCC Scientific Assessment* (Ed. by J.T. Houghton, G.F. Jenkins and J.J. Ephraums), pp. 173–194. WMO/UNEP, Cambridge University Press, Cambridge.

Budyko, M.I., Ronov, A.B. and Yanshin, A.L. (1987). *History of the Earth's Atmosphere.* (English translation.) Springer-Verlag, Berlin.

Cess, R.D. (1989). Gauging water vapour feedback. *Nature* **342**, 736–737.

Cess, R.D. and Potter, G.L. (1988). A methodology for understanding and intercomparing atmospheric climate feedback processes in general circulation models. *J. Geophy. Res.* **93**, 8305–8314.

Cess, R.D., Potter, G.L., Blanchet, J.P., Boer, G.J., Ghan, S.J., Kiehl, J.T. *et al.* (1989). Interpretation of cloud climate feedback as produced by 14 atmospheric general circulation models. *Science* **245**, 513–516.

Charlson, R.J., Lovelock, J.E., Andreae, M.O. and Warren, S.G. (1987). Oceanic phytoplankton, atmospheric sulphur, cloud albedo and climate. *Nature* **326**, 655–661.

Cubasch, U. and Cess, R. (1990). Processes and modelling. In: *Climate Change, the IPCC Scientific Assessment* (Ed. by J.T. Houghton, G.J. Jenkins and J.J. Ephraums), pp. 69–92. WMO/UNEP, Cambridge University Press, Cambridge.

Davis, M.B. (1989). Lags in vegetation response to greenhouse warming. *Clim. Change* **15**, 75–82.

Dickinson, R.E. (1984). Modelling evapotranspiration for three-dimensional global climate models. *Geophys. Mono.* **29**, American Geophysical Union.

Dickinson, R.E. and Henderson-Sellers, A. (1988). Modelling tropical deforestation: a study of GCM land-surface parametrizations. *Quat. J. R. Met. Soc.* **114**, 439–462.

Dickinson, R.E., Henderson-Sellers, A., Kennedy, P.J. and Wilson, M.F. (1986). Biosphere–Atmosphere Transfer Scheme (BATS) for the NCAR Community Climate Model. NCAR Technical Note NCAR/TN-275+STR.

Dickinson, R.E., Meehl, G.A. and Washington, W.M. (1987). Ice-albedo feedback in a CO_2-doubling simulation. *Clim. Change* **10**, 241–248.

Entekhabi, D. and Eagleson, P.S. (1989). Land surface hydrology parameterization for atmospheric general circulation models including subgrid scale spatial variability. *J. Climate* **2**, 816–831.

Folland, C.K., Parker, D.E. and Kates, F.E. (1984). Worldwide marine temperature fluctuations 1856–1981. *Nature* **310**, 670–673.

Gates, W.L. (1985). The use of general circulation models in the analysis of ecosystem impacts of climatic change. *Clim. Change* **7**, 267–284.

Ghan, S.J., Taylor, K.E., Penner, J.E. and Erikson, D.J. (1990). Model test of CCN-cloud albedo climate forcing. *Geophys. Res. Lett.* **17**, 607–610.

Gleick, P.H. (1989). Climate change, hydrology, and water resources. *Rev. Geophys.* **27**, 329–344.

Hansen, J.E., Lacis, A., Rind, D., Russell, L., Stone, P., Fung, I., Ruedy, R. and Lerner, J. (1984). Climate sensitivity analysis of feedback mechanisms. In: *Climate Processes and Climate Sensitivity.* Geophys. Monogr. Ser. 29, pp. 130–163. AGU, Washington, DC.

Harvey, L.D.D. (1989). Transient climatic response to an increase of greenhouse gases. *Clim. Change* **15**, 15–30.

Heinze, C. and Maier-Reimer, E. (1989). Glacial pCO_2 reduction by the world ocean experiments with the Hamburg carbon cycle model. In: *3. World Conference on Analysis and Evaluation of Atmospheric CO₂ Data, Present and Past*. WMO Environmental Pollution Monitoring and Research Program, Rep. No. 59, pp. 9–14.

Henderson-Sellers, A. (1986). Cloud changes in a warmer Europe. *Clim. Change* **8**, 25–52.

Henderson-Sellers, A. and Robinson, P.J. (1986). *Contemporary Climatology*. Longman Scientific and Technical, London.

Ingram, W.J., Wilson, C.A. and Mitchell, J.F.B. (1989). Modelling climate change: an assessment of sea-ice and surface albedo feedbacks. *J. Geophy. Res.* **94**, 8609–8622.

Jones, P.D., Wigley, T.M.L. and Kelly, P.M. (1982). Variations in surface air temperature: Part 1. Northern Hemisphere, 1881–1980. *Mon. Wea. Rev.* **112**, 59–70.

Jones, P.D., Raper, S.C.B., Santer, B., Cherry, B.S.G., Goodess, C.M., Kelly, P.M. et al. (1985). *A Grid Point Surface Air Temperature Data Set for the Northern Hemisphere*. DoE/EV/10098–2, US Dept of Energy, Office of Energy Research, Carbon Dioxide Research Division, Washington, DC.

Jones, P.D., Wigley, T.M.L. and Wright, P.B. (1986). Global temperature variations between 1861 and 1984. *Nature* **322**, 430–434.

Karl, T.R., Tarpley, J.D., Quayle, R.G., Diaz, H.F., Robinson, D.A. and Bradley, R.S. (1989). The recent climate record: what it can and cannot tell us. *Rev. Geophys.* **27**, 405–430.

Kim, J.W., Chang, J.-T., Baker, N.L., Wilks, D.S. and Gates, W.L. (1984). The statistical problem of climate inversion: determination of the relationship between local and large-scale climate. *Mon. Wea. Rev.* **112**, 2069–2077.

Lamb, P.J. (1987). On the development of regional climatic scenarios for policy-oriented climatic-impact assessment. *Bull. Am. Met. Soc.* **68**, 1116–1123.

Lashof, D.A. (1989). The dynamic greenhouse: feedback processes that may influence future concentrations of atmospheric trace gases and climatic change. *Clim. Change* **14**, 213–242.

Lashof, D.A. and Ahuja, D.R. (1990). Relative contributions of greenhouse gas emissions to global warming. *Nature* **344**, 529–531.

Lean, J. and Warrilow, D.A. (1989). Simulation of the regional climatic impact of Amazon deforestation. *Nature* **342**, 411–413.

Lough, J.M., Wigley, T.M.L. and Palutikof, J.P. (1983). Climate impact scenarios for Europe in a warmer world. *J. Clim. App. Met.* **22**, 1673–1684.

MacCracken, M.C. and Luther, F.M. (Eds) (1985). *Projecting the Climatic Effects of Increasing Carbon Dioxide*. DoE/ER-0237, US Dept of Energy, Office of Energy Research, Carbon Dioxide Research Division, Washington, DC.

Manabe, S. and Wetherald, R.T. (1987). Large-scale changes of soil wetness induced by an increase in the atmospheric carbon dioxide. *J. Atmos. Sci.* **44**, 1211–1235.

Meehl, G.A. and Washington, W.M. (1988). A comparison of soil-moisture sensitivity in two global climate models. *J. Atmos. Sci.* **45**, 1476–1492.

Mitchell, J.F.B. and Warrilow, D.A. (1987). Summer dryness in northern mid-latitudes due to increased CO₂. *Nature* **330**, 238–240.

Mitchell, J.F.B., Grahame, N.S. and Needham, K.J. (1988). Climate simulations for 9000 years before present: seasonal variations and effect of the Laurentide ice sheet. *J. Geophys. Res.* **93**, 8283–8304.

Mitchell, J.F.B., Senior, C.A. and Ingram, W.J. (1989). CO₂ and climate: a missing feedback? *Nature* **341**, 132–134.

Mitchell, J.F.B., Manabe, S., Meleshko, V. and Tokioka, T. (1990). Equilibrium climate change — and its implications for the future. In: *Climate Change, the IPCC*

Scientific Assessment (Ed. by J.T. Houghton, G.J. Jenkins and J.J. Ephraums), pp. 131–172. WMO/UNEP, Cambridge University Press, Cambridge.

Overpeck, J.T., Rind, D. and Goldberg, R. (1990). Climate-induced changes in forest disturbance and vegetation. *Nature* **343**, 51–53.

Palutikof, J.P., Wigley, T.M.L. and Lough, J.M. (1984). *Seasonal Climate Scenarios for Europe and North America in a High-CO_2, Warmer World.* DoE/EV/10098–5, TR012, US Dept of Energy, Office of Energy Research, Carbon Dioxide Research Division, Washington, DC.

Palutikof, J.P., Guo, X. and Barthelmie, R.J. (1990). Wind energy and the greenhouse effect. In: *Wind Energy Conversion 1990* (Ed. by T.D. Davies, J.A. Halliday and J.P. Palutikof). Mechanical Engineering Publications, London.

Parry, M.L., Carter, T.R. and Konijn, N.T. (Eds) (1988a). *The Impact of Climatic Variations on Agriculture. Volume 1: Assessment in Cool Temperate and Cold Regions.* Kluwer Academic Publishers, Dordrecht.

Parry, M.L., Carter, T.R. and Konijn, N.T. (Eds) (1988b). *The Impact of Climatic Variations on Agriculture. Volume 2: Assessment in Semi-arid Regions.* Kluwer Academic Publishers, Dordrecht.

Platt, C.M.R. (1989). The role of cloud microphysics in high-cloud feedback effects on climate change. *Nature* **341**, 428–429.

Ramanathan, V., Cess, R.D., Harrison, E.F., Minnis, P., Barkstrom, B.R., Ahmad, E. and Hartmann, D. (1989). Cloud-radiative forcing and climate: results from the earth radiation budget experiment. *Science* **243**, 57–63.

Rind, D. (1988). Dependence of warm and cold climate depiction on climate model resolution. *J. Climate* **1**, 965–997.

Rowntree, P.R. (1988). Land Surface Parametrizations—Basic Concepts and Review of Schemes. Dynamical Climatology DCTN 72, Meteorological Office, Bracknell.

Santer, B. (1988). *Regional Validation of General Circulation Models.* Climatic Research Unit Research Publication, CRURP 9, Norwich.

Santer, B. and Wigley, T.M.L. (1990). Regional validation of means, variances, and spatial patterns in general circulation model control runs. *J. Geophys. Res.* **95**, 829–850.

Schlesinger, M.E. and Mitchell, J.F.B. (1987). Climate model simulations of the equilibrium climatic response to increased carbon dioxide. *Rev. Geophys.* **25**, 760–798.

Schlesinger, M.E. and Zhao, Z.C. (1989). Seasonal climatic change introduced by doubled CO_2 as simulated by the OSU atmospheric GCM/mixed-layer ocean model. *J. Climate* **2**, 429–495.

Schwartz, S.E. (1988). Are global cloud albedo and climate controlled by marine phytoplankton? *Nature* **336**, 441–445.

Sellers, P.J., Mintz, Y., Sud, Y.C. and Dalcher, A. (1986). A simple biosphere model (SiB) for use within general circulation models. *J. Atmos. Sci.* **43**, 505–531.

Shukla, J., Nobre, C. and Sellers, P. (1990). Amazon deforestation and climate change. *Nature* **247**, 1322–1325.

Slade, D.H. (1990). A survey of informed opinion regarding the nature and reality of a "global greenhouse warming"—an editorial. *Clim. Change* **16**, 1–4.

Slingo, A. (1990). Sensitivity of the earth's radiation budget to changes in low clouds. *Nature* **343**, 49–51.

Stouffer, R.J., Manabe, S. and Bryan, K. (1989). Interhemispheric asymmetry in climate response to a gradual increase of atmospheric CO_2. *Nature* **342**, 660–662.

Warrick, R.A., Gifford, R.M. and Parry, M.L. (1986a). CO_2, climatic change and agriculture. In: *The Greenhouse Effect, Climatic Change and Ecosystems* (Ed. by B. Bolin, B.R. Böös, J. Jäger and R.A. Warrick), pp. 393–473. Wiley, Chichester.

Warrick, R.A., Shugart, H.H., Antonovsky, M.J. and others. (1986b). The effects of increased CO_2 and climatic change on terrestrial ecosystems. In: *The Greenhouse Effect, Climatic Change and Ecosystems* (Ed. by B. Bolin, B.R. Döös, J. Jäger and R.A. Warrick), pp. 363–392. Wiley, Chichester.

Warrilow, D.A. and Buckley, E. (1989). The impact of land surface processes on the moisture budget of a climate model. *Ann. Geophys.* 7, 439–449.

Washington, W.M. and Meehl, G.A. (1984). Seasonal cycle experiments on the climate sensitivity due to a doubling of CO_2 with an atmospheric general circulation model coupled to a simple mixed-layer ocean model. *J. Geophys. Res.* 89, 9475–9503.

Washington, W.M. and Meehl, G.A. (1989). Climate sensitivity due to increased CO_2: experiments with a coupled atmosphere and ocean general circulation model. *Clim. Dynam.* 4, 1–38.

Webb, T. and Wigley, T.M.L. (1985) What past climate can indicate about a warmer world. In: *Projecting the Climatic Effects of Increasing Carbon Dioxide* (Ed. by M.C. MacCracken and F.M. Luther). DoE/ER-0237, pp. 237–257. US Dept of Energy, Office of Energy Research, Carbon Dioxide Research Division, Washington, DC.

Wetherald, R.T. and Manabe, S. (1986). An investigation of cloud cover change in response to thermal forcing. *Clim. Change* 8, 5–23.

Wigley, T.M.L. and Raper, S.C.B. (1990). Natural variability of the climate system and detection of the greenhouse effect. *Nature* 344, 324–326.

Wigley, T.M.L. and Santer, B.D. (1990). Statistical comparison of spatial fields in model validation, perturbation, and predictability experiments. *J. Geophys. Res.* 95, 851–865.

Wigley, T.M.L., Jones, P.D. and Kelly, P.M. (1986). Empirical climate studies: Warm world scenarios and the detection of climatic change induced by radiatively active gases. In: *The Greenhouse Effect, Climatic Change and Ecosystems* (Ed. by B. Bolin, B.R. Böös, J. Jäger and R.A. Warrick), pp. 271–322. Wiley, Chichester.

Wigley, T.M.L., Jones, P.D., Briffa, K.R. and Smith, G. (1990). Obtaining sub-grid-scale information from coarse-resolution general circulation model output. *J. Geophys. Res.* 95, 1943–1953.

Wigley, T.M.L., Santer, B.D., Schlesinger, M.E. and Mitchell, J.F.B. (1992). Developing climate scenarios from equilibrium GCM results. *Clim. Change* (submitted).

Wilks, D.S. (1989). Statistical specifications of local surface weather elements from large-scale information. *Theor. Appl. Climatol.* 40, 119–134.

Wilson, M.F. and Henderson-Sellers, A. (1985). A global archive of land cover and soils data for use in general circulation models. *J. Climat.* 5, 119–143.

Wilson, C.A. and Mitchell, J.F.B. (1987). A doubled CO_2 sensitivity experiment with a global climate model including a simple ocean. *J. Geophys. Res.* 92, 13315–13343.

The Potential Effect of Climate Changes on Agriculture and Land Use

I. ABSTRACT

This chapter reviews current knowledge of the likely effects of possible greenhouse-gas induced changes of climate on agriculture and land use. Attention is first given to the types of climate change, particularly reductions in soil water availability, likely to be most critical for agriculture. Secondly, a number of types of effects on agriculture are considered: "direct" effects of elevated CO_2, shifts of thermal and moisture limits to cropping, effects on

ADVANCES IN ECOLOGICAL RESEARCH VOL. 22
ISBN 0–12–013922–7

drought, heat stress and other extremes, effects on pests, weeds and diseases, and effects on soil fertility. Thirdly, a summary is presented of likely overall effects on crop and livestock production. The conclusion is that, while global levels of food production can probably be maintained in the face of climate change the cost of this could be substantial, and that significant geographic shifts of land use will occur.

II. INTRODUCTION

The purpose of this paper is three-fold: firstly, to consider those types of climate change most critical for agriculture; secondly, to summarize present knowledge about the potential socio-economic impact of changes of climate on world agriculture; finally, to establish research priorities for future assessments of impact.

III. ASSESSMENTS OF CRITICAL TYPES OF CLIMATIC CHANGE

A recent review by the Intergovernmental Panel on Climate Change (IPCC, 1990) reported that the potentially most important changes of climate for agriculture upon which there is some agreement by general circulation models (GCMs) include: changes in climatic extremes, warming in the high latitudes, poleward advance of monsoon rainfall, and reduced soil water availability (particularly in mid latitudes in midsummer, and at low latitudes) (Parry, 1990; Parry and Duinker, 1990).

A. Climatic Extremes

It is not clear whether changes in the variability of temperature will occur as a result of climate change. However, even if variability remains unaltered, an increase in average temperatures would result in the increased frequency of temperatures above particular thresholds. Changes in the frequency and distribution of precipitation are less predictable, but the combination of elevated temperatures and drought or flood probably constitutes the greatest risk to agriculture in many regions from global climate change (Parry and Duinker, 1990).

B. Warming in High Latitudes

There is relatively strong agreement between GCM predictions that greenhouse-gas induced warming will be greater at higher latitudes (IPCC, 1990). This will reduce temperature constraints on high latitude agriculture, increase the competition for land here and result in the northward retreat of the southern margin of the boreal forest (Parry and Duinker, 1990). Warming at low latitudes, although less pronounced, is also likely to have a significant impact on agriculture.

C. Poleward Advance of Monsoon Rainfall

In a warmer world the intertropical convergence zones (ITCZs) would be likely to advance further poleward as a result of an enhanced ocean–continent pressure gradient. If this were to occur—and it must be emphasized that this conclusion is based on GCM equilibrium experiments and that the transient response could be different—then total rainfall could increase in some regions of monsoon Africa, monsoon Asia and Australia, though there is currently little agreement on which regions these might be (IPCC, 1990). Rainfall could also be more intense in its occurrence, so flooding and erosion could increase.

D. Reduced Soil Water Availability

Probably the most important consequences for agriculture would stem from higher potential evapotranspiration, primarily due to the higher temperatures of the air and the land surface. Even in the tropics, where temperature increases are expected to be smaller than elsewhere and where precipitation might increase, the increased rate of loss of moisture from plants and soil would be considerable (Rind et al., 1989; Parry, 1990). It may be somewhat reduced by greater air humidity and increased cloudiness during the rainy seasons, but could be pronounced in the dry seasons.

There are three ways in which increases in greenhouse gases (GHG) may be important for agriculture. Firstly, increased atmospheric carbon dioxide (CO_2) concentrations can have a direct effect on the growth rate of crop plants and weeds. Secondly, GHG-induced changes of climate may alter levels of temperature, rainfall and sunshine and this can influence plant and animal productivity. Finally, rises in sea level may lead to loss of farmland by inundation and to increasing salinity of groundwater in coastal areas. These three types of potential impact will be considered in turn.

IV. EFFECTS OF CO_2 ENRICHMENT

A. Effects on Photosynthesis

CO_2 is vital for photosynthesis, and the evidence is that increases in CO_2 concentration would increase the rate of plant growth (Cure, 1985; Cure and Acock, 1986). There are, however, important differences between the photosynthetic mechanisms of different crop plants and hence in their response to increasing CO_2. Plant species with the C3 photosynthetic pathway (e.g. wheat, rice and soybean) tend to respond more positively to increased CO_2 because it suppresses rates of photorespiration.

However, C4 plants (e.g. maize, sorghum, sugarcane and millet) are less responsive to increased CO_2 levels. Since these are largely tropical crops, and most widely grown in Africa, there is thus the suggestion that CO_2 enrichment will benefit temperate and humid tropical agriculture more than that in the semi-arid tropics. Thus, if the effects of climate changes on agriculture in some parts of the semi-arid tropics are negative, then these may not be partially compensated by the beneficial effects of CO_2 enrichment as they might in other regions. In addition we should note that, although C4 crops account for only about 20% of the world's food production, maize alone accounts for 14% of all production and about 75% of all traded grain. It is the major grain used to make up food deficits in famine-prone regions, and any reduction in its output could affect access to food in these areas (Parry, 1990).

C3 crops in temperate and subtropical regions could also benefit from reduced weed infestation. Fourteen of the world's 17 most troublesome terrestrial weed species are C4 plants in C3 crops. The difference in response to increased CO_2 may make such weeds less competitive. In contrast, C3 weeds in C4 crops, particularly in tropical regions, could become more of a problem, although the final outcome will depend on the relative response of crops and weeds to climate changes as well (Parry, 1990).

Many of the pasture and forage grasses of the world are C4 plants, including important prairie grasses in North America and central Asia and in the tropics and subtropics. The carrying capacity of the world's major rangelands is thus unlikely to benefit substantially from CO_2 enrichment (Parry, 1990). Much, of course, will depend on the parallel effects of climate changes on the yield potential of these different crops.

The actual amount of increase in usable yield rather than of total plant matter that might occur as a result of increased photosynthetic rate is also problematic. In controlled environment studies, where temperature, nutrients and moisture are optimal, the yield increase can be substantial,

Table 1

Mean predicted growth and yield increases for various groupings of C3 species for a doubling of atmospheric CO_2 concentration from 330 p.p.m.v. to 660 p.p.m.v. The errors indicated are 95% confidence limits

	Footnote	Immature crops		Mature crops	
		No. of records	% increase of biomass	No. of records	% increase of marketable yield
Fibre crops	a	5	124	2	104
Fruit crops	b	15	40	12	21
Grain crops	c	6	20	15	36
Leaf crops	d	5	37	9	19
Pulses	e	18	43	13	17
Root crops	f	10	49	—	—
C3 weeds	g	10	34	—	—
Trees	h	14	26	—	—
Av. of all C3		(83)	40 ± 7	(51)	26 ± 9

Source: Warrick et al., 1986.
Footnotes: The species represented are:
a. Cotton (*Gossypium hirsutum*).
b. Cucumber (*Cucumis sativus*), eggplant (*Solanum melongena*), okra (*Abelmoschus esculentus*), pepper (*Capsicum annuum*), tomato (*Lycopersicum esculentum*).
c. Barley (*Hordeum vulgare*), rice (*Oryza sativa*), sunflower (*Helianthus annuus*), wheat (*Triticum aestivum*).
d. Cabbage (*Brassica oleracea*), white clover (*Trifolium repens*), fescue (*Festuca elatior*), lettuce (*Lactuca sativa*), Swiss chard (*Beta vulgaris*).
e. Bean (*Phaseolus vulgaris*), pea (*Pisum sativum*), soybean (*Glycine max*).
f. Sugar beet (*Beta vulgaris*), radish (*Raphanus sativus*).
g. *Crotalaria spectabilis*, *Desmodium paniculatum*, jimson weed (*Datura stramonium*), pigweed (*Amaranthus retroflexus*), ragweed (*Ambrosia artemisiifolia*), sicklepod (*Cassia obtusifolia*), velvet leaf (*Abutilon theophasti*).
h. Cotton (*Gossypium deltoides*).

averaging 36% for C3 cereals such as wheat, rice, barley and sunflower under a doubling of ambient CO_2 concentration (Table 1). Few studies have yet been published, however, of the effects of increasing CO_2 in combination with changes of temperature and rainfall.

Little is also known about possible changes in yield quality under increased CO_2. The nitrogen content of plants is likely to decrease, while the carbon content increases, implying reduced protein levels and reduced nutritional levels for livestock and humans. This, however, may also reduce the nutritional value of plants for pests, so that they need to consume more to obtain their required protein intake.

B. Effects on Water Use by Plants

Just as important may be the effect that increased CO_2 has on the closure of stomata. This tends to reduce the water requirements of plants by reducing transpiration (per unit leaf area) thus improving what is termed water use efficiency (the ratio of crop biomass accumulation to the water used in evapotranspiration). A doubling of ambient CO_2 concentration causes about a 40% decrease in stomatal aperture in both C3 and C4 plants which may reduce transpiration by 23–46% (Cure and Acock, 1986; Morison, 1987). This might well help plants in environments where moisture currently limits growth, such as in semi-arid regions, but there remain many uncertainties, such as to what extent the greater leaf area of plants (resulting from increased CO_2) will balance the reduced transpiration per unit leaf area (Allen et al., 1985; Gifford, 1988).

In summary, we can expect that a doubling of atmospheric CO_2 concentrations from 330 to 660 p.p.m.v. might cause a 10–50% increase in growth and yield of C3 crops (such as wheat, rice and soybean) and a 0–10% increase for C4 crops (such as maize and sugarcane) (Warrick et al., 1986). Much depends, however, on the prevailing growing conditions. Our present knowledge is based on experiments mainly in field chambers and has not yet included extensive study of response in the field under suboptimal conditions. Thus, although there are indications that, overall, the effects of increased CO_2 could be distinctly beneficial and could partly compensate for some of the negative effects of CO_2-induced changes of climate, we cannot at present be sure that this will be so.

V. POTENTIAL EFFECTS OF CHANGES OF CLIMATE

A. Changes in Thermal Limits to Agriculture

Increases in temperature can be expected to lengthen the growing season in areas where agricultural potential is currently limited by insufficient warmth, resulting in a poleward shift of thermal limits of agriculture. The consequent extension of potential will be most pronounced in the northern hemisphere because of the greater extent here of temperate agriculture at higher latitudes. There may, however, be important regional variations in our ability to exploit this shift. For example, the greater potential for exploitation of northern soils in Siberia than on the Canadian Shield may mean relatively greater increases in potential in northern Asia than in northern N. America (Parry, 1990).

A number of estimations have been made concerning the northward shift in productive potential in mid-latitude northern hemisphere countries. These relate to changes in the climatic limits for specific crops under a variety of climatic scenarios, and are therefore not readily compatible (William and Oakes, 1978; Newman, 1980; Blasing and Solomon, 1983; Rosenzweig, 1985; Parry and Carter, 1988; Parry et al., 1989). They suggest, however, that a 1°C increase in mean annual temperature would tend to advance the thermal limit of cereal cropping in the mid-latitude northern hemisphere by about 150–200 km, and to raise the altitudinal limit to arable agriculture by about 150–200 m. One such study, a logical development of those considered above, has mapped the shift of growing areas of different cultivars or types of the same crop that might occur under an altered climate (Rosenzweig, 1985). Wheat growing regions in North America were characterized according to their present-day temperature and rainfall regimes, and then re-mapped for the equilibrium climate based on a 2 × CO_2 experiment with the Goddard Institute of Space Studies (GISS) GCM. Results indicated a substantial northward extension of winter wheat into Canada from its current location on the US Great Plains, a switch from hard to soft wheat in the Pacific Northwest due to increased precipitation, and an expansion of areas in autumn-sown spring wheat in the southern USA due to higher winter temperatures (Fig. 1). In Mexico, wheat-growing regions would remain the same but greater high-temperature stress may occur.

A similar magnitude of shift of cropping limits has been estimated for Europe. In this region the major climatic determinant of successful ripening of grain maize (i.e. maize grown for its grain rather than as green fodder) is the warmth of the growing season. An effective temperature sum (ETS) of 850 degree-days above a base temperature of 10°C corresponds closely with the actual limit of its cultivation today (Carter et al., 1991). This boundary extends from the southwestern tip of England through northern central Europe and central Russia to just south of Moscow. Much of the fertile north European plain is therefore currently too cool for grain maize to mature in all but the warmest years.

However, under the 2 × CO_2 equilibrium climates projected by a number of GCM experiments this limit is displaced 200–350 km further north. Figure 2 illustrates the location of the thermal limit to grain maize for 2 × CO_2 climates projected by three GCMs—GISS, Goddard Fluid Dynamics Laboratory (GFDL) and Oregon State University (OSU). The similarity between the figures indicates the level of agreement between the models regarding temperature increases in the summer half of the year. The entire northern European plain is estimated to be within the grain maize limit under a 2 × CO_2 climate, particularly the western part of northern Europe (UK, northern Germany, Denmark) where maritime influence creates a greater sensitivity to warming because greater CO_2-induced temperature increases are expected in winter than in summer. It is worth noting, however, that there is very little agreement between model estimates of precipitation, which can be a critical

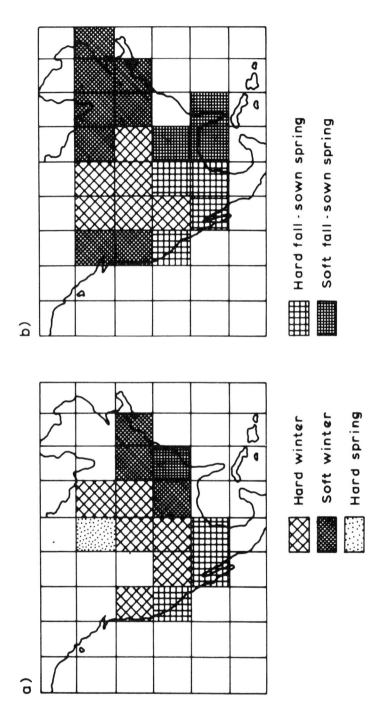

Fig. 1. Simulated North American wheat regions using (a) GISS GCM control, and (b) doubled CO_2 runs. (Source: Rosenzweig, 1985.)

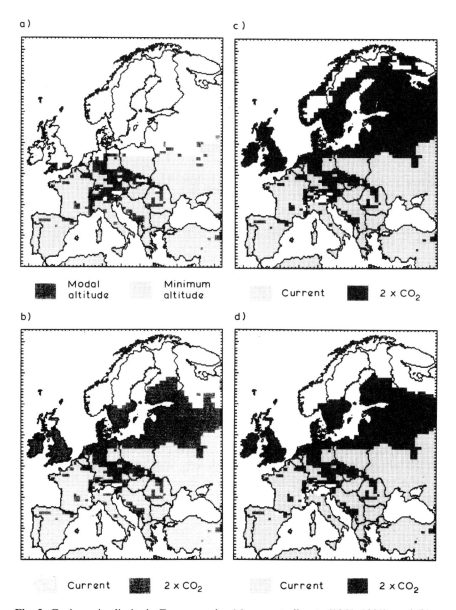

Fig. 2. Grain maize limits in Europe under (a) current climate (1951–1980), and (b) GISS, (c) GFDL, and (d) OSU equilibrium $2 \times CO_2$ climates. (Source: Carter *et al.*, 1991.)

factor for many crops in Europe and is also important for maize. Consequently, we are at present only able to draw a very imperfect picture of how potential growing regions may shift.

Outside North America and Europe, little study has been made of the spatial shift of crop potential. One exception is Japan, where an estimate has been made of the extension of area in which rice could safely be cultivated without severe risk of crop loss due to frost. On the island of Hokkaido, in the north of Japan, the "safely cultivable" area for irrigated rice is estimated to more than double under the GISS 2 × CO_2 climate, assuming there remains adequate precipitation and the crop is fully irrigated (Yoshino et al., 1988). Corresponding poleward shifts of crop potential in the southern hemisphere, for example for cereals, fruit and vegetables in New Zealand, have also been estimated (Salinger et al., 1990). Once more, however, it must be emphasized that these are estimations only of altered potential.

Since it is important to consider those changes of climate that may occur within the next 30 years as well as those within the next century, more recent impact assessments have begun to evaluate potential effects of time-dependent changes in temperature (Carter et al., 1991). Figure 3 illustrates the thermal limit for grain maize in Europe for decadal "time slices" of temperature based on results from experiments for the GISS GCM. The value of this approach is that it can help elucidate the rate of shift of agricultural potential that may occur as a result of global warming.

In this case the transient response data are for Scenario A (one of four conducted at GISS) which assumes a continued rise in emissions of trace gases at growth rates typical of the 1970s and 1980s (i.e. without effective policies of emissions control) representing an exponential increase of 1·5% per annum (Hansen et al., 1990). Under this scenario the indication is that the rate of northward shift of the grain maize limit could approximate 150 km decade^{-1} between the 1990s and 2030, and perhaps 240 km decade^{-1} from 2030 to 2060. Broadly similar rates of shift are implied for many crops throughout the middle and high latitudes, and it remains to be seen whether rates of adaptation in agriculture can match them.

The use of synthetic climatic scenarios, in combination with the transient ones considered above, can serve to relate different rates of possible climatic change to different rates of shift in agricultural potential. In the UK, for example, the effects of warming suggest a poleward shift of limits for grain maize and silage maize by about 300 km for each °C in mean annual temperature (Parry et al., 1989). Under a "business as usual" emissions scenario the temperature increases are currently estimated to be +0·5°C by 2000–2010, about 1·5°C by 2020–2050 and about 3°C by 2050–2100 (IPCC, 1990). This suggests a rate of shift of about 100–150 km decade^{-1}. If emissions were reduced such that rates of warming were (say) cut by a third, then the shift would be reduced to 50–100 km decade^{-1}. Tolerable rates of

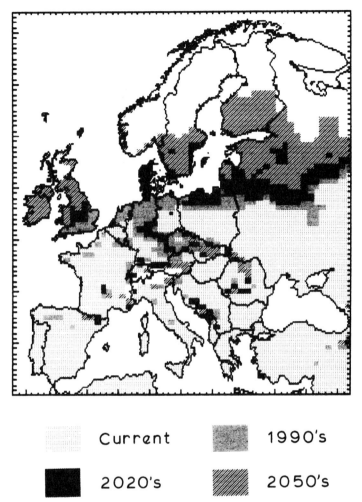

Current   1990's

2020's   2050's

Fig. 3. Grain maize limit under the GISS transient response Scenario A in the 1990s, 2020s, and 2050s (relative to the limit for the current climate). (Source: Carter *et al.*, 1991.)

shift in climatic resources can thus be used as a guide to target rates of tolerable climatic change.

The shifts of crop potential described above are examples of a world-wide relocation of climatic zones that could occur as a result of CO_2 -induced changes of climate, particularly a poleward shift of thermal zones. An illustration of the scale of these shifts is given in Fig. 4 which maps regions that have a present-day climate analogous to the future climate assumed under the GISS 2 × CO_2 scenario. For example, Iceland's climate estimated

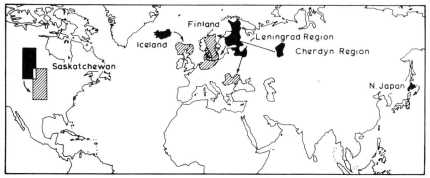

Fig. 4. Present-day analogues of the GISS 2 × CO$_2$ climate estimated for selected regions in the IIASA/UNEP study: Saskatchewan, Iceland, Finland, Leningrad and Cherdyn regions (USSR) and Hokkaido and Tohoku districts (Japan). (Source: Parry and Carter, 1988.)

under the GISS 2 × CO$_2$ scenario is similar to that of northern Britain today. This serves to illustrate not only the magnitude of possible changes in agricultural potential, but also the adaptive responses likely to be required to re-tune agriculture to altered climatic resources (Parry and Carter, 1988). For example, perhaps the combination of barley growing and cattle rearing and fattening, which are successful enterprises in northern Britain today, would be appropriate for Iceland in the future. Due to differences in latitude, however, there are important differences in day length between such regions, and the analogy is far from perfect.

B. Shifts of Moisture Limits to Agriculture

There is much less agreement between GCM-based projections concerning GHG-induced changes in precipitation than there is about temperature—not only concerning changes of magnitude, but also of spatial pattern and distribution through the year. For this reason it is difficult to identify potential shifts in the moisture limits to agriculture. This is particularly so because relatively small changes in the seasonal distribution of rainfall can have disproportionately large effects on the viability of agriculture in tropical areas, largely through changes in growing period when moisture is sufficient and thus through the timing of critical episodes such as planting, etc. However, recent surveys for the IPCC have made a preliminary identification of those regions where there is some agreement amongst 2 × CO$_2$ experiments with GCMs concerning an overall reduction in crop-water availability (Parry, 1990; Parry and Duinker, 1990). It should be emphasized that coincidence of results for these regions is not statistically significant. The regions are:

Table 2
Response of spring-wheat yield (as percentages of the long-term mean) to variations in air temperature (ΔT) and rainfall (ΔR) during the growing season (Cherdyn, forest zone)

	$\Delta T(°C)$		
ΔR(mm)	$-1\cdot0$	0	$+1\cdot0$
-20	93	97	99
0	95	100	103
$+20$	97	101	107

Source: Pitovranov *et al.*, 1988.

(1) Decreases of soil water in December, January and February:
Africa: northeast Africa, Southern Africa
Asia: western Arabian Peninsula; Southeast Asia
Australasia: eastern Australia
N. America: southern USA
S. America: Argentine Pampas
(2) Decreases in soil water in June, July and August:
Africa: north Africa; west Africa
Europe: parts of western Europe
Asia: north and central China; parts of Soviet central Asia and Siberia
N. America: southern USA and Central America
S. America: eastern Brazil
Australasia: western Australia

C. Effects on Yields

Whether crops respond to higher temperatures with an increase or decrease in yield depends on whether they are determinate or indeterminate, and whether their yield is currently strongly limited by insufficient warmth. In cold regions very near the present-day limit to arable agriculture any temperature increase, even as much as the 7–9°C indicated for high latitudes under a doubling of CO_2 can be expected to enhance yields of cereal crops. For example, near the current northern limit of spring wheat production in the European region of the USSR yields are estimated to increase about 3% per°C, assuming no concurrent change in rainfall (Table 2). In Finland, the marketable yield of barley increases 3–5% per°C, and in Iceland hay yields increase about 15% per°C (Kettunen, *et al.*, 1988; Bergthorssen *et al.*, 1988).

Table 3
Response of spring wheat yield (as percentages of the long-term mean) to variations in air temperature (ΔT) and precipitation (ΔP) during the growing season (Palassovka, Volgograd region)

ΔP(mm)	$\Delta T(°C)$				
	-1.0	-0.5	0	$+0.5$	$+1.0$
-40	79	79	76	76	76
-20	92	92	89	89	89
0	104	103	100	100	99
$+20$	115	114	110	109	108
$+40$	125	124	120	118	117

Source: Nikonov et al., 1988.

Away from current temperature-constrained regions of farming and in the core areas of present-day cereal production such as in the Corn Belt of N. America, the European lowlands and the Soviet Ukraine, increases in temperature would probably lead to decreased cereal yield due to a shortened period of crop development. In eastern England, for example, a 3°C rise in mean annual temperature is estimated to reduce winter wheat yield by about 10% although the direct effect of a doubling of ambient atmospheric CO_2 might more than compensate for this.

In other mid-latitude regions much would depend on possible changes in rainfall. For example, in the Volgograd region, just east of the Ukraine, spring wheat yields are estimated to fall only a small amount with a 1°C increase in mean temperature during the growing season, though they could increase or decrease substantially if the temperature was accompanied by an increase or decrease of rainfall (Table 3).

Yields of root crops such as sugar beet and potatoes, with an indeterminate growth habit, can be expected to see an increase in yield with increasing temperatures, provided these do not exceed temperatures optimal for crop development (Squire and Unsworth, 1988).

Changes of temperature would also have an effect on moisture available for crop growth, whether or not levels of rainfall remained unchanged. In general, and at mid-latitudes, evaporation increases by about 5% for each °C of mean annual temperature. Thus, if mean temperature were to increase in the east of England by 2°C potential evaporation would increase by about 9% (assuming no change in rainfall). The effect of this would be small in the early part of the growing season, but after mid-July the soil moisture deficit would be considerably larger than at present and, for some crops, this implies substantially increased demand for irrigation (Rowntree et al., 1989).

In most of the tropical and equatorial regions of the world, and across large areas outside the tropics, the yield of agricultural crops is limited more by the amount of water received by and stored in the soil than by air temperature. Even in the high mid-latitudes such as in southern Scandinavia too little rain can restrict growth of cereal crops during the summer when evapotranspiration exceeds rainfall. In all these areas the amount of dry matter a crop produces is roughly proportional to the amount of water it transpires (Monteith, 1981). This, in turn, is affected by the quantity of rainfall but not in a straightforward manner: it also depends on how much of the rainfall is retained in the soil, how much is lost through evaporation from the soil surface, and how much remains in the soil that the crop cannot extract.

Relatively few studies have been made of the combined effects of possible changes in temperature and rainfall on crop yields, and those that have are based on a variety of different methods. However, a recent review of results from about 10 studies in North America and Europe noted that warming is generally detrimental to yields of wheat and maize in these mid-latitude core cropping regions. With no change in precipitation (or radiation) slight warming (+1°C) might decrease average yields by about 5 ± 4%; and a 2°C warming might reduce average yields by about 10 ± 7%. In addition, reduced precipitation might also decrease yields of wheat and maize in these breadbasket regions. A combination of increased temperatures (+2°C) and reduced precipitation could lower average yields by over a fifth.

D. Effects of Drought, Heat Stress and Other Extremes

Probably most important for agriculture, but about which least is known, are the possible changes in climatic extremes, such as the magnitude and frequency of drought, storms, heat waves and severe frosts (Rind et al., 1989). Some modelling evidence suggests that hurricane intensities will increase with climatic warming (Emanuel, 1987). This has important implications for agriculture in low latitudes, particularly in coastal regions.

Since crop yields often exhibit a non-linear response to heat or cold stress, changes in the probability of extreme temperature events can be significant (Parry, 1976; Mearns et al., 1984). In addition, even assuming no change in the standard deviation of temperature maxima and minima, we should note that the frequency of hot and cold days can be markedly altered by changes in mean monthly temperature. To illustrate, under a $2 \times CO_2$ equilibrium climate the number of days in which temperatures would fall below freezing would decrease from a current average of 39 to 20 in Atlanta, Georgia (USA), while the number of days above 90°F (32°C) would increase from 17 to 53 (EPA, 1989). The frequency and extent of area over which losses of agricultural output could result from heat stress, particularly in tropical regions, is therefore likely to increase significantly. Unfortunately, no studies have yet been made of this. However, the apparently small increases in mean annual temperatures in tropical regions ($\sim$1–2°C under a 2 $\times$ CO_2 climate) could sufficiently increase heat stress on temperate crops such as wheat so that these are no longer suited to such areas. Important wheat producing areas such as N. India could be affected in this way (Parry and Duinker, 1990).

An important additional effect, especially in temperate mid-latitudes, is likely to be the reduction of winter chilling (vernalization). Many temperate crops require a period of low temperatures in winter to either initiate or accelerate the flowering process. Low vernalization results in low flower bud initiation and, ultimately, reduced yields. A 1°C warming would reduce effective winter chilling by between 10 and 30%, thus contributing to a poleward shift of temperate crops (Salinger, 1989).

There is a distinct possibility that, as a result of high rates of evapotranspiration, some regions in the tropics and subtropics could be characterized by a higher frequency of drought or a similar frequency of more intense drought than at present. Current uncertainties about how regional patterns of rainfall will alter mean that no useful prediction of this can at present be made. However, it is clear in some regions that relatively small decreases in water availability can readily produce drought conditions. In India, for example, lower-than-average rainfall in 1987 reduced food grains production from 152 to 134 million tonnes (mt), lowering food buffer stocks from 23 to 9 mt. Changes in the risk and intensity of drought, especially in currently drought-prone regions, represent potentially the most serious impact of climatic change on agriculture both at the global and the regional level.

E. Effects on the Distribution of Agricultural Pests and Diseases

Studies suggest that temperature increases may extend the geographic range of some insect pests currently limited by temperature (EPA, 1989; Hill and Dymock, 1989). As with crops, such effects would probably be greatest at higher latitudes. The number of generations per year produced by multivoltine (i.e. multigenerational) pests would increase, with earlier establishment of pest populations in the growing season and increased abundance during more susceptible stages of growth. An important unknown, however, is the effect that changes in precipitation amount and air humidity may have on the insect pests themselves and on their predators, parasites and diseases. Climate change may significantly influence interspecific interactions between pests and their predators and parasites.

Under a warmer climate at mid-latitudes there would be an increase in the overwintering range and population density of a number of important agricultural pests, such as the potato leafhopper which is a serious pest of soybeans and other crops in the USA (EPA, 1989). Assuming planting dates did not change, warmer temperatures would lead to invasions earlier in the growing season and probably lead to greater damage to crops. In the US Corn Belt increased damage to soybeans is also expected due to earlier infestation by the corn earworm.

Examination of the effect of climatic warming on the distribution of livestock diseases suggests that those at present limited to tropical countries, such as Rift Valley fever and African swine fever, may spread into the mid-latitudes. For example, the horn fly, which currently causes losses of $730·3 million in the US beef and dairy cattle industries, might extend its range under a warmer climate leading to reduced gain in beef cattle and a significant reduction in milk production (Drummond, 1987; EPA, 1989).

In cool temperate regions, where insect pests and diseases are not generally serious at present, damage is likely to increase under warmer conditions. In Iceland, for example, potato blight currently does little damage to potato crops, being limited by the low summer temperatures. However, under a $2 \times CO_2$ climate that may be 4°C warmer than at present, crop losses to disease may increase to 15% (Bergthorsson et al., 1988).

Under warmer and more humid conditions cereals would be more prone to diseases such as septoria. In addition, increases in population levels of disease vectors may well lead to increased epidemics of the diseases they carry. To illustrate, increases in infestations of the bird cherry aphid (*Rhopalosiphum padi*) or grain aphid (*Sitobian avenae*) could lead to increased incidence of barley yellow dwarf virus in cereals.

VI. EFFECTS OF SEA-LEVEL RISE ON AGRICULTURE

GHG-induced warming is expected to lead to rises in sea level as a result of thermal expansion of the oceans and partial melting of glaciers and ice caps, and this in turn is expected to affect agriculture, mainly through the inundation of low-lying farmland but also through the increased salinity of coastal groundwater. The current projection of sea-level rise above present levels is 20 cm ± 10 cm by c. 2030, and 30 cm ± 15 cm by 2050 (Warrick and Oerlemans, 1990).

Preliminary surveys of proneness to inundation have been based on a study of existing contoured topographic maps, in conjunction with knowledge of the local "wave climate" that varies between different coastlines. They have identified 27 countries as being especially vulnerable to sea-level rise, on the basis of the extent of land liable to inundation, the population at risk and the capability to take protective measures (UNEP, 1989). It should be emphasized, however, that these surveys assume a much larger rise in sea levels than is at present estimated to occur within the next century under current trends of increase of GHG concentrations. On an ascending scale of vulnerability (1–10) experts identified the following most vulnerable countries or regions: 10, Bangladesh; 9, Egypt, Thailand; 8, China; 7, western Denmark; 6, Louisiana; 4, Indonesia.

The most severe impacts are likely to stem directly from inundation. Southeast Asia would be most affected because of the extreme vulnerability of several large and heavily populated deltaic regions. For example, with a 1·5-m sea-level rise, about 15% of all land (and about 20% of all farmland) in Bangladesh would be inundated and a further 6% would become more prone to frequent flooding (UNEP, 1989). Altogether 21% of agricultural production could be lost. In Egypt, it is estimated that 17% of national agricultural production and 20% of all farmland, especially the most productive farmland, would be lost as a result of a 1·5-m sea-level rise. Island nations, particularly low-lying coral atolls, have the most to lose. The Maldive Islands in the Indian Ocean would have one half of their land area inundated with a 2-m rise in sea level (UNEP, 1989).

In addition to direct farmland loss from inundation, it is likely that agriculture would experience increased costs from saltwater intrusion into surface water and groundwater in coastal regions. Deeper tidal penetration would increase the risk of flooding and rates of abstraction of groundwater might need to be reduced to prevent recharge of aquifers with sea water.

Further indirect impacts would be likely as a result of the need to relocate both farming populations and production in other regions. In Bangladesh, for example, about 20% of the nation's population would be displaced as a result of the farmland loss estimated for a 1·5-m sea-level rise. It is important to emphasize, however, that the IPCC estimates of sea-level rise are much lower than this (about 0·5 m by 2090 under the IPCC Business-As-Usual case).

VII. SUMMARY OF POTENTIAL EFFECTS OF CLIMATE CHANGE AND SEA-LEVEL RISE

Potential impacts on yields vary greatly according to types of climate change and types of agriculture. In general, there is much uncertainty about how agricultural potential may be affected.

In the northern mid-latitudes where summer drying may reduce productive potential (e.g. in the US Great Plains and Corn Belt, Canadian Prairies, southern Europe, south European USSR) yield potential is estimated to fall by ∼ 10–30% under the GISS equilibrium 2 × CO_2 climate (Parry, 1990).

In the USA a warming of 3·8–6·3°C, with soil moisture reduced by 10% (which is consistent with the GISS and GFDL 2 × CO_2 climate), is estimated to decrease potential yields of maize, allowing for the limited beneficial fertilizing effect of enhanced CO_2 on this C4 plant by ∼4–17% in California, 16–25% on the Great Plains (assuming irrigation), and by 5–14% in the southeast (also assuming irrigation), (Fig. 5). In the Great Lakes region there could be a small increase in potential yields, depending on available moisture.

However, towards the northern edge of current corn producing regions (e.g. the northern edge of the Canadian Prairies, northern Europe, northern USSR and Japan, southern Chile and Argentina) warming may enhance productive potential, particularly when combined with beneficial direct CO_2 effects. Much of this potential may not, however, be exploitable owing to limits placed by inappropriate soils and difficult terrain, and on balance it seems that the advantages of warming at higher latitudes would not compensate for reduced potential in current major cereal producing regions. A summary of estimated impacts for some northern case studies is given in Fig. 6.

Agriculture in Scandinavia stands to gain more from global warming than perhaps any other region of the world. For example, in Finland, where the equilibrium 2 × CO_2 climate is projected to be about 4°C warmer and also wetter than at present, yields of adapted cultivars of spring wheat are estimated to increase by about 10% in the south, up to 20% in the centre and even more in the north. Yields of barley and oats are raised by 9–18%, depending on the region in Finland (Kettunen et al., 1988).

In northern Japan, where temperature is projected to increase by 3–3·5°C and precipitation by 5% (the GISS 2 × CO_2 climate), rice yields are estimated to increase in the north (Hokkaido) by ∼5%, and in the north-central region (Tohoku) by ∼2% (Yoshino et al., 1988). The average increase estimated for the country as a whole is ∼3%. Cultivation limits for rice would rise about 500 m in elevation and advance ∼100 km north in Hokkaido.

In the European USSR (Leningrad region), with May–October temperatures 2–3°C warmer and annual precipitation about 100 mm higher (the GISS 2 × CO_2 climate), yields of winter wheat and maize are likely to increase but those of temperate crops such as barley, oats, potatoes and green vegetables are likely to decrease (Nikonov et al., 1988). In the Perm region, just to the west of the Ural mountains at about 60°N, spring wheat yields are expected to decrease slightly under the warmer growing season, but this may be more than compensated by the direct effects of CO_2, the combined climate and direct CO_2 effects perhaps allowing a 20% increase in yields (Nikonov et al., 1988).

Effects at lower latitudes are much more difficult to estimate because production potential is largely a function of the amount and distribution of precipitation and because there is little agreement about how precipitation may be affected by GHG warming. Because of these uncertainties the tendency has been to assert that worthwhile study must await improved projection of changes in precipitation. Consequently very few estimates are currently available of how yields might respond to a range of possible changes of climate in low-latitude regions. The only comprehensive national estimates available are for Australia where increases in cereal and grassland productivity might occur (except in western Australia) if warming is accompanied by increase in summer rainfall (Pearman, 1988).

The impacts described above relate to possible changes in potential productivity or yield. It should be emphasized that such potential effects are those estimated assuming present-day management and technology. They are not the estimated future actual effects, which will depend on how farmers and governments respond to altered potential through changes in management and technology. The likely effects on actual agricultural output and on other measures of economic performance such as profitability and employment levels are considered in the next section.

VIII. EFFECTS ON PRODUCTION AND LAND USE

To date (1991) six national case studies have been made of the potential impact of climatic changes on agricultural production (in Canada, Iceland, Finland, the USSR, Japan and the United States) (Parry et al., 1988; Smit, 1989; EPA, 1989). These studies are based on results from model experiments of yield responses to altered climate and the effects that altered yields might have on production.

Other countries have conducted national reviews of effects of climate change, basing these on existing knowledge rather than on new research. The most comprehensive of these are for Australia and New Zealand (Pearman, 1988; Salinger et al., 1990). Brief surveys have also been completed in the UK and West Germany (SCEGB, 1989; Department of the Environment, 1991). Several other national assessments are currently in progress but not yet complete.

Fig. 5. Estimated crop yields under the GISS 2 × CO_2 scenario for present-day and for adjusted crop varieties and management in Finland, N. USSR and N. Japan. 1 = present variety; 2, 3, 4 = varieties with thermal requirements 50, 100 and 120 GDD (growing-degree days) higher than present; 5, 6 = newly introduced middle-maturing and late-maturing rice varieties, transplanting date 25 days earlier than present; 7 = present variety with technology trend projected to 2035; 8 = present variety, fertilizer applications 50% above those in 7; 9 = present variety, drainage activity 2 km km^{-2} above that in 7; 10 = combination of 8 and 9; 11 = includes "direct" effects of CO_2. (Source: Parry and Carter, 1988.)

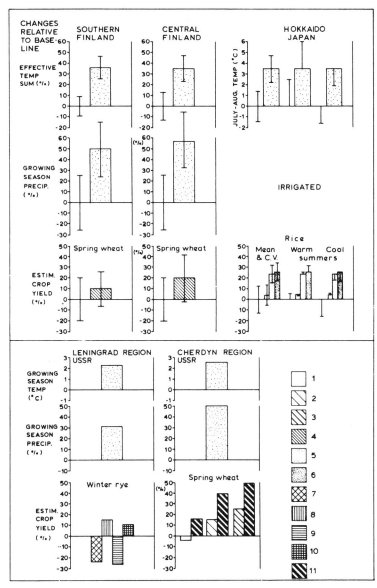

In summary, it seems that overall output from the major present-day grain producing regions could well decrease under the warming and possible drying expected in these regions. In the USA grain production may be reduced by 10–20% and, while production would still be sufficient for domestic needs, the amount for export would probably decline. Production may also decrease in the Canadian prairies and in the southern Soviet Union.

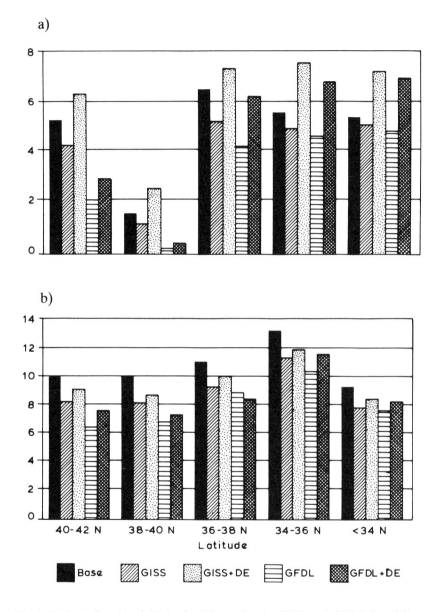

Fig. 6. Estimated maize yields in the USA under the GISS and GFDL 2 × CO_2 climates with and without the direct effects (DE) of CO_2: (a) dryland and (b) irrigated. Estimations for direct effects assume CO_2 concentrations of 660 p.p.m., which are 100 p.p.m. above IPCC estimates and are thus somewhat exaggerated. (Source: Rosenzweig, 1985.)

In Europe production of grains might increase in the UK and the Low Countries if rainfall increases sufficiently, but may fall in southern Europe substantially if there are significant decreases in rainfall as currently estimated in most GCM $2 \times CO_2$ experiments (Parry, 1990). Output could increase in Australia if there is a sufficient increase in summer rainfall to compensate for higher temperatures.

Production could increase in regions currently near the low-temperature limit of grain growing: in the northern hemisphere in the northern Prairies, Scandinavia, north European Soviet Union, and in the southern hemisphere in southern New Zealand, and southern parts of Argentina and Chile. But it is reasonably clear that, because of the limited area unconstrained by inappropriate soils and terrain, increased high-latitude output will probably not compensate for reduced output at mid-latitudes.

IX. EFFECTS ON FOREST LAND USE

Major forest-type zones and species range may shift significantly as a result of climate change. For example, if warming were to induce a northward shift of the boreal forest in northern regions of America, Europe and Asia, it is possible that extensive grazing, livestock rearing and cultivation of quick-maturing crop (farming types currently located at the southern limit of the boreal zone) would be encouraged to shift northwards to exploit regions vacated by forestry. A geographic shift of agriculture in these marginal regions would thus be the combined result of changes in potential for farming and changes in potential for forestry, with the outcome perhaps determined by the comparative advantage of one use over the other; and this might further be influenced by future policies of conservation.

An illustration of the possible extent of poleward shift of the boreal zone in the northern hemisphere is given in Fig. 7. This is based on an estimation of the levels of effective temperature sum above a threshold temperature of 5°C that currently define the northern and southern limits of the boreal (600 and 1300 degree-days respectively) (Kauppi and Posch, 1988). Under the warming projected for a $2 \times CO_2$ climate (in this instance based on experiments with the GISS GCM) these limits are relocated about 500–1000 km further north than at present. If taken as a proxy limit of northern agriculture, this indicates a substantial extension of agricultural potential, although much of this may be severely limited by inappropriate soils and terrain, particularly in northern America and Europe.

Similarly substantial shifts can be expected to occur for vegetation zones throughout the world. An illustration of the possible scale of these shifts, in this instance estimated for Europe for a 5°C increase in mean annual temperature and a 10% increase in precipitation is given in Fig. 7. On

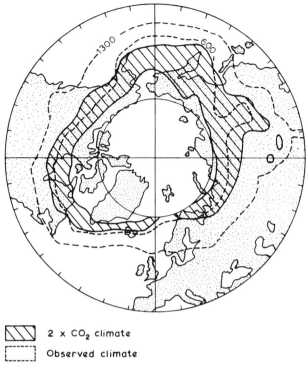

▨	2 x CO₂ climate
⬚	Observed climate

Fig. 7. Calculated boreal zone for the GISS $2 \times CO_2$ climate scenario relative to the calculated present-day zone. (Source: Kauppi and Posch, 1988.)

average, the major vegetation zones shift northwards by 1000 km, the largest changes being in the boreal and Mediterranean regions (De Groot, 1987).

X. CONCLUSIONS

Our assessment of possible effects has, up to this point, assumed that technology and management in agriculture do not alter significantly in response to climatic change, and thus do not alter the magnitude and nature of the impacts that may stem from that change. It is certain, however, that agriculture will adjust and, although these adjustments will be constrained by economic and political factors, it is likely that they will have an important bearing on future impacts. Three broad types of adjustment may be anticipated: changes in land use (e.g. changes in farmed area, crop type and crop location), and changes in technology, (e.g. changes in irrigation, fertilizer use, control of pests/diseases, soil management, farm infrastructure, and changes in crop and livestock husbandry). For a discussion of these adjustments the reader is referred to a number of recent reviews (Parry and Duinker, 1990; Parry 1990).

On balance, the evidence is that food production at the global level can, in the face of estimated changes of climate, be sustained at levels that would occur without a change of climate, but the cost of achieving this is unclear. It could be very large. Increases in productive potential at high mid-latitudes and high latitudes, while being of regional importance, are not likely to open up large new areas for production. The gains in productive potential here due to climatic warming would be unlikely to balance possible large-scale reductions in potential in some major grain-exporting regions at mid-latitude. Moreover, there may well occur severe negative impacts of climate change on food supply at the regional level, particularly in regions of high present-day vulnerability least able to adjust technically to such effects.

The average global increase in overall production costs could thus be small (perhaps a few per cent of world agricultural GDP). Much depends however, on how beneficial are the so-called "direct" effects of increased CO_2 on crop yield. If plant productivity is substantially enhanced *and* more moisture is available in some major production areas, then world productive potential of staple cereals could increase relative to demand with food prices reduced as a result. If, on the contrary, there is little beneficial direct CO_2 effect *and* climate changes are negative for agricultural potential in all or most of the major food-exporting areas, then the average costs of world agricultural production could increase significantly, these increased costs amounting to perhaps over 10% of world agricultural GDP.

Although we know little, at present, about how the frequency of extreme weather events may alter as a result of climatic change, the potential impact of concurrent drought or heat stress in the major food-exporting regions of the world could be severe. In addition, relatively small decreases in rainfall or increases in evapotranspiration could markedly increase both the risk and the intensity of drought in currently drought-prone (and often food-deficient) regions. Change in drought-risk represents potentially the most serious impact of climatic change on agriculture both at the regional and the global level. The regions most at risk from impact from climatic change are probably those currently most vulnerable to climatic variability. Frequently these are low-income regions with a limited ability to adapt through technological change.

This chapter has emphasized the inadequacy of our present knowledge. It is clear that more information on potential impacts would help us identify the full range of potentially useful responses and assist in determining which of these may be most valuable.

Some priorities for future research may be summarized as follows:

(1) Improved knowledge is needed of effects of changes in climate on crop yields and livestock productivity in different regions and under varying types of management; on soil-nutrient depletion; on hydrological conditions as they effect irrigation-water availability; on pests, diseases and soil microbes, and their vectors and on rates of soil erosion and salinization.

(2) Further information is needed on the range of potentially effective technical adjustments at the farm and village level (e.g. irrigation, crop selection, fertilizing, etc.); on the economic, environmental and political constraints on such adjustments, and on the range of potentially effective policy responses at regional, national and international levels (e.g. re-allocations of land use, plant breeding, improved agricultural extension schemes, large-scale water transfers).

Less than a dozen detailed regional assessments have been made of the potential impact of climatic changes on agriculture. There is therefore no adequate basis for predicting likely effects on food production at the regional or world scale. A comprehensive, international research effort is required, now, to redeem the situation.

REFERENCES

Allen, L.H. Jr, Jones, P. and Jones, J.W. (1985). Rising atmospheric CO_2 and evapotranspiration. In: *Advances in Evapotranspiration*. Proceedings of the National Conference on Advance in Evapotranspiration, pp. 13–27. American Society of Agricultural Engineers, St Joseph, Michigan.

Bergthorsson, P., Bjornsson, H., Dyrmundsson, O., Gudmundsson, B., Helgadottir, A. and Jonmundsson, J.V. (1988). The effects of climatic variations on agriculture in Iceland. In: *The Impact of Climatic Variations on Agriculture, Vol. 1 Cool Temperate and Cold Regions* (Ed. by M.L. Parry, T.R. Carter and N.T. Konijn), pp. 382–509. Kluwer, Dordrecht.

Blasing, T.J. and Solomon, A.M. (1983). Response of North American Corn Belt to Climatic Warming. Prepared for the US Department of Energy, Office of Energy Research, Carbon Dioxide Research Division, Washington, DC, DOE/N88–004.

Carter, T.R., Parry, M.L. and Porter, J.H (1991). Climatic change and future agroclimatic potential in Europe. *Int. J. Climatol.* **11**, 251–269.

Cure, J.D. (1985). Carbon dioxide doubling responses: a crop survey. In: *Direct Effects of Increasing Carbon Dioxide on Vegetation*, (Ed. by B.R. Strain and J.D. Cure), pp. 100–116. US DOE/ER–0238, Washington, USA.

Cure, J.D. and Acock, B. (1986). Crop responses to carbon dioxide doubling: a literature survey. *Agric. Forest Meteorol.* **38**, 127–145.

De Groot, R.S. (1987). Assessment of Potential Shifts in Europe's Natural Vegetation Due to Climatic Change and Implications for Conservation. Young Scientists' Summer Program 1987: Final Report. International Institute for Applied Systems Analysis, Laxenburg, Austria.

Department of the Environment (1991). *Potential Impacts of Climate Change in the UK*. HMSO, London.

Drummond, R.O. (1987). Economic aspects of ectoparasites of cattle in North America. In: *Symposium, The Economic Impact of Parasitism in Cattle*, pp. 9–24. XXIII World Veterinary Congress, Montreal.

Emanuel, K.A. (1987). The dependence of hurricane intensity on climate: mathematical simulation of the effects of tropical sea surface temperatures. *Nature* **326**, 483–485.

Environmental Protection Agency (EPA) (1989). The Potential Effects of Global Climate Change on the United States. Report to Congress.

Gifford, R.M. (1988). Direct effect of higher carbon dioxide levels concentrations on vegetation. In: *Greenhouse: Planning for Climate Change* (Ed. by G.I. Pearman), pp. 506–519. CSIRO, Melbourne, Australia.

Hansen, J., Fung, I., Lacis, A., Rind, D., Lebedeff, S., Ruedy, R. *et al.* (1990). Global climate changes as forecast by Goddard Institute for Space Studies three-dimensional model. *J. Geophysic. Res.* **93–D8**, 9341–9364.

Hill, M.G. and Dymock, J.J. (1989). Impact of Climate Change: Agricultural/Horticultural Systems. DSIR Entomology Division, submission to New Zealand Climate Change Programme, Department of Scientific and Industrial Research, New Zealand.

Intergovernmental Panel on Climate Change (IPCC) (1990). *Scientific Assessment of Climate Change: Policymakers Summary*. WMO and UNEP, Geneva and Nairobi.

Kauppi, P. and Posch, M. (1988). A case study of the effects of CO_2 induced climatic warming on forest growth and the forest sector: A. Productivity reactions of northern boreal forests. In: *The Impact of Climatic Variations on Agriculture, Vol. 1, Assessments in Cool Temperate and Cold Regions* (Ed. by M.L. Parry, T.R. Carter and N.T. Konijn), pp. 183–195. Kluwer, Dordrecht.

Kettunen, L., Mukula, J., Pohjonen, V., Rantanen, O. and Varjo, U. (1988). The effects of climatic variations on agriculture in Finland. In: *The Impact of Climatic Variations on Agriculture: Vol. 1, Assessments in Cool Temperate and Cold Regions* (Ed. by M.L. Parry, T.R. Carter and N.T. Konijn), pp. 511–614. Kluwer, Dordrecht.

Mearns, L.O., Katz, R.W. and Schneider, S.H. (1984). Extreme high temperature events: changes in their probabilities with changes in mean temperatures. *J. Clim. and Appl. Meteorol.* **23**, 1601–1613.

Monteith, J.L. (1981). Climatic variation and the growth of crops. *Q. J. R. Meteorol. Soc.* **107**, 749–774.

Morison, J.I.L. (1987). Intercellular CO_2 concentration and stomatal response to CO_2. In: *Stomatal Function* (Ed. by E. Zeiger, I.R. Cowan and G.D. Farquhar), pp. 229–251. Stanford University Press, Stanford.

Newman, J.E. (1980). Climate change impacts on the growing season of the North American Corn Belt. *Biometeorol.* **7** (2), 128–142.

Nikonov, A.A., Petrova, L.N., Stolyarova, H.M., Levedev, V. Yu, Siptits, S.O., Milyitin, N.N. *et al.* (1988). The effects of climatic variations on agriculture in the semi-arid zone of European USSR: A. The Stavropol Territory. In: *The Impact of Climatic Variations on Agriculture, Vol. 2, Assessments in Semi-Arid Regions* (Ed. by M.L. Parry, T.R. Carter and N.T. Konijn), pp. 587–627. Kluwer, Dordrecht.

Parry, M.L. (1976). The significance of the variability of summer warmth in upland Britain. *Weather* **31**, 212–217.

Parry, M.L. (1990). *Climate Change and World Agriculture*. Earthscan, London.

Parry, M.L. and Carter, T.R. (1988). The assessments of the effects of climatic variations on agriculture: aims, methods and summary of results. In: *The Impact of Climatic Variations on Agriculture, Vol. 1, Assessments in Cool Temperate and*

Cold Regions (Ed. by M.L. Parry, T.R. Carter and N.T. Konijn), pp. 11–95. Kluwer, Dordrecht.

Parry, M.L. and Duinker, P.N. (1990). *The Potential Effects of Climatic Change on Agriculture and Forestry*, Working Group II Report, Intergovernmental Panel on Climate Change. WMO and UNEP, Geneva and Nairobi.

Parry, M.L., Carter, T.R. and Konijn, N.T. (1988). *The Impact of Climatic Variations on Agriculture*, Vol. 1 and 2. Kluwer, Doordrecht.

Parry, M.L., Carter, T.R. and Porter, J.H. (1989). The greenhouse effect and the future of UK agriculture. *Journal of the Royal Agricultural Society of England*, **150**, 120–131.

Pearman, G.I. (Ed.) (1988). *Greenhouse: Planning for Climate Change*. CSIRO, Melbourne, Australia.

Pitovranov, S.E. Zakimets, V., Kiselev, V.I. and Simtenko, O.D. (1988). The effects of climatic variations on agriculture in the subarctic zone of the USSR. In: *The Impact of Climatic Variations on Agriculture, Vol. 1, Assessments in Cool Temperate and Cold Regions* (Ed. by M.L. Parry, T.R. Carter and N.T. Konijn), pp. 617–724. Kluwer, Dordrecht.

Rind, D., Goldberg, R. and Ruedy, R. (1989). Change in climate variability in the 21st century. *Climatic Change* **14**, 5–37.

Rind, D., Goldberg, R., Hansen, J. Rosenzweig, C. and Ruedy, R. (1990). Potential evapotranspiration and the likelihood of future drought. *Journal of Geophysical Research* **95**, 9983–10004.

Rosenzweig, C. (1985). Potential CO_2- induced climate effects on North American wheat-producing regions. *Climatic Change* **7**, 367–389.

Rowntree, P.R., Callander, B.A. and Cochrane, J. (1989). Modelling climate change and some potential effects on agriculture in the UK. *Journal of the Royal Society of England*, **149**, 120–126.

Salinger, M.J. (1989). The Effects of Greenhouse Gas Warming on Forestry and Agriculture. *WMO commission of Agrometeorology*. WMO, Geneva.

Salinger, M.J., Williams, W.M., Williams, J.M. and Martin, R.J. (1990). *Carbon Dioxide and Climate Change: Impacts on Agriculture*. New Zealand Meteorological Service; DSIR Grasslands Division; MAFTech.

SCEGB (1989),. Study Commission of Eleventh German Bundestag, *Protecting the Earth's Atmosphere: An International Challenge*. Bonn University, Bonn.

Smit, B. (1989). Climate warming and Canada's comparative position in agriculture. *Climate Change Digest*, CCD 89–01, Environment Canada.

Squire, G.R. and Unsworth, M.H. (1988). Effects of CO_2 and Climatic Change on Agriculture. Contract Report to the Department of the Environment, Department of Physiology and Environmental Science, University of Nottingham, Sutton Bonnington, UK.

UNEP, (1989). Criteria for Assessing Vulnerability to Sea Level Rise: a Global Inventory to High Risk Areas. United Nations Environment Programme and the Government of the Netherlands, Draft Report.

Warrick, R.A. and Oerlemans, J. (1990). Sea level rise. In: *IPCC Scientific Assessment of Climate Change* (Ed. by J.T. Houghton, G.J. Jenkins and J.J. Ephraums), pp. 257–282. WMO and UNEP, Geneva and Nairobi.

Warrick, R.A. and Gifford, R. with Parry, M.L. (1986). CO_2, climatic change and agriculture. In: *The Greenhouse Effect, Climatic Change and Ecosystems* (Ed. by B. Bolin, B.R. Doos, J. Jager and R.A. Warrick) pp. 393–473. SCOPE 29, John Wiley, Chichester.

Williams, G.D.V. and Oakes, W.T. (1978). Climatic resources for maturing barley and wheat in Canada. In: *Essays on Meteorology and Climatology: in Honour of*

Richard W. Longley, Studies in Geography Mono. 3 (Ed. by Haye, K.D. and Reinelt, E.R.), pp. 367–385. University of Alberta, Edmonton.
Yoshino, M.M., Horie, T., Seino, H., Tsujii, H., Uchijima, T. and Uchijima, Z. (1988). The effect of climatic variations on agriculture in Japan. In: *The Impact of Climatic Variations on Agriculture: Vol. 1, Assessments in Cool, Temperate and Cold Regions* (Ed. by M.L. Parry, T.R. Carter and N.T. Konijn), pp. 723–868. Kluwer, Dordrecht.

Modeling the Potential Response of Vegetation to Global Climate Change

T.M. SMITH, H.H. SHUGART, G.B. BONAN and J.B. SMITH

I. INTRODUCTION

The question, "What is the potential response of vegetation to anthropogenically induced changes in the global climate pattern?" is not a single question, but a class of questions. It encompasses a wide array of patterns and processes ranging from photosynthesis at the leaf level to demographic processes influencing the distribution (both temporal and spatial) of species at a continental to global scale. Although plant response to varying environmental conditions has been investigated experimentally (both in the field and laboratory) from the leaf to the population and community level, the ability to extrapolate these observations directly to higher levels of organization operating over much larger time and space scales is limited. Paleobotanical studies continue to provide observations of large-scale responses of vegetation to global climate change in the past, however, they are not analogous

ADVANCES IN ECOLOGICAL RESEARCH VOL. 22
ISBN 0–12–013922–7

in either the cause or potential time-scale of climate change. This limitation of direct observation of vegetation response to changing climate conditions of the magnitude predicted by general circulation models of the earth atmosphere (GCMs) makes it necessary to synthesize the current understanding of the relationship between plant pattern and climate within a conceptual framework which will allow for the prediction of plant patterns under novel environmental conditions. This is the realm of ecological modeling.

Models provide a means of formalizing a set of assumptions/hypotheses linking pattern and process, allowing for extrapolation beyond the range of observed phenomena. There have been a wide array of models developed to explore the response of vegetation to environmental variation. These models range from purely statistical to models which simulate basic physiological and demographic processes. The purpose of this chapter is not to provide an exhaustive review of models relating climate and plant pattern; rather we will examine a specific set of models which we are currently using to investigate the question of plant response to climate change at a global scale. The focus is on developing a methodology for predicting changes in the large-scale distribution of vegetation (i.e. global distribution of biomes or ecosystem complexes) under changing global climate patterns.

II. CLIMATE–VEGETATION CLASSIFICATION

Perhaps the simplest of models for relating vegetation pattern to climate at a global scale is the approach of climate–vegetation classification. Assuming that the broad-scale patterns of vegetation (e.g. biomes) are essentially at equilibrium with present climate conditions, one can relate the distribution of vegetation or plant types with biologically important features of the climate. Global bioclimatic classification schemes (Grisebach, 1938; von Humbolt, 1867; Koppen, 1900, 1918, 1936; Prentice, 1990; Thornthwaite, 1931, 1933, 1948; Holdbridge, 1947, 1959, 1967; Troll and Paffen, 1964; Box, 1978, 1981) are essentially climate classifications defined by the large-scale distribution of vegetation. Although similar in concept, the wide variation in both the terminology used to describe categories of vegetation and the climate variables identified as important in influencing plant pattern make comparisons among models difficult. Bioclimatic classification models have a history of application in simulating the distribution of vegetation under changed climate conditions, both for past climatic conditions associated with the last glacial maximum (Manabe and Stouffer, 1980; Hansen *et al.*, 1984; Prentice and Fung, 1990) and predictions of future climate patterns under conditions of doubled CO_2 (Emanuel *et al.*, 1985; Prentice and Fung, 1990; Smith *et al.*, 1991).

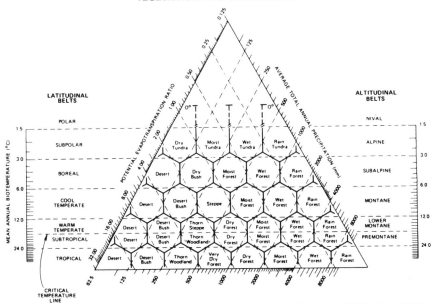

Fig. 1. Holdridge climate–vegetation classification scheme. (Source: Holdridge, 1967.)
A detailed description of model is provided in the text.

III. APPLICATION OF HOLDRIDGE LIFE-ZONE CLASSIFICATION TO CLIMATE CHANGE AT A GLOBAL SCALE

The Holdridge life zone classification system is a bioclimatic model relating
the distribution of natural vegetation associations to climate indices. The
features of the Holdridge classification are summarized in Fig. 1. The life
zones are depicted by a series of hexagons formed by intersecting intervals of
climate variables on logarithmic axes in a triangular coordinate system. Two
variables, average biotemperature and average annual precipitation, determine the classification. Average biotemperature is the average temperature
over a year with the unit temperature values (i.e. daily, weekly or monthly)
that are used in computing the average set to 0°C if they are less than or equal
to 0°C.

In the Holdridge diagram (Fig. 1), identical axes for average annual
precipitation form two sides of an equilateral triangle. A logarithmic axis for
the potential evapotranspiration (PET) ratio (effective humidity) forms the
third side of the triangle, and an axis for mean annual biotemperature is
oriented perpendicular to the base. By marking equal intervals on these
logarithmic axes, hexagons are formed that designate the Holdridge life
zones. Each life zone is named to indicate a vegetation association.

Table 1
General circulation models used to construct climate change scenarios

GCM	Change in mean global: Temperature (°C)	Precipitation (%)
Oregon State University (OSU)[a]	2·84	7·8
Geophysical Fluid Dynamics Laboratory (GFDL)[b]	4·00	8·7
Goddard Institute for Space Studies (GISS)[c]	4·20	11·0
United Kingdom Meteorological Office (UKMO)[d]	5·20	15·0

[a]Schlesinger and Zhao, 1988.
[b]Manabe and Wetherald, 1987.
[c]Hansen *et al.*, 1988.
[d]Mitchell, 1983.

The potential evapotranspiration is the amount of water that would be released to the atmosphere under natural conditions with sufficient, but not excessive, available water throughout the growing season. The potential evapotranspiration ratio is the quotient of PET and average annual precipitation. Holdridge (1959) assumes, on the basis of studies of several ecosystems, that PET is proportional to biotemperature. The PET ratio in the Holdridge diagram is therefore dependent on the two primary variables, annual precipitation and biotemperature.

One additional division in the Holdridge system is based on the occurrence of killing frost. This division is along a critical temperature line that divides hexagons between 12 and 24°C into warm temperate and subtropical zones. This line is adjusted to reflect regional conditions. The complete Holdridge classification at this level includes 37 life zones.

Expected current distributions of Holdridge life zones were mapped using a climate data base of mean monthly precipitation and temperature at a 0·5° (latitude and longitude) resolution (Leemans and Cramer, 1990). Simulations of current and $2 \times CO_2$ climates from four GCMs (Table 1) were used to construct climate change scenarios. Changes in mean monthly precipitation and temperature were calculated for each GCM scenario for each computational grid element by taking the difference between simulated current and $2 \times CO_2$ climates. These data from each GCM were interpolated to 0·5° resolution and changes in monthly precipitation and temperature were then applied to the global climate data base to provide a change scenario. The altered data bases corresponding to each of the four GCM scenarios were then used to reclassify the grid cells (0·5°) using the Holdridge system.

Table 2
Percentage of terrestrial land area
showing shift in Holdridge life zone
under changed climate

Scenario	% Land area
GFDL	48·0
GISS	44·3
OSU	39·4
UKMO	55·0

Maps of the distribution of life zones under current climate and climate change scenarios based on the OSU and UKMO GCMs (Table 1) are presented in Fig. 2 in the colour plate section. All four climate change scenarios investigated show a significant shift in the distribution of life zones (Table 2). The OSU and UKMO scenarios presented in Fig. 2 represent the extremes of the four scenarios investigated (Table 1). The UKMO-based scenario exhibits a large warming at the northern latitudes resulting in the northward shift of the boreal zone into the area now occupied by tundra. In addition, there is extensive drying in the tropics resulting in the transition from mesic forest to woodland on both the South American and African continents. In contrast, OSU shows a lesser degree of warming at the northern latitudes and an expansion of the tropical forested region as a result of increased precipitation. Despite these marked differences there are a number of consistent patterns across scenarios. All four scenarios show a considerable drying in the mid-latitudes of the northern hemisphere. This drying (combination of both increased temperatures and decreased precipitation) results in the transition from currently forested areas to shrublands and grasslands.

A summary of changes in the distribution of major biome types under the four climate change scenarios is shown in Table 3. All scenarios show a decrease in tundra and desert with corresponding increases in the extent of grassland and forest. The total areal coverage of forest increased in all cases. However, the scenarios differed in the extent to which this increase is attributable to mesic and xeric components (Table 3). All four scenarios show an increase in the extent of dry forest, however, the GFDL and UKMO scenarios show a decrease in the extent of mesic forest (sum of moist, wet and rain forest life zones). These overall changes are more easily interpreted when viewed in the context of changes in the forest types forming these two larger categories of forest.

Changes in areal coverage of the five major mesic forest types as defined by Holdridge are presented in Fig. 3. The types are defined by summing over the moist, wet and rain forest cells for each temperature zone (see Fig. 1). The histograms present the areas for each forest type under current climate and

T. M. SMITH *ET AL.*

Table 3

Changes in the areal coverage of major biome-types[a] under current and changed climate conditions

	Area (km² × 10³)				
	Current	OSU	GFDL	GISS	UKMO
Tundra	939	−302	−515	−314	−573
Desert	3699	−619	−630	−962	−980
Grassland	1923	380	969	694	810
Dry forest	1816	4	608	487	1296
Mesic forest	5172	561	−402	120	−519

[a]Tundra: Polar dry tundra, polar moist tundra, polar wet tundra, polar rain tundra.

Desert: polar desert, boreal desert, cool temperate desert, warm temperate desert, subtropical desert, subtropical desert bush, tropical desert, tropical desert bush.

Grassland: cool temperate steppe, warm temperate thorn steppe, subtropical thorn steppe, tropical thorn steppe, tropical very dry forest.

Dry forest: warm temperate dry forest, subtropical dry forest, tropical dry forest.

Mesic forest: moist, wet and rain forest for boreal, cool temperate, warm temperate, subtropical and tropical temperature zones.

the four climate change scenarios. These predicted changes in coverage under the scenarios can be examined in terms of their components. The portion of the histogram shaded black represents the area of that forest type that remained stable (i.e. did not change type) under the given scenario. The portion designated by grey represents the area which was covered by some other mesic forest type under current climate conditions, but is predicted to change to that forest type under the new climate conditions. The unshaded portion is that area which is predicted to change to that forest type but is currently covered by a non-mesic forest type (i.e. dry forest, grassland, desert or tundra).

There is a large increase in tropical forest under all scenarios, ranging from 87% for the UKMO scenario to 159% for OSU. This increase is largely due to the expansion of the tropical zone (defined in terms of biotemperature: see Fig. 1) under the warmer climate conditions as predicted by the GCMs, resulting in the reclassification of areas currently occupied by subtropical and warm temperate forest. Of the areas predicted to change to tropical forest under the climate change scenarios, 74–98% are currently occupied by another mesic forest type. In general, the global warming results in a poleward shift in all forest zones with the subsequent expansion of the boreal forest zone into the region currently occupied by tundra. This poleward shift can be seen in Fig. 3, where the majority of the area involved in the spatial

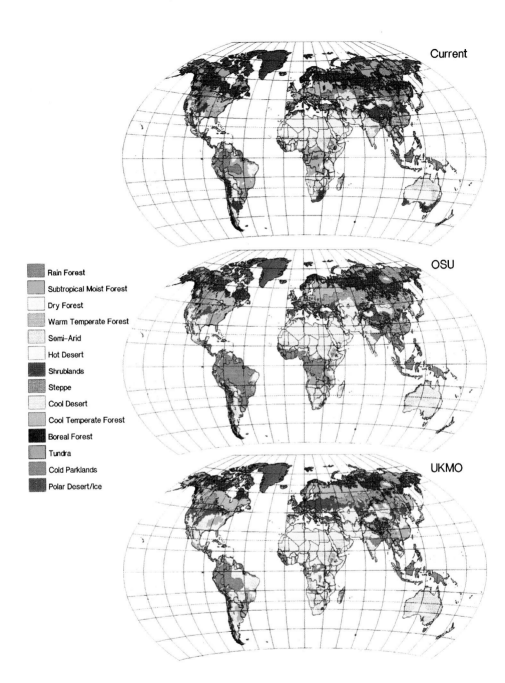

Current

OSU

UKMO

Rain Forest

Subtropical Moist Forest

Dry Forest

Warm Temperate Forest

Semi–Arid

Hot Desert

Shrublands

Steppe

Cool Desert

Cool Temperate Forest

Boreal Forest

Tundra

Cold Parklands

Polar Desert/Ice

Fig. 2. Maps of the distribution of Holdridge life zones under current climate and climate change scenarios based on UKMO and OSU general circulation models (see Table 1 for reference to scenarios).

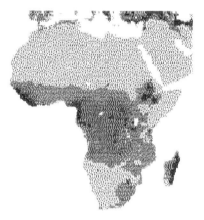

MAP of LAI under Current climate scenario.

Map of LAI under OSU 2x CO_2 climate change scenario.

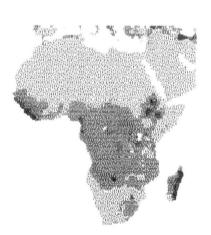

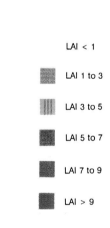

LAI < 1

LAI 1 to 3

LAI 3 to 5

LAI 5 to 7

LAI 7 to 9

LAI > 9

Map of LAI under UKMO 2x CO_2 climate change scenario

Fig. 4. Predicted patterns of leaf area index (LAI) for Africa under current climate and climate change scenarios based on UKMO and OSU general circulation models. Predictions based on model of Woodward (1987).

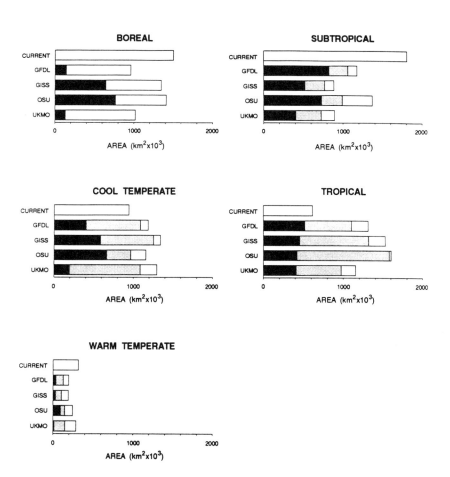

Fig. 3. Areal coverage of mesic forest types under current climate and four climate change scenarios. Black: Area which is currently occupied by that forest type and remains so under the changed climate conditions (i.e. stable). Grey: Area which is predicted to change to that forest type but is currently occupied by some other forest type. Unshaded: Area which is predicted to change to that forest type but is not currently occupied by mesic forest (i.e. tundra, grassland, etc.)

shifts in forest type (with the exception of boreal forest) are from areas currently occupied by other mesic forest types (i.e. grey area of histograms).

IV. APPLICATION OF A PLANT ENERGY BALANCE MODEL TO PREDICTING CHANGES IN LEAF AREA UNDER CHANGING CLIMATE CONDITIONS

The Holdridge life zone system presented above is essentially a correlation between climate and vegetation patterns. As a correlative method, it can be used to interpolate within the range of observed climate conditions (i.e. current climate patterns), however, it cannot provide a means of extrapolating the response of vegetation to novel climate conditions (i.e. combinations of temperature and precipitation outside currently observed global patterns). Extrapolation outside currently observed climate conditions requires a more mechanistic/process-oriented approach relating plant response to environmental conditions. One such approach is based on the assumption that an equilibrium exists between climate, soil water-holding capacity and maximum leaf area in water-limited ecosystems. Given this assumption, it is then possible to solve for maximum sustainable leaf area under a given set of site conditions and climate (Woodward, 1987; Nemani and Running, 1989).

Woodward (1987) developed an approach for predicting leaf area and associated physiognomy using a model of plant energy balance. The Penman–Monteith (Penman, 1948; Monteith, 1973) evapotranspiration model is a biophysical model used to simulate the physiology of water use at the canopy level. Using a set of parameters describing the environment of the canopy, combined with a functional relationship between leaf environment and stomatal conductance, the model predicts evapotranspiration for a given leaf area and climatic conditions. By solving the model iteratively for varying values of leaf area, Woodward used the model to solve for the maximum leaf area which could be sustained under the climatic conditions at any given location.

This approach provides a process-based alternative to the vegetation–climate classification models described above. By using this approach to predict leaf area under current and changed climate conditions, it is possible to predict changes in the leaf area which can be sustained under the changed climate conditions. The comparison of current and changed leaf area index (LAI) can then be used to infer potential changes in the composition and structure of vegetation which may relate to the predicted shifts in LAI.

The expected current distribution of LAI for the continent of Africa was mapped at a 0·5° resolution using a combination of data from the global climate database (Leemans and Cramer, 1990) referred to above in the application of the Holdridge classification system and Muller (1982). Mean monthly statistics were used to simulate daily climate using standard meteorological algorithms (see Woodward, 1987 for detailed discussion). Application of the GCM-based climate change scenarios followed the same procedures outlined above for the analyses based on the Holdridge system.

Maps of the current and changed distribution of LAI for the continent of Africa under the OSU- and UKMO-based scenarios are presented in Fig. 4 in the colour plate section. The climate change under the UKMO scenario results in a significant drying and subsequent reduction in LAI in the tropical forest region of central Africa. Much of the current mesic/rainforest region of central-west Africa is predicted to decline in leaf area to values corresponding to dry woodland/savannah. In contrast, there is an expansion of the tropical forest zone in this region under the OSU scenario.

The resulting patterns of LAI under the two climate change scenarios are in general agreement with the shifts in life zone distribution (i.e. implied physiognomic structure and associated LAI) predicted by the Holdridge system. This agreement between the bioclimatic and the energy-balance models suggests that the predictions of an increase in the area associated with mesic/rainforest in tropical Africa for the OSU scenario using the Holdridge classification system are at least consistent with limitations on plant water balance.

V. MODELING TEMPORAL DYNAMICS

The equilibrium-based approaches discussed above implicitly assume a time-scale sufficient for vegetation migration and eventual equilibrium of vegetation to the new "changed" climate patterns. In contrast, a dynamic view requires the explicit consideration of plant demographic processes to capture the transient nature of vegetation response to changing environmental conditions. There are numerous models of vegetation dynamics which simulate the demographics of plant populations (see Shugart and West, 1980 for review), however, we will focus on a single class of models known as gap models.

VI. INDIVIDUAL-BASED FOREST GAP MODELS

Forest gap models simulate the establishment, growth, and mortality of trees at an annual time-step on a plot of defined size (Shugart, 1984). Although the models differ in their inclusion of processes which may be important in the dynamics of the particular site being simulated (e.g. hurricane disturbance, flooding), all forest gap models share a common set of characteristics and demographic processes.

Each individual plant is modeled as a unique entity with respect to the processes of establishment, growth, and mortality. This allows the model to track species- and size-specific demographic behaviors. The model structure includes two features important to a dynamic description of vegetation: (a) the response of the individual plant to the prevailing environmental conditions, and (b) how the individual modifies those environmental conditions (i.e. the feedback between vegetation structure/composition and the environment).

A. Response of Individual to Environment

The growth of an individual is calculated using a function that is species-specific and predicts, under optimal conditions, an expected growth increment given an individual's current size (Fig. 5). This optimum increment is then modified by a set of environmental response functions, and the realized increment is added to the individual.

Environmental responses are modeled via a "constrained potential" paradigm. In this, an individual has a maximum potential behavior under optimal conditions (i.e. maximum diameter increment, survivorship, or establishment rate). This optimum is then reduced according to the environmental context of the plot (e.g. shading, drought), to yield the realized behavior under ambient conditions. The functions describing species response to environmental resources tend to be generic curves that scale between 0·0 and 1·0, and species are often categorized into a small number of functional types (Fig. 6).

The horizontal position of each individual on the plot is not considered. The assumption is made that each individual on the plot is influenced by (and influences) the growth of all other individuals on the plot. Given this assumption of "horizontal homogeneity", the size of the simulated plot is critical (Shugart and West, 1979). The spatial scale at which these models operate is an area corresponding to the zone of influence of a single individual of maximum size. This allows for an individual growing on the plot to achieve maximum size while at the same time allowing for the death of a large individual to influence the environment on the plot significantly (i.e. gap formation).

Competition in the model is indirect and depends on the relative performance of the individuals under the environmental conditions on the model plot. These environmental conditions may be influenced by the trees themselves (e.g. a tree's leaf area influences light available beneath it), or may be modeled as extrinsic (e.g. temperature). Competitive success thus depends on the environmental conditions on the plot, which species are present, and the relative sizes of the individuals; each of these varies through time in the model.

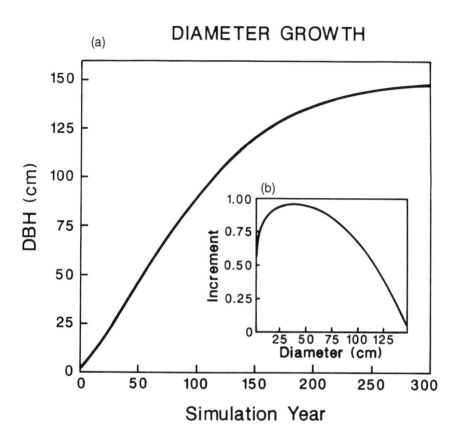

Fig. 5. Example of optimal growth function used to simulate diameter growth of individual trees in forest gap models. Size growth expressed as a function of (a) time, and (b) current size.

The death of individuals is modeled as a stochastic process. Most gap models have two components of mortality: age-related and stress induced. The age-related component applies equally to all individuals of a species and depends on the maximum expected longevity for the species. Stress is defined with respect to a minimal growth increment; individuals failing to meet this minimal condition are subjected to an elevated mortality rate.

Establishment and regeneration are largely stochastic, with maximum potential establishment rates constrained by the same environmental factors that modify growth.

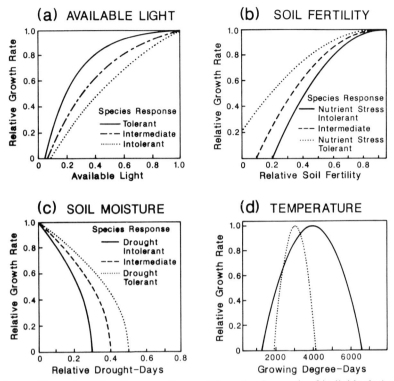

Fig. 6. Example of functions used to modify optimal growth of individuals (see Fig. 5) in forest gap models.

B. Effect of Individual on Environment

Although horizontal homogeneity on the plot is assumed, the vertical structure of the canopy is modeled explicitly. The sizes of individuals (height and leaf area, which are allometric on diameter) are used to construct a vertical leaf-area profile. Using a light extinction equation, the vertical profile of available light is then calculated so that the light environment for each individual can be defined.

All gap models simulate individual tree influence on light availability at height intervals on the plot. Plant influences on other environmental factors are incorporated to varying degrees in different versions of the model; these include soil moisture, fertility, temperature, as well as disturbances such as fires, and herbivory.

Plant influence on soil moisture has been incorporated using a variety of soil hydrology/evapotranspiration models (e.g. Priestly–Taylor, Penman–Monteith) which allow for the distribution of roots and leaf area of the plot to influence the soil moisture profile (Pastor and Post, 1985; Bonan, 1989; Martin, 1990). In return, soil moisture integrated over the growing season is used to modify plant growth (e.g. root biomass and leaf area).

Soil nutrient availability on the plot is influenced by the plants in terms of the quantity and quality of litter input into the soil organic layer for decomposition (Aber et al., 1982; Weinstein et al., 1982; Pastor and Post, 1985). Species are classified into functional categories based on leaf characteristics (e.g. C/N, lignin content) which relate to their decomposability. Nitrogen dynamics (and other mineral nutrients) of the plot are simulated as a function of the prevailing environmental conditions at the soil surface and the quantity and quality of litter material.

Soil temperature is influenced by the collective leaf area of individuals on the plot, which reduces incoming radiation to the soil surface, as well as by the insulating effect of litter material on the plot. In turn, soil temperatures are needed as input for calculating decomposition rates and the development of a permafrost layer at northern latitudes. These processes, together with direct temperature effects on growth, provide a feedback to plant processes (Bonan, 1989, 1991).

In actuality, the modeling framework outlined above has two components or modules; one involving plant demographics, and the other involving the processes, both climatic and biophysical, which define the environment on each plot. An extensive discussion of the biophysical models is beyond the scope of this chapter, but a complete discussion of these models and their incorporation into the gap model structure can be found in the references provided above.

In general, these biophysical models require the transfer of certain parameters from the demographic module describing the vegetation of the plot and in return define the environmental conditions as a function of the geology and soils, prevailing climate and the modifying influence of the plant structure.

The gap model approach has been applied to a wide array of forested systems ranging from boreal (Bonan, 1988; Leemans and Prentice, 1987) to subtropical rainforest (Shugart et al., 1980; Doyle, 1981) and has recently been extended to communities dominated by herbaceous vegetation (Coffin and Lauenroth, 1989).

VII. APPLICATION OF GAP MODELS TO PREDICT FOREST RESPONSE TO CLIMATE CHANGE

Gap models have been applied to examine the response of forested systems to climate changes, both in the reconstruction of prehistoric Quaternary forests (Solomon et al., 1980, 1981; Solomon and Shugart, 1984; Solomon and

Webb, 1985; Bonan and Hayden, 1990), as well as to project possible consequences of future climate change (Solomon *et al.*, 1984; Solomon, 1986; Pastor and Post, 1988; Urban and Shugart, 1989; Bonan *et al.*, 1990; Overpeck *et al.*, 1990). In contrast to the two modeling approaches discussed in the earlier sections, the gap model approach is high resolution in that it can predict species composition, vegetation structure and associated productivity and standing biomass through time. However, it is limited in spatial scale, that is, it is limited in the spatial extent to which the results can be extrapolated. The reason for this limitation is that the information required to parameterize/initialize a model which can address changes in these state variables (e.g. species composition and productivity through time) relate to site-specific features such as topographic position, soil characteristics, land-use history, history of disturbance, and present vegetation structure, all of which may vary over short distances. The application of fine resolution models to provide total coverage over broad regions would be virtually impossible due to both computational and data limitations. As an alternative, sampling approaches have been put forward to provide large-scale coverage over broad environmental gradients (Solomon, 1986; Bonan 1990a,b).

The large-scale bioclimatic models discussed earlier (e.g. Holdridge life zone classification system) can be used to develop a systematic process of site selection based on the implied shifts in the distribution of major ecosystem comlexes. As an example we will use the North American boreal zone.

The shift in the North American boreal forest zone for the UKMO- and OSU-based climate change scenarios shown in Fig. 2 involves an array of transitions between life zones and can be used to locate sites of potential importance for application of stand level models.

A forest gap model (Bonan, 1988) was applied to a series of sites in the North American boreal zone (Fig. 7) to explore the implied transitions based on the Holdridge life zone system. The model simulates plant demographics following the approach outlined above for gap models and includes biophysical models for solar radiation, permafrost, decomposition and soil hydrology/evapotranspiration (Bonan, 1989). Sites were selected based on: (a) representation of differing life zone transitions based on the Holdridge analyses, and (b) availability of data for model validation under current climate. Site descriptions, parameterization and validation of the model for the locations shown in Fig. 7 are discussed in Bonan (1988, 1989).

For each site, stand structure and composition were simulated under current climate conditions, defined by monthly statistics (mean and standard deviation) for precipitation, temperature and cloudiness. The resulting stand description after 500 years of simulation was used to initialize simulations for control (continued current climate statistics) and changed climate conditions based on the four GCMs discussed earlier. Changes in monthly temperature

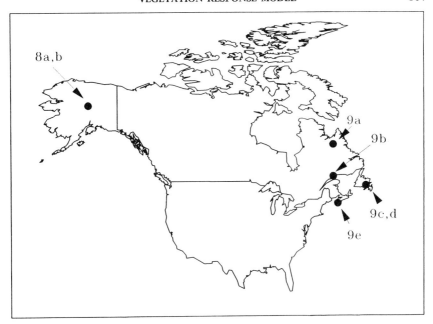

Fig. 7. Location of sites for boreal forest simulations using forest gap model.

and precipitation for each site were gleaned from the global databases for current and changed climate for the 0·5° cell corresponding to the site location. To provide a transient climate scenario based on the temerature and precipitation changes from the equilibrium GCM simulations, the changes were assumed to occur over a 70-year period (years 0–70 of simulation). Annual values for each month were then determined by linearly interpolating between current and 2 × CO_2 values over this period. Climate statistics were then assumed to remain constant after this period for the remainder of the simulation. Only a local seed pool was considered in light of the relatively short duration of the simulations relative to the timescale of observed patterns of species migration (Davis, 1984, 1989; Davis and Botkin, 1985).

The simulation of a site on a south-facing slope in central Alaska is presented in Fig. 8a. The site is currently dominated by white spruce (*Picea glauca*), birch (*Betula papyrifera*) and aspen (*Populus tremuloides*). Under the warmer, drier conditions predicted under all scenarios, the stand declines as a function of moisture stress. The site would eventually be dominated by herbaceous ground cover. The differences in the timing of the decline among the scenarios reflect the varying degree of warming predicted. This result is in agreement with the transition from boreal forest to cool temperate steppe (grassland) predicted for the area using the Holdridge life zone system.

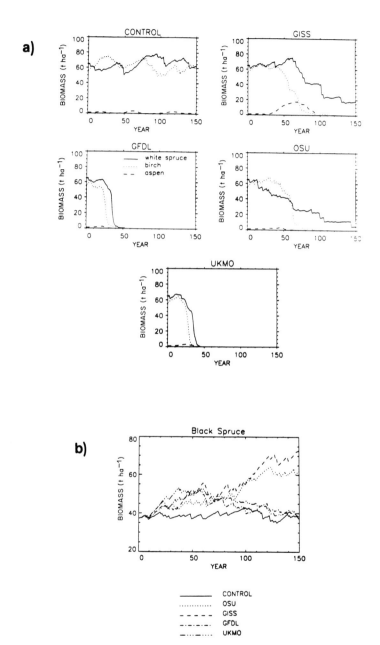

Fig. 8. Gap model simulations of changes in biomass for central Alaska site: (a) south-facing slope, and (b) north-facing slope.

In contrast to the predicted decline in forest cover on the south-facing slope, simulation of a stand in the same location on a north-facing slope (Fig. 8b) shows an increase in biomass of the dominant species, black spruce (*Picea mariana*). This difference is due to the link between the stand water balance and solar radiation. The lower input of solar radiation to the north-facing slope (in comparison to the south-facing) reduces the moisture stress associated with increasing temperature.

Simulation of a site in the boreal forest–tundra transition zone of northern Quebec, Canada is shown in Fig. 9a. The site is currently an open-canopy woodland of black spruce with relatively low productivity and standing biomass. With the warming for this region predicted under all scenarios, the stand increases in biomass to a closed-canopy forest. This transition would be associated with a decline in the present ground cover of moss and lichen. The occurrence of fire would facilitate this transition in that the establishment of seedlings on the site is currently hindered by ground cover.

In contrast to the northern Quebec site, the predicted dynamics of a black spruce stand in central Quebec (Fig. 9b) vary among the scenarios. The OSU and GISS scenarios show little change from the pattern predicted under current climate (i.e. control). These results reflect the lower warming predicted for the region by these two GCMs. In contrast, both the GFDL and UKMO scenarios predict a complete dieback of the stand, with the rates of decline directly related to the degree of warming. These declines are not a function of moisture stress, but result from temperatures rising above those currently associated with the southern limit to the distribution of black spruce. The area is predicted to change to a cool temperate forest under the Holdridge life zone system, and the establishment of species characteristic of that forest type would be dependent on a range of processes operating at the landscape level (e.g. rates of dispersal and establishment).

This result points out a limitation in our current understanding of the distribution of forest zones at a continental scale. The predicted decline in the boreal forest at its southern boundary under increased temperatures has an implied assumption that the southern distributional limits of the component species are physiologically determined (e.g. exceed temperature tolerance for germination, establishment or growth). In contrast, if competition between boreal and northern hardwood (i.e. cool temperate forest) species is a determining factor in the southern distributional limit of boreal forest, then the predicted decline may not occur. In fact, the increased temperatures may result in increased productivity of these stands and the transition to cool temperate forest would involve the eventual invasion of northern hardwood species into established boreal forest. The rate of this transition would be related to the ability of the species to invade established stands; influenced by rates of disturbances such as fire (Overpeck *et al.*, 1990).

110   T. M. SMITH *ET AL.*

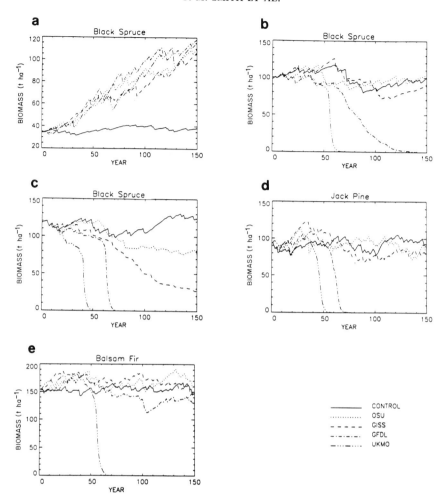

Fig. 9. Gap model simulations of changes in biomass for Canadian boreal sites: (a) Schefferville, northern Quebec, (b) central Quebec, (c) Newfoundland, (d) Newfoundland, and (e) New Brunswick.

Similar results to those observed for central Quebec are predicted for black spruce in Newfoundland (Fig. 9c). Stand biomass declines under all four scenarios with the rates of decline a function of the degree of warming. As with the central Quebec site, the area is predicted to change to a cool temperate forest and the decline in black spruce is a function of the temperatures rising above those currently associated with its southern limit.

Simulation of a balsam fir (*Abies balsamea*) stand on sandy soils in the same area shows slightly different results (Fig. 9d). Only the scenario with the largest temperature increase (UKMO) results in a significant decline in stand biomass. The difference between the two forest types (i.e. black spruce and balsam fir) for this location is a function of the differences between the two species in their growth response to temperature.

Similar results to those observed for balsam fir are predicted for Jack pine (*Pinus banksiana*) in New Brunswick. The two warmest scenarios for the region (UKMO and GFDL) show a decline in stand biomass, while the OSU and GISS scenarios differ little from current climate conditions (i.e. control).

VIII. DISCUSSION

The global analysis of vegetation response to climate change based on the Holdridge life zone system is essentially an equilibrium solution to a dynamic process of vegetation change. The approach has many implicit assumptions about the stability of vegetation associations/types under changed climate conditions, and the ability of vegetation to track (migrate with) spatial changes in climate patterns. As such, the results are best viewed as changes in the climate associated with a given vegetation zone rather than changes in vegetation *per se*. Despite these limitations, the method does provide a means of interpreting the large-scale changes in climate pattern as they relate to the current correlation between climate and vegetation distribution.

Like the Holdridge system, the energy-balance based approach of Woodward (1987) represents an equililibrium solution to a dynamic process. The predicted patterns of LAI under current and changed climate are equilibrium solutions to the corresponding spatial changes in climate patterns from the two GCM scenarios. The model does not explore the temporal dynamics of vegetation response to changed climate conditions bounded by these two solutions, rather it assumes that vegetation can respond by increasing/decreasing leaf area to equilibrate with these new conditions, even though this may require major shifts in species or even life-form composition.

Despite the differences between these two approaches (i.e. Holdridge and Woodward models), there is a general agreement in the predicted patterns of vegetation response for the continent of Africa under the OSU- and UKMO-based climate change scenarios. The agreement between the models for the decline in tropical rainforest under the warmer and drier conditions predicted by the UKMO scenario is to be expected. The predicted climate corresponds to that of areas currently occupied by dry woodland/savannah. In contrast, the predictions of warmer and wetter for the OSU scenario do not have current climate analogs. The energy-balace model suggests that the increased precipitation for the region is sufficient to meet increased water demand of the current leaf area under the higher temperatures. However, it should be noted that the model does not consider direct temperature effects on photosynthesis and respiration. A linked carbon–energy balance model would be necessary to examine the potential response of the changed climate conditions on net primary productivity and subsequent changes in leaf area and standing biomass.

The general qualitative agreement between the transitions predicted by the Holdridge life zone system and the directions of change predicted for the gap model simulations in the North American boreal zone is not surprising. The climate changes precited by the GCMs for this region are not subtle. If the changes in climate patterns are sufficient to shift ecosystems/biomes in climate space (i.e. Holdridge cells; see Fig. 1), it is not surprising that the processes of establishment, growth and mortality of component species are likewise influenced.

Despite the general agreement between the predicted shifts in Holdridge life zones and the results of the forest gap model simulations, the predictions of the forest gap model varied as a function of site differences for a given location (e.g. north- vs. south-facing slope: Fig. 8). Also, although species within a given geographic region may have similar response to large-scale environmental patterns (e.g. temperature and rainfall), interspecific variation is sufficient to result in differential responses in growth under changing climate conditions (e.g. black spruce and balsam fir in Newfoundland: Fig. 9). These differences between the results of the two modeling approaches point out the basic trade-off between resolution and scale discussed earlier. The forest gap model is able to provide a detailed description of vegetation dynamics. However, it is limited in the degree to which the results can be generalized. This limitation is due to both the species and site-specific attributes required for the simulations. To provide complete spatial coverage over a large area it is necessary to reduce the resolution of the description of both the environment and the vegetation itself. This processes of aggregation or averaging will have a direct effect on the predicted dynamics (Smith and Urban, 1988).

Unlike the equilibrium solutions of the Holdridge life zone system, the gap models provide a link between individual plant response and changing climate conditions. This simulation of demographic processes enables the models to track temporal changes in composition and structure as a function of changing climate conditions. Where the shift in climate is detrimental to plant growth (i.e. lower available plant moisture), the dynamics of stand decline are on the time-scale of the change in climate. This is seen in the timing of stand decline in the south-facing slope in central Alaska (Fig. 8b). In contrast, where the conditions for plant growth improve, the dynamics are influenced by a range of factors, including the ability of the present vegetation to respond to the more favorable conditions. The immigration into the area of species which are suited to the new climate conditions will be dependent on processes operating at the landscape scale, such as patterns of dispersal and establishment. The modeling framework as presented here does not include these landscape-level processes, and the integration of processes operating at these disperate scales is one of the major items on the agenda in current climate change research. The purpose of this chapter has been to examine some of the possible methodologies for evaluating vegetation response to climate change and how those methods could be integrated to address patterns and processes operating at differing spatial and temporal scales. The task is not a simple one, and just as the question of vegetation response to climate change is not a single question but a class of questions, so the task of modeling vegetation response to changing climate patterns requires an array of approaches. The results shown are preliminary and for the purpose of exploring the possible implications of climate changes on the order of magnitude predicted by GCMs. As can be seen from the results, the GCMs vary widely in their predictions. In some cases the differences are not only quantitative, but qualitative (e.g. OSU and UKMO scenarios for tropical Africa). Because of this variation, it is probably premature to rely on the GCMs for the development of predictive scenarios in examining the potential implications of anthropogenically induced climate change on global vegetation patterns. However, the magnitudes of change are significant and point to the need to understand the relationships between plant distribution and climate, and the development of methods for evaluating the potential response of vegetation as the predictive powers of the general circulation models improve.

REFERENCES

Aber, J.D., Hendry, G.R., Francis, A.J., Botkin, D.B. and Melillo, J.M. (1982). Potential effects of acid precipitation on soil nitrogen and productivity of forest ecosystems. In: *Acid Precipitation: Effects on Acid Precipitation* (Ed. by F.M. Ditri), pp. 411–434. Ann Arbor Science, Ann Arbor, Michigan.

Bonan, G.B. (1988). Environmental Processes and Vegetation Patterns in Boreal Forests. Ph.D. dissertation, University of Virginia, Charlottesville, VA.

Bonan, G.B. (1989). A computer model of the solar radiation, soil moisture, and soil thermal regimes in boreal forests. *Ecol. Mod.* **45**; 275–306.

Bonan, G.B. (1990a). Carbon and nitrogen cycling in North American boreal forests. I. Litter quality and soil thermal effects in interior Alaska. *Biogeochem.* **10**; 1–28.

Bonan, G.B. (1990b). Carbon and nitrogen cycling in North American boreal forests. II. Biogeographic patterns. *Can. J. For. Res.* **20**, 1077–1088.

Bonan, G.B. (1991). A biophysical surface energy budget analysis of soil temperature in the boreal forests of interior Alaska. *Water Res. Res.*, in press.

Bonan, G.B. and Hayden, B.P. (1990). Using a forest stand simulation model to examine the ecological and climatic significance of the late-Quaternary pine–spruce pollen zone in eastern Virginia, USA. *Quat. Res.* **33**, 204–218.

Bonan, G.B., Shugart, H.H. and Urban, D.L. (1990). The sensitivity of some high-latitude boreal forests to climatic parameters. *Clim. Change* **16**, 9–29.

Box, E.O. (1978). Ecoclimatic Determination of Terrestrial Vegetation Physiognomy. Ph.D. dissertation, University of North Carolina, Chapel Hill, NC.

114   T. M. SMITH *ET AL.*

Box, E.O. (1981). *Macroclimate and Plant Forms: An Introduction to Predictive Modeling in Phytogeography.* Junk, Hague.

Coffin, D.P. and Lauenroth, W.K. (1989). A gap dynamics simulation model of succession in a semi-arid grassland. *Ecol. Mod.* **49**, 229–266.

Davis, M.B. (1984). Climatic instability, time lags and community disequilibrium. In: *Community Ecology* (Ed. by J. Diamond and T.J. Case), pp. 269–284. Harper and Row, New York.

Davis, M.B. (1989). Lags in vegetation response to greenhouse warming. *Clim. Change* **15**, 75–82.

Davis, M.B. and Botkin, D.B. (1985). Sensitivity of cool-temperate forests and their fossil pollen record to rapid temperature change. *Quat. Res.* **23**, 327–340.

Doyle, T.W. (9181). The role of disturbance in the gap dynamics of a montain rain forest: An application of a tropical forest succession model. In: *Forest Succession: Concepts and Applications* (Ed. by D.C. West, H.H. Shugart and D.B. Botkin), pp. 56–73. Springer-Verlag, New York.

Emanuel, W.R., Shugart, H.H. and Stevenson, M.P. (1985). Climatic change and the broad-scale distribution of terrestrial ecosystem complexes. *Clim. Change* **7**, 29–43.

Grisebach, A. (1838). Ueber den Einfluss des Climas auf die Begranzung der naturlichen Floren. *Linnaea* **12**, 159–200.

Hansen, J., Lacis, A., Rind, D., Russell, G., Stone, P., Fung, I. *et al.* (1984). Climate sensitivity: Analysis of feedback mechanisms. In: *Climate Processes and Climate Sensitivity* (Ed. by J.E. Hansen and T. Takashashi). *Geophys. Monogr. Ser.* **29**, pp. 130–163. Maurice Ewing Ser. **5**.

Hansen, J., Fung, I., Lacis, A., Rind, D., Russell, G., Lebedeff, S. *et al.* (1988). Global climate changes as forecast by the GISS-3-D model. *J. Geophys. Res.* **93**, 9341–9364.

Holdridge, L.R. (1947). Determination of world formulations from simple climatic data. *Science* **105**, 367–368.

Holdridge, L.R. (1959). Simple method for determining potential evapotranspiration from temperature data. *Science* **130**, 572.

Holdridge, L.R. (1967). *Life Zone Ecology.* Tropical Science Center, San Jose.

Humbolt, A. von (1867). *Ideen zu einem Geographie der Pflazen nebst einem naturgemalde der Tropenlander.* Tubingen, FRG.

Koppen, W. (1900). Versuch einer Klassification der Klimate, vorzugsweise nach ihren Beziehungen zur pflanzenwelt. *Geogr. Z.* **6**, 593–611.

Koppen, W. (1918). Klassification der Klimate nach Temperatur, Niedenschlag und Jahreslauf. *Petermanns Geogr. Mitt.* **64**, 193–203.

Koppen, W. (1936). Das Geographische System der Klimate. In: *Handbuch der Klimatologie*, Vol. 1, part C (Ed. by W. Koppen and R. Geiger), p. 46. Gebr Borntraeger, Berlin.

Leemans, R. and Cramer, W. (1990). The IIASA climate database for land area on a grid of 0.5° resolution. WP-41, International Institute for Applied Systems Analysis, Luxenburg.

Leemans, R. and Prentice, I.C. (1987). Description and simulation of tree layer composition and size distributions in a primaeval *Picea–Pinus* forest. *Vegetatio* **69**, 147–156.

Manabe, S. and Stouffer, R.J. (1980). Sensitivity of a global climate to an increase in CO_2 concentration in the atmosphere. *J. Geophy. Res.* **8**, 5529–5554.

Manabe, S. and Wetherald, R.T. (1987). Large scale changes in soil wetness induced by an increase in carbon dioxide. *J. Atm. Sci.* **44**, 1211–1235.

Martin, P. (1990). Forest Succession and Climate Change: Coupling Land-Surface Processes and Ecological Dynamics. Ph.D. dissertation, University of California, Berkley.

Mitchell, J.F.B. (1983). The seasonal response of a general circulation model to changes in CO_2 and sea temperature. *Q.J. Roy. Met. Soc.* **109**, 113–152.

Monteith, J.L. (1973). *Principles of Environmental Physics*. E. Arnold, London.

Muller, J.M. (1982). *Selected Climatic Data for a Global Set of Standard Stations for Vegetation Science*. Dr W. Junk, London.

Nemani, R.R. and Running, S.W. (1989). Testing a theoretical climate–soil–leaf area hydrologic equilibrium of forests using satellite data and ecosystem simulation. *Agr. For. Met.* **44**, 245–260.

Olson, J.S., Watts, J.A. and Allison, L.J. (1983). Carbon in Live Vegetation of Major World Ecosystems. ESD Pub. No. 1997, Oak Ridge National Laboratory, TN.

Overpeck, J.T, Rind, D. and Goldberg, R. (1990). Climate-induced changes in forest disturbance and vegetation. *Nature* **343**, 51–53.

Pastor, J. and Post, W.M. (1985). Development of a Linked Forest Productivity–Soil Process Model. ORNL/TM-9519. Oak Ridge National Laboratory, Oak Ridge, TN.

Pastor, J. and Post, W.M. (1986). Influences of climate, soil moisture, and succession on forest carbon and nitrogen cycles. *Biogeochemistry* **2**, 3–27.

Pastor, J. and Post, W.M. (1988). Response of northern forests to CO_2-induced climate change. *Nature* **334**, 55–58.

Penman, H.L. (1948). Natural evaporation from open water, bare soil and grass. *Proc. R. Soc. (Lond.) Series A* **193**, 120–145.

Post, W.M., Emanuel, W.R., Zinke, P.J. and Stangenberger, A.G. (1982). Soil carbon pools and world life zones. *Nature* **298**, 156–159.

Prentice, K.C. and Fung, I.Y. (1990). Bioclimatic simulations test the sensitivity of terrestrial carbon storage to perturbed climates. *Nature* **346**, 48–51.

Prentice, K.C. (1990). Bioclimatic distribution of vegetation for GCM studies. *J. Geophy. Res.* **95 (D8)**, 11811–11830.

Schlesinger, M. and Zhao, Z. (1988). Seasonal climatic changes induced by doubled CO_2 as simulated by the OSU atmospheric GCM/mixed layer ocean model. Climate Research Institute, Oregon State University, Corvallis, OR.

Sedjo, R.A. and Solomon, A.M. (1989). Climate and forests. In: *Greenhouse Warming: Abatement and Adaptation* (Ed. by N.J. Rosenberg, W.E. Easterling, P.R. Crosson, and J. Darmstadter), pp. 105–119. Resources For The Future, Washington, DC.

Shugart, H.H. (1984). *A Theory of Forest Dynamics*. Springer-Verlag, New York.

Shugart, H.H. and West, D.C. (1979). Size and pattern of simulated forest stands. *For. Sci.* **25**, 120–122.

Shugart, H.H. and West, D.C. (1980). Forest succession models. *BioScience* **30**, 308–313.

Shugart, H.H., Hopkins, M.S. Burgess, I.P. and Mortlock, A.T. (1980). The development of a succession model for subtropical rainforest and its application to assess the effects of timber harvest at Wiangaree State Forest, New South Wales. *J. Env. Man.* **11**, 243–265.

Smith, T.M. and Urban, D.L. (1988). Scale and resolution of forest structural pattern. *Vegetatio* **74**, 143–150.

Smith, T.M., Leemans, R. and Shugart, H.H. (1991). Sensitivity of terrestrial carbon storage to CO_2-induced climate change: Comparison of four scenarios based on general circulation models. *Clim. Change*, in press.

Solomon, A.M. (1986). Transient responses of forests to CO_2-induced climate change: Simulation modeling experiments in eastern North America. *Oecologia* **68**, 567–569.

Solomon, A.M. and Shugart, H.H. (1984). Integrating forest-stand simulations with paleoecological records to examine long-term forest dynamics. In: *State and Change of Forest Ecosystems* (Ed. by G.I. Agren), pp. 333–357. Indicators in Current Research, Report Number 13. Swedish University of Agricultural Science, Upsalla, Sweden.

Solomon, A.M. and Webb, T. III. (1985). Computer-aided reconstruction of late-Quarternary landscape dynamics. *Ann. Rev. Ecol. Syst.* **16**, 63–84.

Solomon, A.M., Delcourt H.R., West, D.C. and Blasings, T.J. (1980). Testing a simulation model for reconstruction of prehistoric forest-stand dynamics. *Quat. Res.* **14**, 275–293.

Solomon, A.M., West, D.C. and Solomon, J.A. (1981). Simulating the role of climate change and species immigration in forest succession. In: *Forest Succession* (Ed. by D.C. West, H.H. Shugart and D.B. Botkin), pp. 154–177. Springer-Verlag, New York.

Solomon, A.M., Tharp, M.L., West, D.C., Taylor, G.E., Webb, J.M. and Trimble, J.L. (1984). Response of Unmanaged Forests to CO_2-induced Climate Change: Available Information, Initial Tests and Data Requirements. Tech. Report TROO9., US DOE Carbon Dioxide Research Division, Washington DC.

Thornthwaite, C.W. (1931). The climates of North America according to a new classification. *Geogr. Rev.* **21**, 633–655.

Thornthwaite, C.W. (1933). The climates of the earth. *Geogr. Rev.* **23**: 433–440.

Thornthwaite, C.W. (1948). An approach toward a rational classification of climate. *Geogr. Rev.* **38**, 55–89.

Troll, C. and Paffen, K.H. (1964). Karte Der Jahreszeitenklimate der Erde. Erkund. *Arch. Wiss Geogr.* **18**: 5–28.

Urban, D.L. and Shugart, H.H. (1989). Forest response to climate change: A simulation study for Southeastern forests. In: *The Potential Effects of Global Climate Change on the United States* (Ed. by J. Smith and D. Tirpak), pp. 3-1 to 3-45. EPA-230-05-89-054, US Environmental Protection Agency, Washington, DC.

Weinstein, D.A., Shugart, H.H. and West, D.C. (1982). The Long-term Nutrient Retention Properties of Forest Ecosystems: A Simulation Investigation. ORNL/TM-8472. Oak Ridge National Laboratory, Oak Ridge, TN.

Woodward, F.I. (1987). *Climate and Plant Distribution.* Cambridge University Press, Cambridge.

Effects of Climatic Change on the Population Dynamics of Crop Pests

M.E. CAMMELL and J.D. KNIGHT

I. SUMMARY

The "greenhouse effect" may lead to a notable rise in global mean temperature and also changes in patterns of precipitation and wind. Such changes in climate and weather could profoundly affect the population dynamics and status of insect pests of crops. These may arise not only as a result of direct effects on the distribution and abundance of pest populations but also via effects on the pests' host plants, competitors and natural enemies.

Climatic factors, notably temperature, directly affect the survival, development, reproduction and movement of individual insects and thus the potential distribution and abundance of a particular pest species. In temperate regions, the distribution and survival of many pests is often limited by low temperatures and particularly by the cold conditions which occur in winter. Therefore, warmer conditions in such regions are likely to extend the

ADVANCES IN ECOLOGICAL RESEARCH VOL. 22
ISBN 0–12–013922–7

potential geographic range of some insect pests and also to increase their abundance where decreased development times will enable extra generations to occur. Thus warmer conditions may increase the importance of some existing pests and also lead to the occurrence of new pests which were previously restricted by unfavourable, low temperatures. However, warmer conditions during the autumn–spring period could adversely affect, at least in the short term, the phenology of some cold-hardy, indigenous, temperate pest species whose life cycles are currently synchronized with prevailing environmental conditions. In lower latitudes, hotter and more arid conditions may also adversely affect some pest species.

Within the climatically favourable range of a particular pest, its population may be constrained by the availability of suitable host plants and the action of natural enemies. Climatic change may affect host plants in several ways; for example drought stress may cause changes in plant chemistry and plant structure which may increase or decrease the plant's suitability as a host. Also, climatic change may affect host plant phenology and the distribution and abundance of both crop and non-crop hosts. The distribution and abundance of natural enemies is also likely to be affected by changes in climate. In temperate climates, warmer conditions will tend to favour many natural enemies but whether they play a more effective role in control will also depend upon the responses of the pest to changed climatic conditions. To provide a realistic assessment of whether a particular pest will or will not become more widespread and abundant requires consideration of the overall pest–host plant–natural enemy relationship and how this is likely to alter under different climatic conditions and over varying time scales.

Crop production and crop protection systems may also change in response to climatic changes and these may affect the status of particular pests. Clearly, climatic change increases the need for developing and implementing soundly based pest management strategies. Although precise climatic change cannot at present be reliably predicted, the potentially considerable and wide-ranging consequences for pest management require that the impact of various climatic scenarios on crop pests should be considered as a matter of urgency.

II. INTRODUCTION

There is increasing evidence that the "greenhouse effect" will result in substantial global warming during the next 50–100 years (Houghton et al., 1990). According to most climate models the globally averaged rise in surface air temperature will then be within the range 1·5 – 5·5°C and is likely to be accompanied by changes in precipitation patterns (Bolin et al., 1986). However, there is considerable uncertainty in terms of the amount and pattern of warming at the regional level and also the effects of such changes upon other climatic components such as precipitation and wind patterns. It is suggested from climate models that warming is likely to be greater in higher latitudes and that such increase will be greater during winter than summer. In the UK this change in climate would be similar in effect to a southward shift in latitude of about 10° (Wilson, 1989). Also, there may be increases in precipitation in high mid-latitudes, particularly in winter, but drier summers in low mid-latitudes (Bolin et al., 1986). Furthermore, although long-term changes in mean temperatures and precipitation will probably follow a relatively slow cumulative pattern, the short-term effects may involve sudden changes in the frequency and magnitude of hitherto unusual or extreme climatic events such as drought and storm (Bolin et al., 1986). It is against this often complex and uncertain background that possible effects on the dynamics of insect pests of crops will be discussed.

A characteristic of many crop pests is that populations show a sudden, often explosive, increase in numbers over a relatively short period of time, which leads to an outbreak and crop damage. Such outbreaks may vary considerably in magnitude, persistence and in their frequency and pattern of occurrence. This variation may occur not only between different species but also between populations of the same species in different seasons and localities. The outbreak phase often ends abruptly with a sudden decline and this may be followed by a return to a relatively stable mean density, although populations of opportunistic pest species in unstable environments tend to remain in a state of flux. These changes in population numbers occur as a result of varying mortality, natality and movement of individuals and it is the relative effects of biotic and abiotic factors on these three components which determine the dynamics of a population. Clearly the dynamics of pest populations are complex and diverse and yet it is fundamental that the applied ecologist should be able to explain and predict the conditions under which an outbreak of a particular pest population will or will not occur.

For many years it has been widely recognized that climate is often the principal factor limiting the geographical distribution of a particular species population and also that variations in climate and of its short-term expression which we call "weather" have profound effects upon pest abundance (Uvarov, 1931). Furthermore, the synchronization of pest life cycles with seasonal changes in climate is often dependent upon the insects' response to certain climatic cues during the life cycle. However, there has been considerable debate in relation to the role of weather in the overall dynamics of pest populations. During the 1940s and 1950s contrasting theories were developed to explain both the numerical changes in, and the apparent stability of, insect populations (Andrewartha and Birch, 1954; Nicholson, 1958). These theories largely centred upon the roles of abiotic factors, notably climate, and of biotic factors such as natural enemies and their relative importance as density-dependent and density-independent mechanisms. A particular theory tended to emphasize certain aspects and therefore theories differed in their interpretation of the importance of particular individual factors. In fact, these differences in emphasis highlighted the complexity and diversity of population dynamics and the need for a more unified approach.

Since the early 1960s the study of population dynamics has moved increasingly towards a more synoptic or holistic approach which embraces both population and environment (Clark et al., 1967; Southwood, 1975, 1977). Such an approach has been made possible by the development of more sophisticated techniques, notably computer modelling, which enables the ecologist to analyse and evaluate a complexity of factors and interactions which may occur both within the population and also between the population and its environment. Thus, in recent years there has been considerable progress in our fundamental understanding of the nature of pest population dynamics and of the causes of pest outbreaks (Barbosa and Schultz, 1987). Currently, it is widely recognized that climate usually sets the limits of the arena within which everything else reacts (Huffaker et al., 1984). Thus, biotic factors come into play only if the abiotic environment remains favourable for a sufficient time for population pressures to build and then these become increasingly important. Furthermore, climatic factors not only have many direct effects on pest populations but also important are climatic effects which act indirectly through their effects on the pests' natural enemies and host plants.

Clearly the dynamics of pest populations are complex. However, it is also equally important that ecologists concerned with bioclimatic studies should recognize that climate and weather are also complex and dynamic. As Southwood (1968) aptly remarked "the problems of the meteorologist and the population ecologist have much in common". Thus, bioclimatic studies involve a whole continuum of events in time and space, from varying microenvironmental effects on the individual organism to those widespread climatic patterns and processes which affect the entire pest species population. To quote Wellington and Trimble (1984) "if we are to understand fully the impact of weather on an insect, we must be able to project our imaginations up and down-scale well beyond the limits our own size imposes".

Since the early 1900s bioclimatic data have been used extensively as the basis for pest forecasting systems. Important early contributions were made by Bodenheimer (1925) who described the potential distribution of the Mediterranean fruit fly, Ceratitis capitata using appropriate physiological data and mean monthly temperatures at different localities, and Cook (1924) who plotted monthly temperature and precipitation tolerances for the pale western cutworm, Agrotis orthogonia and projected these to predict potential outbreak areas for the pest. Also particularly noteworthy were the studies of Glenn (1922) and Shelford (1927, 1929) who used the physiological time-scale (day degrees) as the basis for predicting the time of spring emergence of the Codling moth, Cydia pomonella and thus improved the timing of chemical controls. In 1931, Uvarov (1931) published his classical work, *Insects and Climate*, with the principal aim of providing a basis for practical bioclimatic studies of economically important insects: this comprehensive and critical review can still be read with considerable benefit.

Since Uvarov's review there have been numerous studies on biometeorology and its application in pest forecasting and a wide range of experimental and analytical techniques has been developed (Messenger, 1959, 1974). During the past 10 years in particular, there has been considerable interest in developing forecasting systems for a multitude of purposes, ranging from those offering short-term tactical advice on the need for, and timing of, chemical controls to those longer-term strategic forecasts concerned with, for example, prediction of potential risks from exotic pests in new localities (Norton and Way, 1983; Sutherst and Maywald, 1985; Cammell and Way, 1987). Key individual meteorological factors that account for most of the variability form the basis of many practical forecasting models and key factor analysis (Varley and Gradwell, 1960, 1968) may also offer a first step towards developing other models. Continuing improvements in the measurement of biological and climatic factors and the application of computer technology will undoubtedly make the implementation of increasingly more sophisticated predictive models possible. Predicting the effects of a range of climatic scenarios on the many facets of population dynamics offers a formidable but exciting challenge to those concerned with pest forecasting.

Much progress has been made in using climatic data to predict pest distribution, survival and abundance under varying climatic conditions. In this chapter we examine the possibilities and limitations of using such information to predict possible effects of climatic change on pest populations. First, we examine the direct effects of varying temperature and moisture on the survival, development and reproduction of the individual insect and also of temperature and wind on insect movement. Secondly, we consider how these climatic factors directly affect distribution and abundance at the level of the species population. Thirdly, we examine the indirect effects of climate on pest populations through its effects upon the pests' natural enemies and host plants. Fourthly, we discuss climatic change in relation to its possible varying effects upon pest–host plant–natural enemy relationships and how such variation may affect the distribution and abundance of pest populations, using aphids as an example. Finally, the implications of climatic change for pest management are considered. In this context we are principally concerned with evaluating the status of pests, which depends not only on the numbers of the organism present but also on socio-economic sensitivity and biological tolerance of the host to pest attack (Norton and Conway, 1977; Southwood, 1977). Figure 1 shows the major pathways by which climate may affect the distribution, abundance and status of pests.

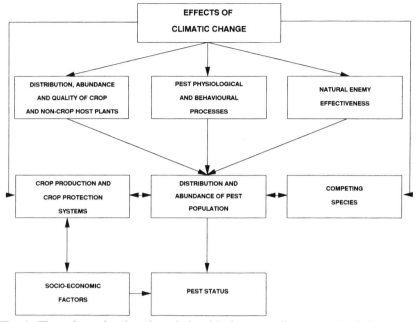

Fig. 1. Flow chart showing the relationship between climate, pests, their natural enemies and host plants.

III. DIRECT EFFECTS OF CLIMATE ON PEST POPULATIONS

The assessment of the direct effects of climate on populations involves carefully designed and complementary laboratory and field studies. Laboratory studies are used to provide precise information on survival, growth, development, fecundity, birth and death rates under a wide range of experimentally controlled environmental conditions. Such data is then used to calculate various life-history parameters for a particular species and to predict numerical changes in the field under varying climatic conditions. Field studies are used to evaluate this data critically in relation to the heterogeneous population in its natural environment. Additional information and modifications arising from field studies are then used to design further laboratory experiments, which in turn may suggest further field studies.

A. Effects on the Physiological and Behavioural Processes of Individual Organisms

Two important climatic factors affecting the physiology of insects are temperature and moisture. Furthermore, their effects are often interactive and are therefore conveniently considered together.

Insects are ectothermic with their physiological processes considerably influenced by ambient temperatures. Typically there are upper and lower lethal temperature limits between which the organism survives and within the survival range there is a more restricted range for growth and development and an even further restricted range for reproduction. Relative humidity tends to modify the effects of temperature.

1. Lethal Limits and Survival

The upper temperature at which death occurs depends on the species, the duration of exposure and its interaction with other factors, notably humidity. For many species the lethal temperature for short-term exposures is within the range 40–50°C (Chapman, 1988) although more prolonged exposure to temperatures slightly below these may cause cessation of feeding and development and subsequent death. At low humidities relatively large insects may exhibit notable evaporative cooling, allowing them to survive higher temperatures in the short term whereas over longer periods of time greater evaporation in dry air may result in the insect dying from desiccation (Wigglesworth, 1972).

The lower lethal temperature varies considerably in different species (Salt, 1961). Many species from warm environments soon die even at temperatures well above freezing-point, probably because of adverse changes in their normal metabolic processes which result in accumulation of toxic products. Also, some species may remain alive for a considerable time at low temperatures but eventually starve to death. For example, *Locusta* sp. is unable to feed below about 20°C (Chapman, 1988). In contrast, many insects in cool and cold environments are able to survive temperatures considerably below 0°C as a result of supercooling and freezing tolerance (Salt, 1961). Cold hardiness is particularly pronounced in diapausing and hibernating stages of insects where they may need to survive long periods below 0°C (Andrewartha and Birch, 1954; Wigglesworth, 1972).

The limits of insect survival may also vary depending upon the stage of development. For example, in many pests of stored products, eggs and pupae are usually more susceptible to extreme temperatures compared with adults. Larvae usually increase in resistance as they grow except at each moulting period when they become more susceptible, and fully grown larvae are usually the most resistant stage to temperature extremes (Howe, 1963). Also, in the aphids *Myzus persicae* and *Sitobion avenae* the nymphs are more cold-hardy than adults but early nymphal instars more resistant than later instars (Bale *et al.*, 1987).

In practice, defining precise and meaningful lethal temperature limits is often difficult. The major problem is in relating results obtained in the

124   M. E. CAMMELL AND J. D. KNIGHT

laboratory to field conditions. In the laboratory, probit analysis (Finney, 1971) has been used to assess the survival rates of batches of insects exposed for different periods at differing temperatures. However, responses to temperatures are not static but vary according to the previous experience of the insect. Thus acclimatization and adaptation may occur and lead to differences in the tolerance of particular populations to both high and low temperatures and their lethal effects. Work by Carter (1972) on the green spruce aphid, *Elatobium abietinum* demonstrates the important differences which may occur in the cold-hardiness of insects in relation to their previous experiences. Thus about 50% of aphids reared at 0°C for 5 days survived −15°C whereas about 50% of those reared at 15°C died at −9°C (Fig. 2). Consequently, in relatively mild winters the aphids do not become cold-hardened through exposure to low temperatures and so may be particularly susceptible to subsequent sudden cold snaps. In daily fluctuating regimes, tolerance to a given extreme maximum may be affected by the daily minimum (Meats, 1984). Minima too close or too far from an extreme maximum may result in that maximum causing greater mortality than it would if it were preceded by a minimum of intermediate distance (Meats 1984, 1989a). In the field, insects may avoid exposure to otherwise lethal ambient temperatures by various behavioural responses, such as when air temperatures are dangerously high the insects may move to cooler parts of the habitat.

Therefore realistic studies of lethal temperatures must consider the possible variation between populations arising from acclimatization and adaptation in varying environments and also ensure that experimental conditions relate accurately to those experienced by an individual insect in its micro-environment. Particularly important in the micro-environment are the often considerable insulating effects of snow which protect overwintering stages of some insects from rapid temperature changes or very low temperatures (Bakke, 1968; Speight and Wainhouse, 1989). Furthermore, insects in the soil and within plant tissues may also be buffered from extreme air temperatures. There may also be considerable differences in the temperatures experienced by insects which are in sun-exposed sites as compared with those in shaded sites (Way and Banks, 1964; Speight and Wainhouse, 1989).

During the life cycles of many pest species there are periods when climatic conditions permit survival but are unfavourable for insect growth, development and reproduction. These periods may occur irregularly and insects typically respond to these unpredictable changes by either becoming quiescent or dispersing to more favourable areas. Alternatively, these periods may occur regularly and arise through widespread seasonal fluctuations in temperature and moisture. Such changes are predictable and insects respond by undergoing certain physiological and behavioural changes which lead to

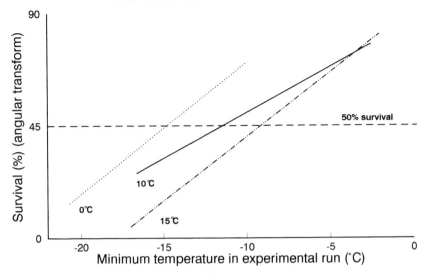

Fig. 2. Regressions of survival of the green spruce aphid on minimum temperature when exposed to low temperature for 6 h. Before exposure, aphids were pretreated at 0, 10, 15°C for at least 5 days. (Source: Speight and Wainhouse (1989) after Carter, 1972.) (Crown Copyright, Published with kind permission of the Forestry Commission.)

dormancy and thus enhance survival during an approaching unfavourable period (Tauber and Tauber, 1976; Tauber *et al.*, 1984). Dormancy and particularly diapause are also important in synchronizing the phenology of pest life cycles because their termination and initiation set the beginning and end of population growth.

Diapause has been recorded in all life stages but for each species the diapausing stage is genetically fixed. Also the species-specific characteristics that comprise a particular diapause are primarily under genetic control (Tauber *et al.*, 1984). However, diapause induction, maintenance and termination are profoundly influenced by environmental factors, notably photoperiod and temperature. Temperature influences diapause in several ways. First, it often interacts with photoperiod to induce diapause. In species that develop without diapause in long day-lengths, low temperatures promote diapause and high temperatures tend to suppress it, whereas in short-day insects or those that aestivate, high temperatures often have the reverse effect. Thermoperiods may also enhance or diminish the effect of diapause-inducing photoperiods and may even induce diapause under non-photoperiodic conditions. Thermoperiodic effects on diapause have been demonstrated in a number of pest species, including the European corn borer, *Ostrinia nubilalis*, pink bollworm, *Pectinophora gossypiella* and the cabbage white butterfly, *Pieris brassica* (Beck, 1983). Secondly, temperature is important in maintaining diapause in many species. The temperature range for rapid diapause development is usually much below that for non-diapause development and low thermal thresholds during diapause ensure that diapause is maintained and development is prevented by warm conditions that may occur before or during winter (Tauber and Tauber, 1976). Thirdly, temperature may also be an active diapause-terminating stimulus.

The effects of moisture on diapause are in general of only minor import- ance. Water absorption may be required to terminate overwintering egg diapause but evidence is debatable. Moisture may also influence diapause development in some insect larvae such as *O. nubilalis* (Beck, 1980).

In conclusion, possible changes in the magnitude and pattern of tempera- tures whilst photoperiods remain the same may have varying effects on the diapause of different pest species because individual species respond differ- ently to temperature, photoperiod and their interactions. Also, as discussed later, seasonal synchrony between the pest species and its parasitoids may be affected where the host pest and the parasitoids respond differently to changed environmental conditions during diapause.

2. Development

The number of days required to complete development depends upon the insect's temperature-dependent rate curve and the temperature regime it experiences. Characteristically, as discussed earlier, there is a threshold below which no development occurs. Above this threshold development increases gradually and then there is a range of temperatures over which development increases linearly to reach a maximum or optimum rate before declining rapidly close to the upper lethal limit. Thus, developmental time for a particular species can be expressed more meaningfully in terms of physiologi- cal time in which a constant number of heat units (i.e. thermal constant) above the lower developmental threshold are required to complete develop- ment.

Temperature optima for rates of development for pest species usually lie in the range from 22 to 38°C with the lower limits being in the range of 12 – 22°C below the optimum (Taylor, 1981). Optima may also vary between different developmental stages. For example the optimum for eggs of *Lymantria* sp. is 22°C whereas for larvae it is 28°C (Zwolfer, 1935). The minimum time periods taken to complete development from birth to adulthood may also vary considerably between individual species. For example, the pea aphid, *Acyrthosiphon pisum* takes less than 7 days at its optimum temperature of 27°C whereas the cutworm, *Noctua pronuba* requires over 50 days at its optimum of 29°C (Taylor, 1981).

Humidity may also affect the developmental rates of many species. Optimum conditions for many lie in the range of 60 – 80% relative humidity and insects will often respond to slight differences in humidity by moving to the preferred humidity whenever possible. In many species there is a gradual lengthening of development time as relative humidity declines and this may also be accompanied by a lower optimum temperature than occurs at higher humidities, as, for example, in *Dermestes frischii* (Amos, 1968). However, some species, particularly those such as the khapra beetle, *Trogoderma granarium* which live in dry environments, are unaffected by humidity over a very wide range and effects only become evident at extreme temperatures. Interactions between relative humidity and temperature are often conveniently considered in terms of thermohygrograms (Howe and Burges, 1953) (Fig. 3).

The most frequently used method of predicting the development rates of individual life stages is the temperature summation model. The input into such models is the developmental time of various life stages of the insect as determined over a range of constant temperatures in the laboratory and assumes a linear relationship between development rates and temperature. This model has proved reasonably accurate for predicting developmental rates in the field, particularly when temperatures are within the most favourable range for development. However, the model has predictive limitations when temperatures are around the lower and upper thresholds, for the development curve is sigmoid rather than linear and therefore deviations from linearity are most pronounced towards the temperature extremes. In recent years, non-linear models have been developed which are based on the logistic equation, sigmoid curve or polynomial functions (Baker and Miller, 1974; Stinner *et al.*, 1974; Fletcher and Kapatos, 1983; Fletcher and Comins, 1985). Developmental data obtained at constant temperatures may also have certain important limitations when applied to the fluctuating conditions which occur in the natural environment (Ratte, 1985). Thus at relatively low mean temperatures developmental rates are greater under fluctuating temperatures than those obtained at equivalent mean constant temperatures, and the greater the diurnal temperature variation the greater the effect (Baker, 1980).

3. Reproduction

Temperature may considerably influence the total number of eggs laid and the pattern of oviposition of individual adults. First, temperature, through its previous effects on the growth of the immature stages of a species, may influence the size and weight of the adult and in females there is often a close relationship between body mass, the number of ovarioles and egg mass (Ratte, 1985). Secondly, temperatures during the reproductive phase of adults can considerably influence fecundity. In fact, reproduction is affected adversely by extremes of temperature more readily than most other physiological functions. The temperature range for egg production is maximal at temperatures relatively close to the upper limit for reproduction and then

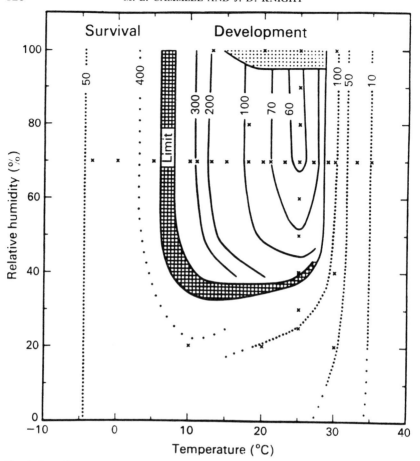

Fig. 3. Development time (days) and physical limits for survival of *Ptinus tectus*. (Source: Howe and Burges, 1953.) (Published with kind permission of CABI.)

falls steeply at higher temperatures and more gradually at lower, with the optimum differing from species to species (Bursell, 1974a). Furthermore, as the temperature limits for reproduction are approached this may not only result in decreased numbers of eggs per female but also in a loss in fertility of an increasing proportion of females (Ratte, 1985). Both total fecundity and the range over which egg production occurs may vary under fluctuating temperatures compared with similar mean constant temperatures. For example, the spotted alfalfa aphid, *Therioaphis maculata* produces offspring between 8 and 32°C under fluctuating temperatures whereas at equivalent constant temperatures they are only produced between 15 and 30°C (Messenger, 1964b).

In general, rates of oviposition increase as relative humidity is raised. However, at the humid end of the range, the situation differs from species to species. Thus in *Locusta* sp. eggs are laid at the highest rate when the relative humidity is 70% and at higher humidities there is a sharp drop whereas in *Cryptolestes* sp. the rate of oviposition increases progressively towards saturation (Bursell, 1974b). Humidity may also affect other behavioural aspects. For example, the calling behaviour and hence subsequent mating success of *O. nubilalis* is affected by humidity (Webster and Cardé, 1982).

4. Movement

The trivial and migratory movements of insects involve both behavioural processes and the effects of various meteorological factors, notably temperature and the wind. Migration is of fundamental importance in the dynamics of many pest populations for it enables species to relinquish habitats that are becoming unfavourable and to exploit those where conditions are more favourable for survival, development and reproduction. The most dramatic example of widespread, long-distance migrations are those of locusts, armyworms, leafhoppers and aphids and these may involve the influence of mesocale weather systems (Johnson, 1969; Rabb and Kennedy, 1979).

There is a temperature threshold below which winged insects are unable to take-off (Taylor, 1963) and a lower one below which they are unable to fly (Cockbain, 1961). The threshold for take-off varies not only between species but also there is variation between individuals of the same species. For example, the proportion of individuals of the aphids, *S. avenae* and *Metopolophium dirhodum* taking-off increases as temperatures increase (Walters and Dixon, 1984). Also Wiktelius (1981) has shown that the emigrants of the bird-cherry aphid, *Rhopalosiphum padi* that leave the primary host, *Prunus padus* in spring have a notably higher minimum temperature threshold than gynoparae that return to the primary host in autumn.

In some pests, for example, the Colorado beetle, *Leptinotarsa decemlineata*, intense insolation is seemingly more important than a high ambient temperature in allowing, or stimulating, take-off (Johnson, 1969). In temperate *Pierid* butterflies, take-off and flight may be severely limited by low temperatures and lack of solar radiation at critical times and this may ultimately lead to much reduced oviposition because the short-lived adults must fly and search out oviposition sites and lay eggs (Kingsolver, 1989). Also there are upper threshold limits; for example the scolytid beetle, *Conophthorus coniperdi* tends not to fly at temperatures above 35°C (Henson, 1962). As mentioned earlier, the temperature thresholds for flight tend to be lower than for take-off. For example, *Dendroctonus pseudotsugae* is able to fly at 13°C but does not take-off below 18 – 20°C (Johnson, 1969). The air speeds of some insects are affected by temperature but others show remarkably constant speeds over a wide range of ambient temperatures.

Few studies have been done on the effects of relative humidity on the take-off and flight of insects. Wellington (1945) concluded that, within the range

25–92%, there was no appreciable effect of relative humidity. However Broadbent (1949) recorded diminished frequency of take-off by the aphids *Brevicoryne brassicae* and *M. persicae* at high humidities.

The major determinants of migration are the behavioural mechanisms of the pest and the nature of the wind. This is particularly evident in weak fliers such as aphids. During take-off, flight and landing, the alate aphid changes its behaviour or "mood" (Moericke, 1955; Harrewijn *et al.*, 1981). Initially there is a strong urge to leave the plant on which it has developed and the aphid flies rapidly upwards in response to predominantly short wavelengths of light from the sky and host plant stimuli are suppressed. Strong winds inhibit take-off but the urge for flight is so great that even strong wind merely delays departure. Flight then settles down to a period of cruising but as it progresses the urge for flight decreases, there is an increasing responsiveness to longer wavelengths of light from vegetation, and the aphid has the desire to land and colonize fresh host plants. However, the direction and distance travelled is considerably influenced by the nature of the wind. During periods of calm weather aphids may be able to undertake short low-level flights. However, in many geographical regions such opportunities only occur rarely and alatae are usually carried to high altitudes by turbulent convection and then enter the general atmospheric circulation within which they have little control over their destination. Disembarking will also depend upon wind speed and normally requires convection currents for the alate aphid to enter the boundary layer where it is able to control its own movements because flight speed exceeds wind speed (Taylor, 1965).

Wind-borne dispersal is seemingly a haphazard means of distribution. Nevertheless, it has proved a highly successful means by which many pest species can scan new hosts over large distances. Clearly, changes in any of the many facets of wind systems could profoundly affect the dynamics of many migratory and dispersive pest species. Models have been recently developed that simulate the effects of many of these systems on airborne movement of insects (Rabb and Kennedy, 1979; Pedgley, 1982; McManus, 1988).

B. Effects on Distribution and Abundance of Pest Populations

In the previous section we briefly described how the interactions of organisms with temperature, moisture and wind determine important characteristics of individuals with regards to survival, development, reproduction and movement. In this section we consider how these characteristics of individuals translate into population-level phenomena in terms of distribution and abundance of the species population.

1. Distribution

The current potential range of a species is the maximum range which can be occupied without intrinsic change in tolerance of the population or extrinsic change in the environment (Munroe, 1984), especially climate. A particular species may not necessarily occupy its entire potential range, for it may not have had time to spread that far or it may be prevented from doing so by lack of suitable host plants, a geographical barrier or quarantine measures. Thus, climatic change may not only change the potential range of a species but it may also influence the rate of spread and degree of accessibility into the hitherto unoccupied areas of the range.

The potential and actual geographical ranges of pest species vary considerably in size. However, many important pest species are relatively widespread geographically (Anon, 1989) and their actual range limits are also diffuse because often such species are able to exploit, through migration, those areas which permit only sporadic occurrence but where, nevertheless, they attain pest status.

In temperate regions, the distribution and survival of many pests is often limited by low temperatures and particularly by the cold conditions that occur in winter. Thus, some insect species which are endemic pests in warmer climates may occasionally invade temperate areas but are then either unable to complete development during the summer because of insufficient thermal units or may complete one or more generations during a favourable summer but are then caught at the onset of winter in a stage of development that is not cold-hardy or able to migrate. In pest species of temperate origin the life cycle usually includes some physiological mechanism such as diapause that permits survival over what would otherwise be a lethally cold winter period. Also local populations of a widely distributed species may show more rapid rates of development and lower developmental thresholds in cooler parts of their range (Campbell et al., 1974). Furthermore, highly migratory pests that are incapable of overwintering may still reach outbreak numbers if they colonize in large numbers and have high reproductive potentials and short life cycles. For example, the lepidopterous pests, *Spodoptera frugiperda* and *Heliothis zea* migrate each year from the southern USA to the northeast via northerly moving weather fronts. Neither species overwinters in the northeast but they routinely reach outbreak proportions (Ferro, 1987).

Solomon and Adamson (1955) studied a wide range of storage pests for their ability to survive winter in unheated buildings in Britain. They showed that there were considerable differences in cold-hardiness and acclimatization not only between different species but also between races and strains of the same species. In general the ability or failure of species to overwinter is related to their original habitats. Thus species such as *Tribolium castaneum*, which originated in hot climates, are most susceptible to winter cold whereas relatively cold-hardy species, such as *Ptinus tectus* originated in temperate regions or have been long established outdoors in such regions. However, some species originating in warm regions show remarkable survival adaptations to temperate climates. For example, *T. granarium* originated from

India and has a temperature range for development of 20–41°C with optimum conditions of 33–37°C. Yet larvae of *T. granarium* which enter a complex facultative diapause in response to low temperatures and deteriorating food conditions are able to survive for many months at temperatures around 2–4°C and for periods of several days at temperatures below freezing (Solomon and Adamson, 1955; Burges, 1962).

High temperatures may also limit pest distribution but effects are often difficult to evaluate because in many regions excessively high temperatures are closely correlated with aridity or extremely low humidities which are also important in limiting distributions. In areas where very hot, dry summers occur regularly, pest species either aestivate or migrate to cooler, moister areas of their range.

Several different approaches have been developed for predicting the potential distribution of particular pest species (Messenger, 1974). One approach which has been used extensively is to compare climatic conditions in known areas of distribution with those of the uninfested areas. The method involves constructing climatographs for different localities by plotting mean temperatures and humidity or rainfall values and then comparing climatographs from areas where the pest is known to occur with those from elsewhere. An early example of the use of the climatograph was that of Cook (1925) who described the potential distribution of the alfalfa weevil, *Hypera postica* in the USA by comparing climatographs of different localities in the USA with the climatographs within the known areas of indigenous distribution of the species in Europe and the near east. Bodenheimer (1938, 1951), combined biological and climatic data in the form of ecoclimatographs to describe the worldwide distribution of *C. capitata* and Bourke (1961) defined the climatic suitability of different areas of Europe for establishment of the Japanese beetle, *Popillia japonica*, which is a serious pest in the Far East and North America. Similar methods have been applied to the Queensland fruit fly, *Dacus tryoni* (Meats, 1989b) and to several stored-products beetles, for example *P. tectus, T. granarium* and the cigarette beetle, *Lasioderma serricorne* (Howe and Burges, 1953; Howe, 1963; Freeman, 1977).

The accuracy with which the potential distribution of a particular pest species can be predicted depends crucially upon the correct choice, measurement and assessment of those individual meteorological factors which consistently prevent the survival and establishment of the insect population. Ideally we require a thorough understanding of the short-term responses of the insect to appropriate climatic factors throughout its life cycle and over a period of several years. Baker (1972) provides a guide to the information needed to assess potential distribution and these comprise: First, measurement of rates of development of each stage in the life cycle in appropriate environments. Secondly, estimation of the annual timing of the life cycle by summing daily increments of development. Thirdly, estimation of the success of each major activity (e.g. mating, oviposition) in the life cycle in relation to weather at appropriate times of the year; and, fourthly, estimation of mortality during each stage in the life cycle. Several models have been developed to simulate pest life cycles (Welch *et al.*, 1978; Baker and Cohen, 1985) and these may be used not only to forecast the phenology of established indigenous pests but also to make predictions about the survival and performance of potential exotic pests under a range of different climatic conditions.

In many cases the development of predictive population models is limited by inadequate and inappropriate biological and climatic data and in such cases the CLIMEX system is a more realistic approach to predicting potential distributions (Sutherst and Maywald, 1985). CLIMEX is a computer-based system which uses a worldwide climatic database and parameters describing the response of individual species to temperature and moisture to generate an ecoclimatic index. This index describes the suitability of any location for permanent colonization by the particular species and has been used, for example, to assess the potential establishment and spread of *C. capitata* and *L. decemlineata* in New Zealand (Worner, 1988). Undoubtedly, CLIMEX has considerable potential for predicting possible changes in pest distributions due to various climatic changes arising from the "greenhouse effect" (Maywald and Sutherst, 1989) and is likely to be increasingly used in such studies.

In conclusion, warmer conditions are likely to extend the geographic range of some insect pests currently limited by low temperatures and such effects are likely to be greatest at higher latitudes. In particular, migrant pests and new pests with high reproductive potentials and short life cycles, are likely to exploit more favourable climatic conditions rapidly but their permanent establishment is still likely to be limited in many temperate areas by the onset of cold conditions and particularly by the perhaps delayed occurrence of the first frosts. It is also important to bear in mind that a relatively small change in the frequency of extremes such as the timing of first-frost date, particularly in marginal areas, may result in a relatively large biological effect upon the pest population. In contrast, warmer conditions during the autum–spring period may adversely effect the phenology of cold-hardy temperate pest species and expose susceptible stages to adverse weather conditions. For example, studies on the forest tent caterpillar, *Malacosoma disstria*, have shown that unusually warm weather in early spring induces premature eclosion and budbreak and if this is followed by a prolonged period of cold there is high mortality of young larvae and disruption of phenological synchrony (Martinat, 1987). Also, hotter and particularly more arid conditions in lower latitudes may further restrict the distribution of pest species.

2. Abundance

Within the potential range of a particular species there are varying zones of actual distribution and abundance. Huffaker and Messenger (1964) divided these into four major zones, namely (a) regions which have the most continuously favourable conditions; (b) intermediate regions of favourability and permanent occupancy; (c) very unstable conditions where permanent occupancy is barely possible and (d) regions of extreme changes in favourability where permanent occupancy is not possible but where temporary occupancy derives from immigrants from other areas. Within each zone there are irregular patches of occupation and the spaces between the patches represent the degree of variation possible in each habitat due to changes in physical conditions and biotic factors such as food and natural enemies (Fig. 4). Huffaker *et al.* (1984) hypothesized that in Zone 1 density-dependent factors are likely to dominate but that conditioning physical factors, notably climate, would play an increasingly important role in Zones 2 and 3 and that in Zone 4 changes in population density would be dominated by physical factors, or associated habitat conditions. This interpretation of the varying abundance of a species population vividly illustrates the complex nature of pest population dynamics and of the varying effects that climate is likely to impose on a particular pest species in different regions of its potential range.

Weather may directly affect population numbers in two major ways. First, and most dramatic, is the impact of a single or isolated catastrophic change in ambient conditions which leads to widespread and immediate mortality. Examples include the impact of unusual winter cold or desiccation, effects of late or early frosts and heavy precipitation (Henson, 1968; Harcourt, 1971; Jones, 1979). If severe weather occurs at a time of relatively high population density it may contribute to stability whereas if it occurs when population numbers are low it may cause local extinction (Huffaker *et al.*, 1984). Secondly, temperature is the main variable which determines rates of growth, development and reproduction. Therefore, periods of warm, favourable, weather can cause rapid increases in pest abundance, particularly in agroecosystems, which lead to pest outbreaks. Also, changes in temperature, humidity and/or wind patterns can trigger outbreaks by directly stimulating mass insect movement which may subsequently lead to dramatic local concentrations of individuals (Wellington and Trimble, 1984). Conversely, unfavourable cold or very hot weather conditions may restrict population growth and development and thus delay or prevent outbreaks.

The development and widespread use of degree-day models to explain outbreaks is ample testimony to the importance of temperature in the growth and development of pest populations. Also the potential for population increase of a particular species has been frequently estimated by using the parameter r, the intrinsic rate of increase, which takes into account rates of development, oviposition, mortality and sex ratio. Howe and colleagues (Howe, 1963) used data from detailed laboratory experiments to calculate intrinsic rates of increase for a range of stored-products pests and used the parameter to compare species in relation to different environmental con-

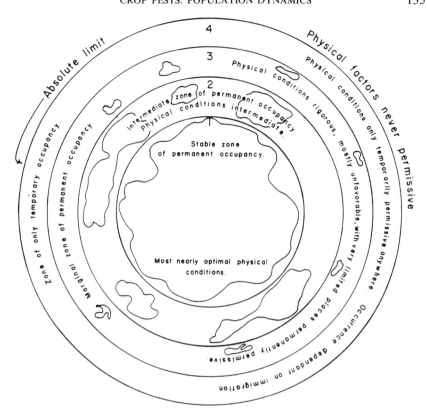

Fig. 4. The effect of conditioning and regulating factors on the geographic distribution of a species population. (Source: Huffaker and Messenger, 1964.)

ditions. Despite the limitations associated with the use of the parameter, Howe (1963) considered that it worked well when comparing the relative status of several grain pests, such as *Sitophilus oryzae, Sitophilus zea-mais, Sitophilus granarius, Rhizopertha dominica* and *T. granarium* in different climatic areas of the world.

There have been several attempts to correlate long-term fluctuations in insect population numbers with various climatic and weather factors and these have met with varying degrees of success (Andrewartha and Birch, 1954, 1984). The theory of "climatic release" postulates that fluctuations in the abundance of many forest pests are under long-term climatic control and that outbreaks are triggered by certain climatic anomalies which favour an increase in fecundity and/or survival and persist over several consecutive generations so that effects may be multiplicative (Graham, 1939; Wellington, 1954; Greenbank, 1956). Thus over a period of a few years there is a continuous increase in density which ultimately leads to the "released" population reaching outbreak numbers. Furthermore, climatic anomalies may be predictable where they are associated with cyclical patterns of weather change. Martinat (1987) summarizes recent studies which attempt to relate outbreaks of several important forest pests to climatic conditions and then critically assesses the climatic release theory in view of recent advances in climatology and data analysis.

In conclusion, warmer conditions, especially in temperate climates, are likely to increase the abundance of some pest species where decreased development times will enable extra generations to occur. More favourable conditions in such areas are also likely to lead to the occurrence of some new pests which were previously restricted by low temperatures. Recent work has shown that there is a link between the local abundance of species at sites where they occur, and the size of their geographic range (Brown, 1984; Gaston and Lawton, 1990). Thus, in species where their range is expanded by climatic change there is likely to be a corresponding increase in abundance within the new range.

So far we have considered the distribution and abundance of pest populations as being constrained only by the direct effects of climate. Of course, in the natural environment populations will be further constrained by the availability of suitable host plants, the action of natural enemies and interference and competition with other insect species. In the following section we consider how host plants and natural enemies and interspecific competition are likely to be affected by climatic change and the way in which these effects will influence the distribution and abundance of pest populations.

IV. INDIRECT EFFECTS OF CLIMATE ON PEST POPULATIONS

A. Effects of Climate Change on Host Plants

The insect–host plant relationship is a long one. The earliest reliable record of an insect feeding on a plant is probably from the Carboniferous period about 300 million years ago. Pest species may have arisen in a number of different ways. First, they may have co-evolved with their host plants and become recognized as pests as the plant was adopted as a crop. Secondly, the insects may be pre-adapted to a crop by evolving on a plant similar to a crop and simply transferring to the crop as it became more abundant. Thirdly, the pest may be opportunistic in exploiting plants by being able to adapt rapidly to a range of new hosts.

A comprehensive review of the direct effects of climatic change on plants is outside the scope of this chapter but a few of the more important changes that may affect the pests will be considered.

1. Host Plant Growth and Greenhouse Gases

The increase in greenhouse gases is the driving force behind the predicted changes in climate. It is estimated that the increases in greenhouse gases will be equivalent to a doubling of CO_2 by about 2030. However because CO_2 accounts for only half the greenhouse-gas forcing, the level of CO_2 will only have actually increased by approximately 50% (Parry, 1989). Nevertheless this increase in CO_2 will have significant effects on the growth of plants. The effects of raised CO_2 levels on crop production will be beneficial in increasing yield as both C3 and C4 plants generally respond to elevated CO_2 levels with increased water use efficiencies (Wong, 1979; Rogers et al., 1983). Also, C3 plants respond with increased photosynthetic rates but with decreased leaf nitrogen levels when grown under enriched CO_2 levels (Rogers et al., 1983). Some C4 crops respond less vigorously to increased CO_2 than others and these include maize, sorghum and sugar cane. It is worth noting that the more troublesome weeds in northern Europe are all C3 species and should benefit from CO_2 enhancement. It is likely that they may become more troublesome if C4 plants such as maize are grown (Parry, 1989).

The changes to plant growth mentioned above (increased carbon but decreased nitrogen levels in the leaves) may lead to insects consuming more plant material to obtain their required nitrogen nutrient levels. This increased consumption has been demonstrated in leaf eating insects by Scriber and Slansky (1981). They showed that in laboratory experiments lepidopteran larvae respond to reduced nitrogen levels in two ways, either with increased feeding or with reduced rates of growth. A further example of the effects of decreased leaf nitrogen on insect feeding has been shown by the increased consumption of leaf tissue by the soybean looper, *Pseudoplusia includens* (Lincoln et al., 1986). Thus in this particular case the increase in productivity of the plant due to increased CO_2 may be offset by the increased feeding of the herbivores. It is quite probable that the same is true of other pests and the benefits due to increased CO_2 levels may be lost to greater pest damage. The increased levels of CO_2 may also affect the production of leaf secondary chemicals such as toxins or attractants and may therefore influence insect herbivore feeding (Lincoln et al., 1986).

2. Drought Stress

A recent review of the role of drought stress in provoking insect outbreaks is given by Mattson and Haack (1987). Stress changes plant amino acid concentrations, and this may lead to pest outbreaks (Brodbeck and Strong, 1987). White (1978) has suggested that these outbreaks can be caused by increases in the concentration of certain amino acids vital to the fecundity or survival of insects which leads to increased numbers. If the climate becomes warmer and drier then this will increase the periods during which the plant is water stressed, even allowing for improved water use efficiency from increased CO_2 levels. In general, plants which are under stress and so growing less than optimally are those which are most susceptible to pest attack (Edwards and Wratten, 1980). In fact a shortage of water, within a certain range above that which inhibits plants growing in the first place, is perhaps the single most important factor which gives rise to conditions suitable for insect outbreaks.

Plant growth will be adversely affected by moisture stress and therefore result in less material being available to herbivores which may limit pest numbers. The texture of leaves may also alter and increased thickness and waxiness could make them less palatable or digestible for the insects. Leaves on stressed plants also change colour which may result in problems for insects that use spectral cues to locate hosts. In contrast, plant surface temperature of droughted plants may typically be 2–4°C higher than well-watered plants which could result in more rapid insect development. Furthermore, foliar concentrations of secondary compounds such as cyanogenic glycosides, glucosinolates and other sulphur compounds, alkaloids, and terpenoids tend to increase in water-stressed plants (Mattson and Haack, 1987). Some of these secondary compounds are also attractants for insect pest species. Water-stressed conifers tend to produce lower concentrations of terpenes but the reduction moves the level from repellant to attractant for many bark beetles.

There is much evidence that plant-sucking insects attack drought-stressed plants because the insect receives an enriched food supply in the sap when water is scarce (Edwards and Wratten, 1980). For example, if a plant is well watered then, in general, aphids are less fecund than if the plant is slightly wilted. However if the plant is highly wilted then the aphids do not do well (van Emden and Wearing, 1965). Thus, an increased tendency for the host to become wilted will result in aphids being more fecund and therefore building up larger colonies and causing increased losses. White (1969) developed an index of plant stress which used data on winter and summer rainfall and this index correlated well with the outbreak of the psyllid, *Cardiospira densitexta* on eucalyptus in Australia. It is possible that similar indices could be calculated for crop plants and their pests to show the possible effects on pest status of increased drought stress.

The ability of insects to perform better on droughted plants is not universal and there are cases where the pest does less well. For example, the maize stem borer, *Chilo* sp. responds to reductions in the water content of maize stems by aestivating and this is generally detrimental to the insect as survival rates are lower than when it remains active (Scheltes, 1978). Another example of an insect doing less well when its host plant is wilted is the cercropia moth, *Hyalophora cercropia*. When the larvae are fed on leaves with reduced water content they grow more slowly and develop into smaller pupae than those fed on leaves containing 70% or more water (Scriber, 1977).

It is likely that the indirect effects of increased temperatures and periods of drought will be generally beneficial to insect pests. The deleterious effects of increased levels of secondary compounds, changes in spectral quality of the host and leaf texture are generally unlikely to counteract the advantages conferred by the increased nutritional status of the plant and the production of compounds that are phagostimulatory.

In recent years there has been a considerable increase in the irrigation of crops (Stewart and Nielsen, 1990) and no doubt it will become even more widespread as evapotranspiration deficits increase as a result of climatic warming. Both increases and decreases in the size of insect populations have been recorded as a result of irrigation. Irrigated crops may provide lush and attractive host plants with improved microclimates for some pests, for example *H. zea* and *Leptoglossus clypealis* (Fereres and Goldhamer, 1990; Grimes and El Zik, 1990), whereas for others it may be detrimental. For example, moist soil conditions cause notable mortality of young larvae of the cutworm, *Agrotis segetum*, and correctly timed irrigation may effectively control this pest. Furthermore, as described later, warm, humid conditions within crop canopies may also favour epizootics of fungal pathogens of both insect pests and their natural enemies.

3. Host Plant Phenology

The predicted increases in mean temperatures will have a number of effects on crop plants over and above determining which plants can be grown where. The changes in climate will affect the rates at which the plants will develop and at what time of year they will produce leaves, flowers and fruit. For a pest attack on a plant to be successful it must often be closely synchronized to critical growth stages in the development of the host plant, consequently anything that may upset the synchrony of the pest and its host is important. Warmer conditions may lead to earlier budburst in perennial plants, such as fruit trees. Budburst would be expected to occur earlier if it were triggered by the accumulation of a certain thermal time. If this were so then budburst would occur, on average, at temperatures lower than it does at present. However, in many tree species there is an increase in the thermal time requirement to initiate budburst associated with decreased chilling. Thus during a mild winter the plant has to accumulate a greater number of day degrees before budburst than during a "normal" winter. This increase prevents very early budburst in warm springs and lessens the risk of frost damage.

Empirical models have suggested that, on average, Cox's apple in Kent would blossom 18–24 days earlier than at present following a 2°C warming, but that *Picea sitchensis* in the Scottish uplands would burst its buds only 5 days earlier (Cannell and Smith, 1986). The implication of earlier budburst is that the insect–host plant relationship may become temporarily uncoupled. An example of this type of uncoupling that may occur is shown by the apple-blossom weevil, *Anthonomus pomorum*, which lays its eggs in the flower buds of apple trees. The larva normally constructs a closed chamber from the petals shortly after hatching. In warm springs the apple trees respond by blossoming earlier and the flowers are too far open for the larva to construct its chamber. This is because the adult weevils are less affected by the warming and so lay their eggs only a little earlier than normal but this is later relative to the state of development of the tree (Uvarov, 1931). However, in the long term some individuals within a pest population are likely to be selected in response to changes in the phenology of their host plants.

Changes in climate may well lead to growers changing their cropping practices and crops such as sugar beet could be sown earlier in the season. At present most bolting-resistant varieties need to receive more than about 35 vernalizing days (days when the maximum temperature < 12°C) before any significant bolting occurs. If a rise of 3°C in average temperatures is assumed the safe sowing date is moved from late March to the end of January (Thomas, 1989). This shift in sowing date may result in the crop suffering lower levels of attack at the vulnerable seedling stage because pest species may still be in diapause and will not complete diapause until a certain photoperiod is experienced.

The necessity of synchronization between the pest and its host is not only necessary at the start of the season but also later in the year, as demonstrated by the relationship between the pine shoot moth, *Evetria buoliana* (Schiff.) on Scots pine in western Europe and North America. Outbreaks of this pest occur frequently in North America where the summers are warm but only rarely in western Europe when the outbreaks are coincident with warm summers. Harris (1960) showed that the survival of the pine shoot moth on Scots pine was related to the amount of resin encountered during the initial attack on the buds. The larvae were unable to establish themselves in very resinous buds. The resin canals in the buds are developed in response to short day-length at the end of summer while the development of the moth is associated with summer temperature. Thus more larvae survive following a warm summer than a cold one because they develop faster and attack the buds before there is well-developed resin protection.

4. Crop Hosts

Insects have the ability to locate and colonize most species of introduced plants remarkably quickly and the more widely a plant becomes established the more insect species it recruits (Strong *et al.*, 1984). The changes in climate are likely to lead to crops being grown in areas of the world where they are not grown at present. For instance, in Britain the temperature limit for the growth and ripening of grain maize would shift from its present position in the extreme south of England to northern Scotland if a 3°C increase in mean temperature occurred (Parry, 1989). If this leads to the large-scale expansion of this crop it is possible that it may be attacked by a number of insects that are presently not regarded as pests. Alternatively, the present limited area of this crop in Britain may have prevented the establishment and build-up of some pest species and therefore, at least initially, pest populations may be relatively small compared to those on crops in areas where it has been grown more extensively for longer periods. Thus, in the long term, pest problems are likely to increase as the crop becomes more widely grown.

If new crops are introduced into an area, then it will be particularly difficult to predict which pest species may occur since many of the indigenous insect species that colonize a newly introduced plant are polyphagous often on plants unrelated to the introduction (Strong *et al.*, 1984). In general, external chewing and sucking herbivores have higher rates of colonization than endophagous gall formers and miners and will therefore make up the vast majority of any new pest species that may arise. It is possible that pests occurring on the crop in the areas where it is already grown could be imported with the crop unless strict quarantine is practised. Any pests that are introduced by accident along with the crop are likely to be particularly successful since they will probably not have any of their native natural enemies to exert control over them.

5. Non-Crop Hosts

The effects of climate change on non-crop plants will also be important where these are an essential link in the pest life cycle. For example heteroecious aphid species have an alternate woody winter host. If the aphids only survive over the winter as eggs on their woody host then they will only overwinter where the host occurs. If the woody host is at the limit of its northern range then an increase in average temperature will result in the northwards migration of the plant, assuming their are no other constraints on its colonization of new areas, since the poleward spread of a species is controlled by the minimum temperature which will be regularly experienced by the plant (Woodward, 1987). A warming of only 1°C by the year 2030 represents a northward shift of present conditions of 300 km or a rate of advancement of suitable conditions of 7·5 km per year. However, the maximum observed migration rate, obtained from the pollen record, of any tree species is only approximately 2 km per year. Thus the rate at which the pest species will be able to establish populations permanently in the new area that is suitable for it will be limited by the rate of spread of the host plant into these new areas. Therefore, there will be a continued need for pests to migrate into the new areas and at the limits of the new extended range relatively fewer migrants are likely to colonize crops successfully.

In summary, new introductions of crops may lead to the emergence of new pests. An accidental introduction of a pest to a new crop must be avoided by the use of strict quarantine, particularly as such an introduction will probably occur without its native natural enemies. Plants may generally do better than at present but this increased productivity could be offset by the increased feeding of pests. Changes in the timing of budburst and sowing dates may lead to the uncoupling of host plant–insect relationships but such uncoupling may only be temporary as some pests are likely to adapt to changed phenologies. The rate of spread of pests into new climatically suitable areas may be delayed by the rate of spread of their alternate and alternative host plants.

B. Effects of Climate on Natural Enemies of Pests

1. Pathogens of Pests

The role of diseases in causing the collapse of insect outbreaks is well known (Tanada, 1959; Anderson and May, 1980). The effect of climatic changes on the different types of pathogen will vary. Many insect pathogenic viruses are inactivated by exposure to sunlight, especially the ultra-violet fraction, but are less affected by temperature, humidity, surface moisture and chemicals (Jacques, 1985). Consequently if there are increased levels of sunlight the number of epidemics of virus diseases is likely to decrease. However, it is possible that there will be increased cloud cover and therefore viruses could remain active in the environment for longer, resulting in higher levels of control. Climate also affects the pathogenicity of the virus. In general, the higher the temperature the greater the virulence and pathogenicity of the virus, resulting in a shortened period of lethal infection. Very high temperatures sometimes increase the resistance of the host insects to viruses (Tanada, 1965) but this would appear to be atypical. Thus it would be expected that an increase in temperatures will lead to more effective control of insects by viruses. Humidity is also important in determining the pathogenicity of the viruses because high humidity can prevent the desiccation of the pathogens and rainfall can assist in their dispersal. Some virus infections do better under conditions of high humidity, resulting in higher rates of mortality of the infected insect (Franz, 1971).

The insect pathogenic bacteria show similar responses to climatic variables as virus diseases. Pathogenicity of *Bacillus thuringiensis* is greatest at temperatures where the growth of the bacterium and metabolic rate of the insect are most rapid, hence temperatures of 32°C are more likely to cause mortality of the host than lower ones (Raun *et al.*, 1966). High humidity also leads to greater mortality due to increased pathogen survival and possibly greater stress on the host.

It is fungal pathogens which are seemingly most affected by climatic variables. In general, the limits for growth are between 5 and 35°C with the optima falling between 20 and 30°C (Roberts and Yendol, 1971). Whilst these are the limits for growth the higher temperatures tend to reduce the viability of spores (Steinhaus, 1960) and can also lead to the insects developing and moulting much more rapidly and escaping infection simply because the hyphae cannot penetrate the integument before it is shed at the next moult. Rainfall or high atmospheric humidity have been observed to be an important requisite for the induction of epizootics of fungal pathogens (Hagen and van den Bosch, 1968; Ferron, 1978). Humidity can be limiting at two periods of an epizootic. First, high humidity is needed by most fungi in order to germinate and cause disease. Secondly the new spores needed to spread and heighten the epizootic are usually produced on the cadavers only at very high humidities (Roberts and Yendol, 1971).

In conclusion, insect pathogens will respond to changes in climate quite markedly. If conditions were to become warmer, drier and sunnier, then the effect of pathogens on the insect population dynamics may be notably smaller than currently occurs. However, if the weather becomes warmer and wetter then some pathogens may exert greater control.

2. Predators and Parasitoids of Pests

The changes in climate will directly effect the population dynamics of these insects as it does those of pest species. Thus natural enemy behaviour and oviposition rates are affected by temperature, humidity and wind (Juillet, 1964; Barbosa and Frongillo, 1979; Jackson and Butler, 1984) and increases in temperature will increase rates of development and reproduction. Clearly, if the natural enemies respond with proportionally greater increases than the pest species then they are likely to exert greater control.

The differential effects of temperature on aphids and their natural enemies clearly demonstrate how temperature can greatly affect the level of control of these pests. For example, at temperatures below about 10°C the pea aphid, *Acyrthosiphon pisum*, can build up populations at greater speed than can be checked by the coccinellid, *Coccinella septempunctata*. However, at temperatures above about 11°C the coccinellid is potentially capable of reducing the aphid population (Dunn, 1952) (Fig. 5). Another aphid predator that also shows a positive response to increased temperatures is the larva of *Syrphus corollae*. Larvae kept at 28°C and 6°C did not consume significantly different numbers of aphids during their development but the rate of consumption at higher temperatures was much greater (i.e. approximately 55 aphids per day

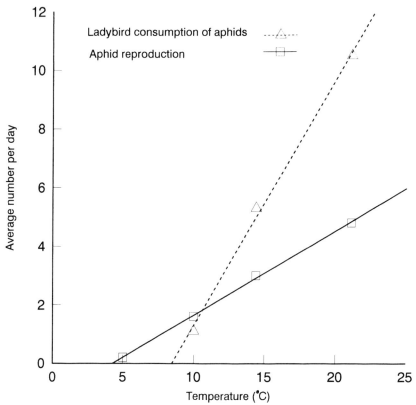

Fig. 5. The average number of aphids consumed and produced each day at various temperatures. (Source: Dunn, 1952.) (Published with kind permission of Horticulture Research International, Warwickshire, UK.)

were consumed by each individual syrphid larva at 28°C, compared with only 8 aphids per day at 6°C) (Benested, 1970).

In laboratory experiments, a braconid parasitoid, *Praon palitans*, of the alfalfa aphid, *Therioaphis maculata*, prevented the aphid from reaching damaging levels when kept at a mean temperature of 21°C. However, at a mean temperature of 12·5°C the parasitoid was only able to persist and did not effectively control the increase of its host (Messenger, 1964a).

Parasitoids have often been used in classical biological control pro-grammes and there are several examples where these have failed because of climatic incompatibility. An interesting example is that of a strain of the parasite, *Trioxys pallidus*, imported from France to California to control the walnut aphid, *Chromaphis juglandicola*, which failed to perform effectively in the hot interior of California but subsequently a second strain of the same species from Iran proved highly effective (van den Bosch and Messenger,

1973). Such studies clearly emphasize the importance of the need for precise knowledge concerning natural enemy–bioclimatic relationships (Messenger, 1970).

Because parasitoids are usually species specific, and may even require hosts in a specific morphological or physiological state, they have to adapt their seasonal cycles very closely to those of their hosts. Diapause plays a large part in maintaining this synchrony (Tauber *et al.*, 1983). In northern temperate regions many insects enter diapause in response to the influences of low autumnal temperatures or, more commonly, shortening periods of daylight (Askew, 1971). They remain in a condition of diapause throughout the winter until another environmental stimulus supervenes to "break" diapause and allow development to continue. Parasitoids show three types of relationship with their hosts during diapause. These are not discrete but have a continuum in the degree to which the parasitoid is reliant upon the hosts for diapause-regulating cues:

(1) Parasitoid diapause is primarily regulated by abiotic cues (temperature and/or photoperiod) and therefore independent of the hosts' diapause.
(2) Simultaneous dependence of the parasitoids' diapause on the hosts' diapause and abiotic cues.
(3) Dependence of parasitoid diapause entirely on the hosts' physiological state (Tauber *et al.*, 1983).

Photoperiod is the most common diapause-inducing stimulus. This may cause problems if the pest species does not enter diapause but continues to reproduce and develop in the autumn because of higher average temperatures, for the pest may then escape from the controlling effect of the natural enemies that are obliged to enter diapause. If the natural enemy relies on a combination of photoperiod and temperature for its cue to enter diapause this disruption is less likely to occur since the entry into diapause may be delayed under warmer conditions. Likewise, if the cue for entering diapause is entirely dependent on the hosts' physiological state then there should be no decoupling of the relationship. An example of the combined effects of both photoperiod and temperature being important in the induction and termination of diapause is shown by the work of Tauber *et al.* (1983) on *Tetrastichus julis*, a parasitoid of the cereal leaf beetle, *Oulema melanopa*. When temperatures are high over the winter period a long photoperiod is required to break diapause, whereas if temperatures are low a short photoperiod is sufficient. Greater numbers of the parasitoid emerge successfully at lower temperatures than at high ones so an increase in mean temperatures may result in fewer parasitoids emerging and exerting less control. Subsequent post-diapause development is controlled by a specific thermal time requirement which, if greater than that of the leaf beetle, could result in the parasitoid emerging after the pest and being unable to locate the correct developmental stage of the host to parasitize.

In conclusion, predicted warmer conditions will tend to favour many natural enemies but whether they play a more effective role will also depend upon the response of the pest to changed climatic conditions. If the general warming is sufficient for natural enemies to be able to control pest populations early in the year then, assuming the crops are not planted earlier, there may be fewer outbreaks. However, if crops are sown earlier in response to warmer conditions then the situation may not be so different from present, with the pests building up correspondingly earlier at the higher temperatures and forming large populations before the natural enemies can exert effective control.

The effects of climate on hyperparasites are likely to be similar to those on all other insect species, consequently the relationships between the hyperparasites and their parasitoid hosts may be altered. There is some discussion over the effects of hyperparasites on the natural control of pest species but obligate hyperparasites will usually increase host pest abundance (Saunders, 1989). Thus if the change in climate favours the hyperparasite rather than the parasite then natural control of a pest may be reduced. Alternatively the changes may favour the parasite and better control will result. However, mathematical models have suggested that hyperparasites can enhance local stability (Hassell and Waage, 1984) and therefore their action may result in better long-term control of some pests by their parasites.

C. Interspecific Competition

A particular crop is usually attacked by a complex of pest species, the species and its relative abundance sometimes varying between different climatic areas. The individual species of the pest complex may occur on the plant at different times or occupy different niches. However, when two or more species overlap in time and space then interspecific competition may occur and this can have profound effects on the dynamics of the competing species. Furthermore, the effects of interspecific competition may vary in relation to different meteorological conditions. For example, work by Park (1954) on the flour beetles, *Tribolium confusum* and *T. castaneum* showed that under controlled laboratory conditions, *T. castaneum* is the more successful competitor in relatively hot, moist conditions whereas *T. confusum* is the more successful of the two competing species under cooler, drier conditions. Clearly, the interactions which occur between species in particular pest complexes could be altered under different climatic scenarios.

V. EFFECTS OF CLIMATE ON PEST–HOST PLANT–NATURAL ENEMY RELATIONSHIPS

As shown earlier, climatic change is likely not only to effect the distribution and abundance of pest populations but also those of their host plants and natural enemies. Therefore, to provide a realistic assessment of whether a particular pest will or will not become more widespread and abundant we need to consider how the pest–host plant–natural enemy relationship is likely to alter under different climatic conditions and over varying time scales. This relationship is examined with particular reference to aphid pests.

Aphids are the most important group of pest insects on crops of the temperate climatic zones and there is considerable data on their biology, ecology, natural enemies and host plants and of the varying effects of weather (Dixon, 1985; Wellings and Dixon, 1987; Minks and Harrewijn, 1989). The life cycles of many aphids are complex and involve cyclical parthenogenesis, viviparity and polymorphism. Many important pest species have hetero-ecious, holocyclic life cycles in which they alternate between a primary woody host where they lay overwintering eggs and secondary herbaceous hosts on which they spend the summer. However, in some aphid species in some parts of their geographic range there is partial or complete anholocycly, in which case they overwinter as parthenogenetic adults and nymphs on secondary hosts.

The impact of cold winter conditions on aphid populations varies consi-derably depending upon whether the aphid overwinters parthenogenetically or as an egg. The peach potato aphid, *Myzus persicae*, the potato aphid, *Macrosiphum euphorbiae* and the grain aphid, *Sitobion avenae* commonly overwinter parthenogenetically in the UK on various weeds and overwinter-ing crops, and severe weather conditions, particularly sustained freezing conditions, have notably adverse effects on overwintering populations of these aphid species (Watson *et al.*, 1975; Turl, 1983; Watson and Carter, 1983; Knight *et al.*, 1986). Severe weather may also adversely affect the survival and quality of the overwintering herbaceous hosts and, in fact, work by Turl (1983) has shown that overwintering success of anholocyclic popula-tions of *M. persicae* may depend as much on the relative cold-hardiness of the preferred herbaceous weed hosts as on the cold-hardiness of the aphids themselves. In contrast, the survival of eggs is unlikely to be significantly influenced by winter temperatures as they are extremely cold-hardy (Way and Banks, 1964; Sømme, 1969).

The numbers of aphids migrating and the size of populations on host plants are often correlated with varying temperature, moisture and wind patterns (Jones, 1979; A'Brook, 1983). In the UK the size of the spring migration of *S. avenae* is inversely related to winter temperatures below 0°C (Watson and Carter, 1983). Also, the time of the beginning and end of the migration of the damson-hop aphid, *Phorodon humuli* from its primary hosts, *Prunus* spp. to hops can be usefully predicted from climatic data. Thus the beginning of migration is associated with temperature in late March and early April and with periods of rainfall in mid-January and mid-April, whilst the end of migration is associated with temperature and sunshine in mid- and late-June and with mid-January rainfall (Thomas *et al.*, 1983). Work by Jones (1979) on cereal aphids highlights the importance of the direct effects of weather on the abundance of aphids on crops. Thus the accumulated temperatures above 5°C up to the emergence of the ears accounted for between 55 and 99% of the variation in the abundance of aphids over a 5-year period. Furthermore, heavy rains and high winds had adverse effects by dislodging aphids from their plants but such effects varied between different cereal aphid species, for *Metopolophium dirhodum* was notably more easily dislodged than *S. avenae*.

The importance of both the direct effects of weather and of indirect effects via the host plant is clearly demonstrated by work on the cowpea aphid, *Aphis craccivora* in southeast Australia (Gutierrez *et al.*, 1974b; Gutierrez, 1987). The aphid breeds on various species of medic, clover and vetch which dry off during the hot dry summers and are therefore unsuitable for aphids. Pasture growth is strongly influenced by rainfall patterns, and its phenology determines that of the aphid populations. Thus cowpea aphid populations are common in a very wet season when pastures are lush whereas in a drought year aphids are relatively scarce. During winter, light ground frosts adversely affect the frost-susceptible aphid but not the plants. Therefore aphids are largely absent in the temperate regions of Australia during winter and summer and the aphid must re-invade each year from the frost-free subtropical areas of eastern Australia. Mean soil moisture and temperature indices have been used to define the limits of climatic favourability for the aphid during different parts of the year and a comprehensive simulation model for the aphid has also been developed (Gutierrez *et al.*, 1974a). Further evidence of the importance of indirect climatic effects via host plants has also been provided by A'Brook (1981, 1983) who showed that numbers of the aphids, *Rhopalosiphum insertum* and *R. padi* trapped in the autumn at Aberystwyth (UK) were associated with high summer rainfall and soil moisture and that seemingly this relationship occurred because cereal aphid numbers are associated with the yields of their summer grass hosts which are in turn related to summer rainfall.

The general role of natural enemies in reducing and regulating populations of aphids has been extensively reviewed and it is widely recognized that their effects may vary considerably under different climatic conditions (Hagen and van den Bosch, 1968; Messenger, 1970; Minks and Harrewijn, 1989). Temperature has the most important climatic influence on predators and parasites whereas relative humidity is very important in the development of epizootics of fungal pathogens.

The development rates of predators and parasites are influenced by temperature as with their aphid hosts but in general their lower thermal threshold is higher (Campbell et al., 1974). Thus at low temperatures their impact on aphid populations tends to be relatively small but as temperatures rise this impact becomes increasingly more important (Messenger and Force, 1963; van Emden, 1966). Temperatures may also affect the functional responses, searching rates and handling times of predators and parasitoids. Hodek (1973) demonstrated the differential effect of temperature on the build-up of populations of the black bean aphid, *Aphis fabae*, and the preying efficiency of adults of *Coccinella septempunctata* under experimental field conditions on sugar beet in Czechoslovakia. During a relatively cool season, coccinillids failed to control aphid populations when the ratio of aphids to beetles was 90 : 1 or greater but in a hotter season the coccinellids eliminated the aphid population even when there were 200 aphids per beetle. Detailed studies in Canada on the relationship between coccinellids and the pea aphid, *A. pisum*, have shown that the predator–prey relationship is considerably influenced by temperature including the rates at which adult beetles enter and depart from crops (Baumgaertner et al., 1981). In fact, factors influencing natural enemy movement may be the key to understanding their role in preventing aphid outbreaks (Wellings and Dixon, 1987). Overall, low temperatures tend to reduce the effectiveness of predators and parasites and there is a temperature band in which aphid populations are able to increase, albeit slowly, in the absence of actively searching natural enemies (Wellings and Dixon, 1987). This transient relatively enemy-free space may be particularly important during periods of the aphid life cycle in autumn and spring when the rates of survival and reproduction of relatively small populations determine the size of subsequent populations.

In conclusion, the impact of natural enemies on aphid populations is extremely variable and is likely to be greatest when a natural enemy synchronizes with early establishment of the aphid colonies and either completely destroys it or severely checks the initial rapid reproduction on which build-up of large colonies depends. Unfortunately, natural enemy populations usually develop most rapidly after the aphid colonies are well established and therefore have a limited effect upon peak numbers, although they are often a major factor in the early collapse of the population. Undoubtedly, in many areas unfavourable climatic conditions impose a major constraint on the effectiveness of natural enemies at critical times in the life cycle of the pest.

In temperate climates, warmer conditions during winter are likely to increase both the survival and development rates of those aphids which overwinter parthenogenetically and particularly of those species which

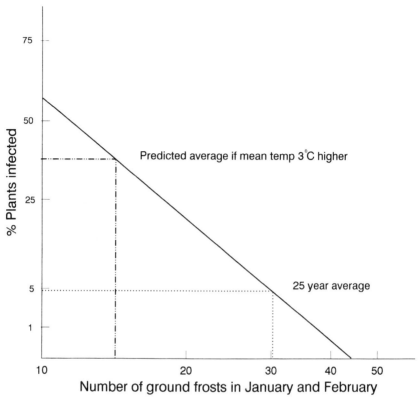

Fig. 6. Relationship between the incidence of virus yellows in sugar beet and winter weather. (Source: Thomas, 1989.) (Published with kind permission of British Sugar plc).

overwinter on cold-susceptible herbaceous hosts. Parthenogenetic overwintering is likely to become more common and widespread and this will lead to larger and earlier migrations of certain pest species, which is likely to increase problems with aphid-transmitted viruses on crops such as potatoes and sugar beet. This will be particularly so where the overwintering host plant is also a reservoir for the pathogen. Thomas (1989) has predicted that milder winters in the UK could lead to a notable increase in the frequency of serious epidemics of virus yellows on sugar beet (Fig. 6), which in turn could lead to increasing pesticide usage and the faster selection of pesticide-resistant aphids.

Aphids which overwinter holocyclically are unlikely to benefit from milder winters but eggs will probably hatch earlier in spring and the populations will then build up more rapidly during warmer conditions in spring, which may also lead to earlier migrations to crops. Of course, milder winter and spring conditions can also favour the survival and activity of some natural enemies and this may lead to the early collapse of aphid populations. Increased early activity of natural enemies may be particularly important in preventing build-up of populations from overwintering eggs if milder conditions improve the synchrony of the natural enemy with its aphid host in such a way that there is notable mortality of immature fundatrices and thus prevents the initial rapid build-up of the aphid population.

The increased likelihood of sowing crops earlier in the year and, in some cases, in the previous autumn is likely to be fully exploited by aphids which will rapidly build up on highly nutritious plants during the more favourable weather conditions of late spring–early summer. However, as crops mature and become less nutritious to aphids, and there is also increasing natural enemy action, then aphid populations on crops will probably reach peak numbers and collapse earlier than currently occurs. Similar effects may also occur on alternative or alternate non-crop hosts. The outcome will be that for some aphid species, there will be an increased problem of survival during prolonged hot, dry conditions in summer on suboptimal host plants and at a time when natural enemies are likely to be abundant. Of course, the introduction of new crops and cropping patterns in response to climate change could extend the period over which susceptible hosts are available and this would probably lead to further aphid outbreaks later in the season. Also the earlier sowing of crops in autumn would have similar effects.

Clearly, aphids with their high reproductive rates, short generation times and ability to colonize rapidly a succession of new host plants, are ideally suited to exploit and adapt to changes brought about by changed climate. In temperate climates, aphids and aphid-borne viruses are likely to become more common, particularly early in the season and their distribution is also likely to extend northwards. However, the long-term population dynamics of many important pest species depends upon year-round conditions both in the crop and non-crop environments and until these are more fully understood our predictions of possible changes in long-term abundance and distribution of aphid pests in response to climatic change must be suitably qualified.

VI. IMPLICATIONS OF CLIMATIC CHANGE FOR PEST MANAGEMENT

Undoubtedly, climatic change increases the need for developing and implementing soundly based pest management strategies. Basically, we need to predict how climatic change in a particular area is likely to change the status and phenology of individual pest species and also the risk from potential pests so that optimum pest control strategies can be devised.

As mentioned earlier, the status of a pest depends upon the numbers of individuals present, socio-economic factors and the biological tolerance of the crop to pest attack, all of which may be affected by climatic change. Pests vary in status from those of high or "key" status, which cause considerable economic crop losses unless controlled, to those of only low status which rarely justify specific control measures. A particular pest species may even vary in status on the same crop between regions or under different cropping systems. Potential pests are either exotic or indigenous in origin. Those of exotic origin have varying pest status on crops in their native environments but are currently prevented from establishing in a new region because of climatic constraints, absence of suitable crop hosts or insufficient time to spread that far. Thus under changed climatic conditions some hitherto potential pests may rapidly attain pest status in a new region. Indigenous insect species with pest potential are currently present in the region concerned but are on non-crop hosts or on crops where they do not cause economic damage. However they may also rapidly assume pest status under more favourable climatic conditions and particularly on susceptible host crops which may be introduced as a result of those more favourable climatic conditions.

It is important when evaluating pest status that we recognize the scale and complexity of the pest problem. First, worldwide there are several thousand known crop pest species and on a single, widespread crop there may be in excess of 100 different species which occur within varying pest complexes in different geographic regions. Secondly, there are a considerable number of phytophagous insect species in the world from which more pest species may emerge. For example, there are some 55 000 known species of the phytophagous beetle family Curculionidae, some of which are known pests. Furthermore, this number probably only represents about 10–20% of the estimated total world species of that family. Thirdly, insects exhibit considerable genetic diversity and phenotypic plasticity which enables them to adapt to changing conditions, and particular pest species may exist as several biotypes which may interact differentially with different crop hosts. For example, there are three biotypes of the greenbug, *Shizaphis graminum* that differ in their ability to live on, and damage, varieties of wheat, barley and sorghum (Lattin and Oman, 1983).

Studies on the population dynamics of pest species and of the likely causes of pest outbreaks (Barbosa and Schultz, 1987; Wallner, 1987) have highlighted both the inherent characteristics of pest species and the conditions under which outbreaks are likely to occur. Such data provide a useful background for assessing the likely risks posed by particular types of insect species under varying conditions, including climatic change. Opportunistic or *r*-strategist species (Southwood, 1977) are ideally adapted to exploit agroecosystems and these types of pest are likely to invade rapidly and spread into new areas which become climatically favourable. In particular those pest species with wide geographic ranges, wide range of habitats and host plants, rapid reproduction and highly migratory behaviour are likely to invade quickly and spread rapidly on crops in new areas, particularly as they are also likely, at least initially, to escape from the constraints imposed by their native natural enemies. Clearly, opportunistic species which are currently absent from areas because of unfavourable climatic conditions should receive priority attention in order to define the particular changes in conditions under which they are likely to cause crop losses in new areas.

Predicting changes in the status of existing pests and of the likely status of potential pests as a result of climatic change can be conveniently considered as follows. Initially, predictions should be made in relation to the likely new potential ranges of particular pests under different climatic scenarios. CLIMEX provides a useful approach in this respect and from such data we will be able to define the maximum likely distribution of a particular pest. The range of the pest may then be further defined into zones of varying potential abundance, the extent to which this is done depending upon the data available and the degree of precision and detail required. Such predictions of the likely changes in the potential distribution and abundance of pests are particularly useful for formulating policy with regards to the possible use of quarantine measures.

Of course, the actual distribution and abundance of a pest within its climatically suitable limits will depend largely upon the availability of suitable host plants and the constraints imposed by natural enemies. In particular, we need to identify those existing crops which are likely to be grown more widely and also those new crops which will probably be introduced into an area as a result of more suitable climatic conditions. Climatic matching of the distribution and abundance of pests and their crop hosts will help to define areas at greatest risk from pest attack and to assess future priorities in terms of pest management. In this respect, it may be useful to examine the status of particular pests on crops in areas that are currently climatically similar to the predicted climate. For example, pests which are normally associated with southern Europe are likely to become increasingly important in northern Europe. On sugar beet these include tortoise beetles, flea beetles and caterpillars of various moths including cutworms (Thomas, 1989). Furthermore, on cereals the pentatomid grain bugs, *Aelia* spp. and *Eurygaster* spp., the beetle, *Zabrus tenebrionides, O. nubilalis* and stalk borers, *Sesamia* spp. are currently important pests in southern Europe (Cavalloro and Sunderland, 1988).

It is important to recognize that pest status also depends upon the effectiveness of available control measures which may be affected by climate. Thus, a potentially very damaging and frequent pest which can be controlled effectively at relatively low cost may result in a lower overall loss of revenue than a pest which is potentially less damaging and infrequent but is not controlled effectively. Climate may affect many aspects of chemical control such as residue persistence, toxic action and treatment dosage and frequency. Also the number of days which are climatically suitable for applying chemical sprays may alter if there is a change in the frequency of weather factors such as strong winds and rain which may prevent application. Furthermore, climate may also affect other control components such as biological and cultural controls. Therefore changes in climate may alter the effectiveness of various control methods and may also change the emphasis placed on any one particular method.

Neither agriculture, nor the tactics and strategies for dealing with pests are static, and changes in crop production and crop protection systems may affect pest status. These changes may arise either in direct response to climatic change or as result of unrelated developments and clearly they should be included in any model which is used to predict how climatic change is likely to alter pest status.

VII. CONCLUSIONS

In this chapter the aim has been to show the various ways in which climate can affect the population dynamics of pest species both directly and indirectly and to indicate what may occur under certain changed climatic conditions. The information gained from numerous studies on insect populations has provided the basis for these current interpretations and has also served to highlight the need for additional relevant information.

The precise climatic changes that may occur in the future cannot be predicted with any degree of certainty. However, because the potential consequences for agriculture and for pest control, in particular, are considerable and wide-ranging, it is important that the impact of various climatic scenarios on pests should be considered as a matter of urgency. The monitoring of a number of indicator pest species should be undertaken. The chosen species should ideally show measurable changes in dynamics as a result of only very small changes in temperature and moisture and would therefore provide model systems on which hypotheses could be tested as more information becomes available in relation to pest–host plant–natural enemy interactions.

ACKNOWLEDGEMENTS

We wish to thank colleagues at Silwood Park, especially Dr G. A. Norton and Professor M. J. Way, for their helpful comments during the preparation of this chapter.

REFERENCES

A'Brook, J. (1981). Some observations in west Wales on the relationships between the number of alate aphids and weather. *Ann. appl. Biol.* **97**, 11–15.

A'Brook, J. (1983). Forecasting the incidence of aphids using weather data. *Bulletin OEPP* **13**, 229–233.

Amos, T.G. (1968). Some laboratory observations on the rates of development, mortality and oviposition of *Dermestes frischii* (Kug.) (Col. Dermestidae). *J. stored Prod. Res* **4**, 103–117.

Anderson, R.M. and May, R.M. (1980). Infectious diseases and population cycles of forest insects. *Science* **210**, 658–661.

Andrewartha, H.G. and Birch, L.C. (1954). *The Distribution and Abundance of Animals.* University of Chicago Press, Chicago.

Andrewartha, H.G. and Birch, L.C. (1984). *The Ecological Web.* University of Chicago Press, Chicago.

Anon. (1989). *Distribution of Pests, Series A (Agricultural).* CAB International Institute of Entomology, London.

Askew, R.R. (1971). *Parasitic Insects.* Heinemann, London.

Baker, C.R.B. (1972). An approach to determining potential pest distribution. *Bull. OEPP* **3**, 5–22.

Baker, C.R.B. (1980). Some problems in using Meteorological data to forecast the timing of insect life cycles. *Bull. OEPP* **10**, 83–91.

Baker, C.R.B. and Cohen, L.I. (1985). Further development of a computer model for simulating pest life cycles. *Bull. OEPP* **15**, 317–324.

Baker, C.R.B. and Miller, G.W. (1974). Some effects of temperature and larval food on the development of *Spodoptera littoralis* (Boisd) (Lep, Noctuidae). *Bull. ent. Res.* **63**, 495–511.

Bakke, A. (1968). Ecological studies on bark beetles (Coleoptera:Scolytidae) associated with Scots pine (*Pinus sylvestris* L.) in Norway with particular reference to the influence of temperature. *Medd. Norske Skog.* **21**, 443–602.

Bale, J.S., Harrington, R., Knight, J.D. and Clough, M.S. (1987). Principles of the cold hardiness and overwintering survival of aphids. *Proc. Crop Prot. in Northern Britain, Dundee 1987*, 97–104. British Crop Protection Council.

Barbosa, P. and Frongillo, E.A. (1979). Influence of light intensity and temperature on the locomotory and flight activity of *Brachymeria intermedia* (Hymenoptera: Chalcididae), a pupal parasitoid of the gypsy moth. *Entomophaga* **22**, 405–411.

Barbosa, P. and Schultz, J.C. (Eds) (1987). *Insect Outbreaks.* Academic Press, London.

Baumgaertner, J.U., Frazer, B.D., Gilbert, N., Gill, B., Gutierrez, A.P., Ives, P.M. *et al.* (1981). Coccinellids (Coleoptera) and aphids (Homoptera). *Can. Ent.* **113**, 975–1048.

Beck, S.D. (1980). *Insect Photoperiodism.* Academic Press, New York.

Beck, S.D. (1983). Insect thermoperiodism. *A. Rev. Ent.* **28**, 91–108.

Benested, E. (1970). Food consumption at various temperature conditions in larvae of *Syrphus corollae* (Fabr.) (Dipt., Syrphidae). *Norsk Ent. Tiddss.* **17**, 87–91.

Bodenheimer, F.S. (1925). On predicting the developmental cycles of insects. 1. *Ceratitis capitata*, Wied. *Bull. Soc. ent. Egypte* **8**, 149–157.

Bodenheimer, F.S. (1938). *Problems of Animal Ecology.* Clarendon Press, Oxford.

Bodenheimer, F.S. (1951). *Citrus Entomology in the Middle East.* Dr W. Junk, The Hague.

Bolin, B, Döös, B.R., Jager, J. and Warrick, R.A. (Eds) (1986). *The Greenhouse Effect, Climatic Change and Ecosystems.* (SCOPE 29). John Wiley, New York.

Bosch, van den, R. and Messenger, P.S. (Eds) (1973). *Biological Control*. Intext Educational Publishers, New York.

Bourke, P.M.A. (1961). Climatic aspects of the possible establishment of the Japanese beetle in Europe. *World Meteorological Organization Technical Note 41*, pp. 1–9.

Broadbent, L. (1949). Factors affecting the activity of alatae of the aphids *Myzus persicae* (Sulzer) and *Brevicoryne brassicae* (L.). *Ann. appl. Biol.* **36**, 40–62.

Brodbeck, B. and Strong, D. (1987). Amino acid nutrition of herbivorous insects and stress to host plants. In: *Insect Outbreaks* (Ed. by P. Barbosa and J.C. Schultz), pp. 347–364. Academic Press, London.

Brown, J.H. (1984). On the relationship between abundance and distribution of species. *Am. Nat.* **124**, 255–279.

Burges, H.D. (1962). Diapause, pest status and control of the khapra beetle, *Trogoderma granarium* Everts. *Ann. appl. Biol.* **50**, 614–617.

Bursell, E. (1974a). Environmental aspects—temperature. In: *The Physiology of Insecta* (2nd Edn) (Ed. by M. Rockstein), pp. 1–41. Academic Press, New York.

Bursell, E. (1974b). Environmental aspects—Humidity. In: *The Physiology of Insecta* (2nd Edn) (Ed. by M. Rockstein), pp. 43–84. Academic Press, New York.

Cammell, M.E. and Way, M.J. (1987). Forecasting and monitoring. In: *Integrated Pest Management* (Ed. by A.J. Burn, T.H. Coaker and P.C. Jepson), pp. 1–26. Academic Press, London.

Campbell, A., Frazer, B.D., Gilbert, N., Gutierrez, A.P. and Mackauer, M. (1974). Temperature requirements of some aphids and their parasites. *J. appl. Ecol.* **11**, 431–438.

Cannell, M.G.R. and Smith, R.I. (1986). Climatic warming, spring budburst and frost damage on trees. *J. appl. Ecol.* **23**, 177–191.

Carter, C.I. (1972). Winter temperatures and survival of the green spruce aphid *Elatobium abietinum* (Walker). *Forestry Commission Forest Record,* **84**. HMSO, London.

Cavalloro, R. and Sunderland K.D. (Eds) (1988). *Integrated Crop Protection in Cereals*. Proceedings of EC Experts Group, Littlehampton. A. A. Balkema, Rotterdam.

Chapman, R.F. (1988). *The Insects : Structure and Function* (3rd Edn). Hodder & Stoughton, Sevenoaks.

Clark, L.R., Geier, P.W., Hughes, R.D. and Morris, R.F. (1967). *The Ecology of Insect Populations in Theory and Practice*. Methuen, London.

Cockbain, A.J. (1961). Low temperature thresholds for flight in *Aphis fabae* Scop. *Ent. exp. appl.* **4**, 211–219.

Cook, W.C. (1924). The distribution of the pale Western Cutworm *Porosagrotis orthogonia*, Morr., a study in the physical ecology. *Ecology* **5**, 60–69.

Cook, W.C. (1925). The distribution of the alfalfa weevil (*Phytonomus posticus*, Gyll.) A study in physical ecology. *J. Agric. Res.* **30**, 479–491.

Dixon, A.F.G. (1985). *Aphid Ecology*. Blackie, Glasgow.

Dunn, J.A. (1952). The effects of temperature on the pea aphid–ladybird relationship. *Second Ann. Rep. Nat. Veg. Res. Sta.* **2**, 21–23.

Edwards, P.J. and Wratten, S.D. (1980). *Ecology of Insect–Plant Interactions*. Edward Arnold, London.

Emden, van, H.F. (1966). The effectiveness of aphidophagous insects in reducing aphid populations. In: *Ecology of Aphidophagous Insects* (Ed. by I. Hodek), pp. 227–235. Dr W. Junk, The Hague.

Emden, van, H.F. and Wearing, C.H. (1965). The role of the aphid host plant in delaying economic damage levels in crops. *Ann. appl. Biol.* **56**, 323–324.

Fereres, E. and Goldhamer, D.A. (1990). Deciduous fruit and nut trees. In: *Irrigation of Agricultural Crops* (Ed. by B.A. Stewart and R.D. Nielsen), pp. 987–1017. Agronomy no. 30, ASA, CSSA, SSSA, Madison, Wisconsin.

Ferro, D.N. (1987). Insect pest outbreaks in agroecosystems. In: *Insect Outbreaks* (Ed. by P. Barbosa and J.C. Schultz), pp. 195–215. Academic Press, London.

Ferron, P. (1978). Biological control of insect pests by entomogenous fungi. *A. Rev. Ent.* **23**, 409–442.

Finney, D.J. (1971). *Probit Analysis*. Cambridge University Press, Cambridge.

Fletcher, B.S. and Comins, H. (1985). The development and use of a computer simulation model to study the population dynamics of *Dacus oleae* and other fruit flies. *Atti XIV Congr. Naz. Ent. Itali.* Palermo-Erice-Bagheria, pp. 561–575.

Fletcher, B.S. and Kapatos, E.T. (1983). An evaluation of different temperature-development rate models for predicting the phenology of the olive fly *Dacus oleae*. In: *Fruit Flies of Economic Importance* (Ed. by R. Cavalloro), pp. 321–329. Balkema, Rotterdam.

Franz, J.M. (1971). Influence of environment and modern trends in crop protection on microbial control. In: *Microbial Control of Insects and Mites* (Ed. by H.D. Burges and N.W. Hussey), pp. 407–444. Academic Press, London.

Freeman, J.A. (1977). Prediction of new storage pest problems. In: *Origins of Pest, Parasite, Disease and Weed Problems* (Ed. by J.M. Cherrett and G.R. Sagar), pp. 265–302. 18th Symposium British Ecological Society. Blackwell, Oxford.

Gaston, K.J. and Lawton, J.H. (1990). Effects of scale and habitat on the relationship between regional distribution and local abundance. *Oikos*, **58**, 329–335.

Glenn, F.W. (1922). Relation of temperature to development of the codling moth. *Econ. ent.* **15**, 193–198.

Graham, S.A. (1939). Forest insect populations. *Ecol. Monographs.* **9**, 301–310.

Greenbank, D.O. (1956). The role of climate and dispersal in the initiation of outbreaks of the spruce budworm in New Brunswick. I. The role of Climate. *Can. J. Zool.* **34**, 453–476.

Grimes, D.W. and El Zik, K.M. (1990). Cotton. In: *Irrigation of Agricultural Crops* (Ed. by B.A. Stewart and R.D. Nielsen), pp. 741–773. Agronomy no. 30, ASA, CSSA, SSSA, Madison, Wisconsin.

Gutierrez, A.P. (1987). Analyzing the effects of climate and weather on pests. In: *Agrometeorology* (Ed. F. Prodi, F. Rossi and G. Cristoferi), pp. 203–223. Commune di Cesena, Cesena, Italy.

Gutierrez, A.P., Havenstein, D.E., Nix, H.A. and Moore, P.A. (1974a). The ecology of *Aphis craccivora* Koch and subterranean clover stunt virus in south-east Australia. II. A model of cowpea aphid populations in temperate pastures. *J. appl. Ecol.*, **11**, 1–20.

Gutierrez, A.P., Nix, H.A., Havenstein, D.E. and Moore, P.A. (1974b). The ecology of *Aphis craccivora* and subterranean clover stunt virus in south-east Australia III. A regional perspective on the phenology and migration of the cowpea aphid. *J. appl. Ecol.* **11**, 21–35.

Hagen, K.S. and Van den Bosch, R. (1968). Impact of pathogens, parasites and predators on aphids. *A. Rev. Ent.* **13**, 325–384.

Harcourt, D.G. (1971). Population dynamics of *Leptinotarsa decimlineata* (Say.) in Eastern Ontario III. Major population processes. *Can. Ent.* **103**, 1049–1061.

Harrewijn, P., Van Hoof, H.A. and Noordink, J.P.W. (1981). Flight behaviour of the aphid *Myzus persicae* during its maiden flight. *Neth. J. Pl. Path.* **87**, 111–117.

Harris, P. (1960). Production of pine resin and its effect on the survival of *Rhyacionia buoliana* (Schiff.)(Lepidoptera: Olethreutidae). *Can. J. Zool.* **38**, 121–130.

Hassell, M.P. and Waage, J.K. (1984). Host-parasitoid population interactions. *A. Rev. Ent.* **29**, 89–114.

Henson, W.R. (1962). Laboratory studies on the adult behaviour of *Conophthorus coniperdi* (Schwarz) (Coleoptera: Scolytidae). III Flight. *Ann. ent. soc. Am.* **55**, 524–530.

Henson, W.R. (1968). Some recent changes in the approach to studies of climatic effects on insect populations. In: *Insect Abundance* (Ed. by T.R.E. Southwood), pp. 37–46. Blackwell, Oxford.

Hodek, I. (1973). *Biology of Coccinellidae*. Academia, Prague.

Houghton, J.T., Jenkins, G.J. and Ephraums, J.J. (1990). *Climate Change: The IPCC Scientific Assessment*. Cambridge University Press, Cambridge.

Howe, R.W. (1963). The prediction of the status of a pest by means of laboratory experiments. *World Rev. Pest Cont.* **2**, 2–12.

Howe, R.W. and Burges, H.D. (1953). Studies on beetles of the family Ptinidae. IX A laboratory study of the biology of *Ptinus tectus* Boield. *Bull. ent. Res.* **44**, 461–516.

Huffaker, C.B. and Messenger, P.S. (1964). The concept and significance of natural control. In: *Biological Control of Insect Pests and Weeds* (Ed. by P. Debach), pp. 74–117. Chapman and Hall, London.

Huffaker, C.B., Berryman, A.A. and Laine, J.E. (1984). Natural control of insect populations. In: *Ecological Entomology* (Ed. by C. B. Huffaker and R. L. Rabb), pp. 359–398. John Wiley, New York.

Jackson, C.G. and Butler, G.D. (1984). Development time of three species of Bracon (Hymenoptera: Braconidae) on the pink bollworm (Lepidoptera: Gelechiidae) in relation to temperature. *Ann. ent soc. Am.* **77**, 539–542.

Jaques, R.P. (1985). Stability of insect viruses in the environment. In: *Viral Insecticides for Biological Control* (Ed. by Maramorosch and K.E. Sherman), pp. 285–360. Academic Press, London.

Johnson, C.G. (1969). *Migration and Dispersal of Insects by Flight*. Methuen, London.

Jones, M.G. (1979). Abundance of aphids on cereals from before 1973 to 1977. *J. appl. Ecol.*, **16**, 1–22.

Juillet, J.A. (1964). Influence of weather on flight activity of parasitic hymenoptera. *Can. J. Zool.* **42**, 1133–1141.

Kingsolver, J.G. (1989). Weather and the population dynamics of insects: integrating physiological and population ecology. *Phys. Zool.* **62**, 314–334.

Knight, J.D., Bale, J.S., Franks, F., Mathias, S.F. and Baust, J.G. (1986). Insect cold hardiness: supercooling points and pre-freeze mortality. *Cryo-let.* **7**, 194–203.

Lattin, J.D. and Oman, P. (1983). Where are the exotic insect threats? In: *Exotic Plant Pests and North American Agriculture* (Ed. by C. L. Wilson and C. L. Graham), pp. 93–137. Academic Press, New York.

Lincoln, D.E., Couvet, D. and Sionit, N. (1986). Response of an insect herbivore to host plants grown in carbon dioxide enriched atmospheres. *Oecologia (Berl.)*, **69**, 556–560.

McManus, M.L. (1988). Weather behaviour and insect dispersal. *Mem. ent. soc. Can.* **146**, 71–94.

Martinat, P.J. (1987). The role of climatic variation and weather in forest insect outbreaks. In: *Insect Outbreaks* (Ed. by P. Barbosa and J.C. Schultz), pp. 241–268. Academic Press, London.

Mattson, W.J. and Haack, R.A. (1987). The role of drought stress in provoking outbreaks of phytophagous insects. In: *Insect Outbreaks* (Ed. by P. Barbosa and J.C. Schultz), pp. 365–407. Academic Press, London.

Maywald, G.F. and Sutherst, R.W. (1989). CLIMEX: Recent developments in a computer program for comparing climates in ecology. *Proc. Simulation Soc. Aust.*, 8th biennial conf., pp. 134–140.

Meats, A. (1984). Thermal constraints to successful development of the Queensland fruit fly in regions of constant and fluctuating temperature. *Ent. exp. appl.* **36**, 55–59.

Meats, A. (1989a). Abiotic mortality factors—temperature. In: *World Crop Pests. Fruit Flies: Their Biology, Natural Enemies and Control, Vol. 3b* (Ed. by A.S. Robinson and G. Hooper), pp. 229–239. Elsevier, Amsterdam.

Meats, A. (1989b). Bioclimatic potential. In: *World Crop Pests. Fruit Flies: Their Biology, Natural Enemies and Control, Vol. 3b* (Ed. by A.S. Robinson and G.Hooper), pp. 241–252. Elsevier, Amsterdam.

Messenger, P.S. (1959). Bioclimatic studies with insects. *A. Rev. Ent.* **4**, 183–206.

Messenger, P.S. (1964a). Use of life tables in a bioclimatic study of an experimental aphid–braconid wasp host parasite system. *Ecology* **45**, 119–131.

Messenger, P.S. (1964b). The influence of rhythmically fluctuating temperatures on the development and reproduction of the spotted alfalfa aphid. *Therioaphis maculata. Econ. ent.* **57**, 71–76.

Messenger, P.S. (1970). Bioclimatic inputs to biological control and pest management programs. In: *Concepts of Pest Management* (Ed. by R.L. Rabb and F.E. Guthrie), pp. 84–102. North Carolina State University, Raleigh.

Messenger, P.S. (1974). Bioclimatology and prediction of population trends. *Proc. FAO Conf. Ecol. Relation to Plant Pest Control*, pp. 21–45. FAO, Rome.

Messenger, P.S. and Force, D.C. (1963). An experimental host–parasite system: *Therioaphis maculata* (Buckton)–*Praon palitans* Muesebeck (Homoptera: Aphididae–Hymenoptera: Braconidae). *Ecology* **44**, 532–540.

Minks, A.K. and Harrewijn, P. (Eds) (1989). *World Crop Pests. Aphids; Their Biology, Natural Enemies and Control*. Elsevier, Amsterdam.

Moericke, V. (1955). Über die lebensgewohnheiten der geflugelten blattläuse (aphidina) unter besonderer beruchsichtigung des verhaltens beim landen. *Zeit. angew. Ent.* **37**, 29–91.

Munroe, E. (1984). Biogeography and evolutionary history: wide-scale and long-term patterns of insects. In: *Ecological Entomology* (Ed. by C.B. Huffaker and R.L. Rabb), pp. 279–304. Wiley, New York.

Nicholson, A.J. (1958). Dynamics of insect populations. *A. Rev. Ent.* **3**, 107–136.

Norton, G.A. and Conway, G.R. (1977). The economic and social context of pest disease and weed problems. In: *Origins of Pest, Parasite, Disease and Weed Problems* (Ed. by J.M. Cherrett and G.R. Sagar), pp. 205–226. 18th Symposium British Ecological Society. Blackwell, Oxford.

Norton, G.A. and Way, M.J. (1983). Forecasting and crop protection decision making—realities and future needs. *10th Int. Congr. Pl. Prot., Brighton*, **1**, 131–138.

Park, T. (1954). Experimental studies on interspecific competition. II. Temperature, humidity and competition in two species of *Tribolium. Phys. Zool.* **27**, 177–238.

Parry, M.L. (1989). The potential impact of agriculture of the "greenhouse effect". In: *The "Greenhouse Effect" and UK Agriculture* (Ed by R.M. Bennett), pp. 27–48. Centre for Agriculture Strategy, Reading University, Reading.

Pedgley, D.E. (1982). *Windborne Pests and Diseases: Meteorology of Airborne Organisms*. Ellis Horwood, Chichester.

Rabb, R.L. and Kennedy, G.G. (Eds) (1979). *Movement of Highly Mobile Insects: Concepts and Methodology in Research*. North Carolina State University, Raleigh.

Ratte, H.T. (1985). Temperature and insect development. In: *Environmental Physiology and Biochemistry of Insects* (Ed. by K.H. Hoffmann), pp. 33–65. Springer-Verlag, Berlin.

Raun, E.S., Sutter, G.R. and Revelo, M.A. (1966). Ecological factors affecting the pathogenicity of *Bacillus thuringiensis* var. *thuringiensis* to the European corn borer and fall armyworm. *J. invert. Path.* **8**, 365–375.

Roberts, D.W. and Yendol, W. G. (1971). Use of fungi for the microbial control of insects. In: *Microbial Control of Insects and Mites* (Ed. by H.D. Burges and N.W. Hussey), pp. 125–149. Academic Press, London.

Rogers, H.H., Thomas, J.F. and Bingham, G.E. (1983). Response of agronomic and forest species to elevated atmospheric carbon dioxide. *Science* **220**, 428–429.

Salt, R.W. (1961). Principles of insect cold hardiness. *A. Rev. Ent.* **6**, 55–74.

Saunders, D.J. (1989). Hyperparasites. In: *Aphids. Their Biology, Natural Enemies and Control* (Ed. by A.K. Minks and P. Harrewijn), pp.189–203. Elsevier, Amsterdam.

Scheltes, P. (1978). The condition of the host plant during aestivation–diapause of the stalk-borers *Chilo partellus* and *Chilo orichalcociliella* (Lepidoptera, Pyralidae) in Kenya. *Ent. exp. appl.* **24**, 679–688.

Scriber, J.M. (1977). Limiting effects of low leaf-water content on the nitrogen utilization, energy budget and larval growth of *Hyalophora cercropia* (Lepidoptera: Saturniidae). *Oecologia (Berl.)* **28**, 269–287.

Scriber, J.M. and Slansky, F. (1981). The nutritional ecology of immature insects. *A. Rev. Ent.* **26**, 183–211.

Shelford, V.E. (1927). An experimental investigation of the relations of the codling moth to weather and climate. *Bull. Ill. Nat. Hist. Surv.* **16**, 307–440.

Shelford, V.E. (1929). *Laboratory and Field Ecology*. Williams & Wilkins, Baltimore.

Solomon, M.E. and Adamson, B.E. (1955). The powers of survival of storage and domestic pests under winter conditions in Britain. *Bull. ent. Res.* **46**, 311–355.

Sømme, L. (1969). Mannitol and glycerol in overwintering aphid eggs. *Norsk Ent. Tidss.* **16**, 107–111.

Southwood, T.R.E. (1968). The Abundance of Animals. Inaugural lecture. Imperial College, London.

Southwood, T.R.E. (1975). The dynamics of insect populations. In: *Insects, Science and Society* (Ed. by D. Pimentel), pp. 155–199. Academic Press, New York.

Southwood, T.R.E. (1977). The relevance of population dynamic theory to pest status. In: *Origins of Pest, Parasite, Disease and Weed Problems* (Ed. by J.M. Cherrett and G.R. Sagar), pp. 35–54. 18th Symposium British Ecological Society. Blackwell, Oxford.

Speight, M.R. and Wainhouse, D. (1989). *Ecology and Management of Forest Insects*. Oxford Science Publications, Clarendon Press, Oxford.

Steinhaus, E.A. (1960). The duration of viability and infectivity of certain insect pathogens. *J. insect Path.* **2**, 225–229.

Stewart, B.A. and Nielsen, R.D. (Eds) (1990). *Irrigation of Agricultural Crops*. Agronomy no. 30, ASA, CSSA, SSSA, Madison, Wisconsin.

Stinner, R.E., Gutierrez, A.P. and Butler, G.D. (1974). An algorithm for temperature dependent growth simulation. *Can. Ent.* **106**, 519–524.

Strong, D.R., Lawton, J.H. and Southwood, T.R.E. (1984). *Insects on Plants*. Blackwell, Oxford.

Sutherst, R.W. and Maywald, G.F. (1985). A computerised system for matching climates in ecology. *Agric. Ecosystems Env.* **13**, 281–299.

Tanada, Y. (1959). Microbial control of insect pests. *A. Rev. Ent.* **4**, 277–302.

Tanada, Y. (1965). Factors affecting the susceptibility of insects to viruses. *Entomophaga* **10**, 139–150.

Tauber, M.J. and Tauber, C.A. (1976). Insect seasonality: Diapause maintenance, termination and post diapause development. *A. Rev. Ent.* **21**, 81–107.

Tauber, M.J., Tauber, C.A., Nechols, J.R. and Obrycki, J.J. (1983). Seasonal activity of parasitoids: control of external, internal and genetic factors. In: *Diapause and Life Cycle Strategies in Insects* (Ed. by V.K. Brown and I. Hodek), pp. 87–125. Dr W. Junk, The Hague.

Tauber, M.J., Tauber, C.A. and Masaki, S. (1984). Adaptations to hazardous seasonal conditions: dormancy, migration and polyphenism. In: *Ecological Entomology* (Ed. by C.B. Huffaker and R.L. Rabb), pp. 149–183. John Wiley, New York.

Taylor, F. (1981). Ecology and evolution of physiological time in insects. *Am. Nat.* **117**, 1–23.

Taylor, L.R. (1963). Analysis of the effect of temperature on insects in flight. *J. anim. Ecol.* **32**, 99–117.

Taylor, L.R. (1965). Flight behaviour and aphid migration. *Proc. N. Cent., Ent. Soc. Am.* **20**, 9–19.

Thomas, G.G., Goldwin, G.K. and Tatchell, G.M. (1983). Associations between weather factors and the spring migrations of the damson-hop aphid, *Phorodon humuli. Ann. appl. Biol.* **102**, 7–17.

Thomas, T.H. (1989). Sugar beet in the greenhouse—a global warming warning. *Beet Review* **57**, 24–26.

Turl, L.A.D. (1983). The effect of winter weather on the survival of aphid populations on weeds in Scotland. *Bull. OEPP* **13**, 139–143.

Uvarov, B.P. (1931). Insects and climate. *Trans. ent. Soc. Lond.* **79**, 1–247.

Varley, G.C. and Gradwell, G.R. (1960). Key factors in population studies. *J. anim. Ecol.* **29**, 399–401.

Varley, G.C. and Gradwell, G.R. (1968). Population models for the winter moth. In: *Insect Abundance* (Ed. by T.R.E. Southwood), pp. 132–142. Blackwell, Oxford.

Wallner, W.E. (1987). Factors affecting insect population dynamics: differences between outbreak and non-outbreak species. *A. Rev. Ent.* **32**, 317–340.

Walters, K.F.A. and Dixon, A.F.G. (1984). The effect of temperature and wind on the flight activity of cereal aphids. *Ann. appl. Biol.* **104**, 17–26.

Watson, M.A., Heathcote, G.D., Lauckner, F.B. and Sowray, P.A. (1975). The use of weather data and counts of aphids in the field to predict the incidence of yellowing viruses of sugar beet crops in England in relation to the use of insecticides. *Ann. appl. Biol.* **81**, 181–198.

Watson, S.J. and Carter, N. (1983). Weather and modelling cereal aphid populations in Norfolk (UK). *Bull. OEPP* **13**, 223–227.

Way, M.J. and Banks, C.J. (1964). Natural mortality of eggs of the black bean aphid, *Aphis fabae* Scop. on the spindle tree, *Euonymus europaeus* L. *Ann. appl. Biol.* **54**, 255–267.

Webster, R.P. and Cardé, R.T. (1982). Influence of relative humidity on calling behaviour of female European corn borer (*Ostrinia nubilalis*). *Ent. exp. appl.*, **32**, 181–185.

Welch, S.M., Croft, B.A., Brunner, J.F. and Michels, M.F. (1978). PETE: an extension phenology modelling system for management of multi-species pest complex. *Env. Ent.* **7**, 487–497.

Wellings, P.W. and Dixon, A.F.G. (1987). The role of weather and natural enemies in determining aphid outbreaks. In: *Insect Outbreaks* (Ed. by P. Barbosa and J.C. Schultz), pp. 313–346. Academic Press, London.

Wellington, W.G. (1945). Conditions governing the distribution of insects in the free atmosphere I. Atmospheric pressure, temperature and humidity. *Can. Ent.* **77**, 7–15.

Wellington, W.G. (1954). Weather and climate in forest entomology. *Meteo. Monographs* **2**, 11–18.

Wellington, W.G. and Trimble, R.M. (1984). Weather. In: *Ecological Entomology* (Ed. by C.B. Huffaker and R.L. Rabb), pp. 399–425. John Wiley, New York.

White, T.C.R. (1969). An index to measure weather-induced stress of trees associated with outbreaks of psyllids in Australia. *Ecology* **50**, 905–909.

White, T.C.R. (1978). The importance of relative shortage of food in animal ecology. *Oecologia (Berl.)* **33**, 71–86.

Wigglesworth, V.B. (1972). *The Principles of Insect Physiology.* Chapman & Hall, London.

Wiktelius, S. (1981). The diurnal flight periodicities and temperature threshold for flight for different migrant forms of *Rhopalosiphum padi*: L. (Hom., Aphididae). *Zeit. angew. Ent* **93**, 449–457.

Wilson, P. (1989). The "greenhouse effect" and animal production in the UK. In: *The "Greenhouse Effect" and UK Agriculture* (Ed. by R.M. Bennett), CAS Paper 19, pp. 53–66. Centre for Agriculture Strategy, Reading University, Reading.

Wong, S.C. (1979). Elevated atmospheric partial pressure of CO_2 and plant growth. *Oecologia (Berl.)* **44**, 68–74.

Woodward, F.I. (1987). *Climate and Plant Distribution.* Cambridge University Press, Cambridge.

Worner, S.P. (1988). Ecoclimatic assessment of potential establishment of exotic pests. *Econ. ent.* **81**, 973–983.

Zwolfer, W. (1935). Die temperaturabhangigkeit der Entwicklung der nonne (*Lymantria monacha* L.) und ihre Bevolkerungswissenschaftliche Auswertung. *Zeit. angew. Ent.* **21**, 333–384.

Responses of Soils to Climate Change

J.M. ANDERSON

I. SUMMARY

Increased trace gas concentrations in the atmosphere are forcing climate change towards warmer and drier conditions at mid-latitudes and longer summer seasons and higher temperatures at high latitudes. Current climate change scenarios suggest that this could result in changes in vegetation and soils in the boreal and tundra regions. Historically changes in land use, particularly in the tropics, and anthropogenic sources of methane, have been the major contributors to climate warming but industrial sources of CO_2 are currently more significant. In a warmer world, however, increased emissions of CO_2 and CH_4 from boreal and tundra soils could have positive feedback effects on climate change. This chapter considers the effects of climate on

ADVANCES IN ECOLOGICAL RESEARCH VOL. 22
ISBN 0–12–013922–7

carbon dynamics in soils through interactions of vegetation, topography, soil physicochemical conditions and parent material. Processes of soil organic matter formation, carbon dating and manipulative experiments are considered which show that the carbon pools formed in cold soils are susceptible to increased mineralization rates if soil temperatures increase. Under current projections of climate-warming, soils at high latitudes could release about 1 Gt C year^{-1} over the next 50–60 years; a comparable flux to current effects of deforestation in the tropics and about 20% of industrial emissions. Warmer, more developed soils, contain larger pools of stabilized carbon formed through interactions with soil minerals but the rate of accumulation associated with changing plant/soil relationships is unlikely to constitute a significant carbon sink on any immediate time-scale. The net shifts in carbon pools in vegetation and soil as a consequence of enrichment by CO_2 and N deposition affecting the quantity, quality and location of plant production are major uncertainties in current projections of global carbon balances.

II. INTRODUCTION

Atmospheric concentrations of carbon dioxide (CO_2), methane (CH_4), nitrous oxide (N_2O) and chlorofluorohydrocarbons (CFCs) have increased dramatically as a consequence of agricultural and industrial development. There is increasingly a consensus that these trace gases are effecting changes in the global heat balance with the result that significant changes in climate are predicted over the next century (Houghton et al., 1990). The resulting geographical shifts in climate patterns, and increased concentrations of atmospheric CO_2, are likely to have profound consequences for the distribution, structure and functioning of ecosystems (Melillo et al., 1990). These changes will be amplified or damped by physical feedback processes including geophysical effects (water vapour, albedo), the consequences of changes in temperature (atmospheric circulation, ocean currents and gas solubilities), biogeochemical effects (mobilization of methane hydrates, production and decomposition on land and in oceans) and changes in human activity (Lashof, 1989).

This chapter concerns the interactions between climate and soils. The current status of soils as a source of CO_2 and other trace gases has been very comprehensively reviewed (Schlesinger, 1977, 1984; Palm et al., 1986; Bolin 1986; Houghton et al., 1987; Cicerone and Oremland, 1988; Watson et al., 1990; Bouman, 1990a; Mosier et al., 1991) and will only be considered briefly here. Rather, emphasis is placed on the effects of climate change on soils through interactions with vegetation and parent materials (Fig. 1) to provide an insight into the dynamic processes involved.

A comprehensive synthesis of the response of soils to global warming is provided by Scharpenseel *et al.* (1990) which was unavailable when this chapter was prepared.

III. BIOTIC AND INDUSTRIAL SOURCES OF TRACE GASES

Records show that land and sea temperatures have increased by 0·3–0·6°C over the last century with the warmest years on record occurring in the last decade. It is estimated that over this period CO_2 has contributed about 56% of the "greenhouse warming". The other trace gases, although present in

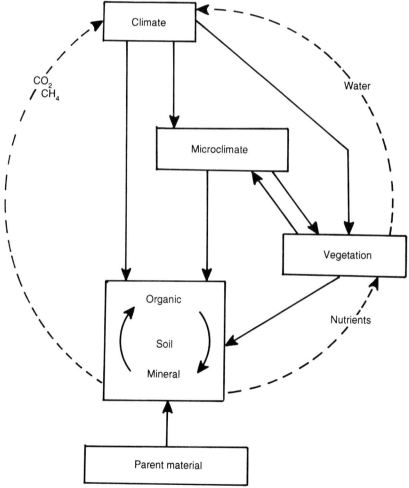

Fig. 1. Interrelationships of factors influencing the development of soil and associated carbon pools. Feedbacks to climate are shown as broken lines.

166 J. M. ANDERSON

Table 1

Characteristics of major trace gases in the atmosphere affecting global warming (data from Houghton et al., 1990)

	CO_2	CH_4	N_2O	CFCs
Concentration	p.p.m.v.	p.p.m.v.	p.p.b.v.	p.p.b.v.
Pre-industrial (1750–1800)	280	0·8	288	na
Present day (1990)	353	1·72	310	0·2–0·3
Annual % increase	0·5	0·9	0·25	4
Residence time (years)	100	10	150	65–130
Radiative absorption relative to CO_2	1	32	150	> 10 000

na = not applicable.

lower concentrations, have higher radiative absorption potentials than CO_2 (Table 1) and 44% of radiative forcing is attributable to CH_4 (11%), N_2O (6%), CFCs (24%) and stratospheric water (3%). Recently, Wang et al. (1991) have challenged the expression of relative warming effects of trace gases as CO_2 equivalents (Lashof and Ahuja, 1990) suggesting that CO_2 has different climate-forcing effects. None the less, with the exception of CFCs, which are entirely of industrial origin, CO_2 and CH_4 are currently the most important biogenic greenhouse gases implicated in climate change. Water vapour also has a powerful positive feedback effect on climate warming in response to trace gas concentrations (Ramanathan et al., 1989; Rind et al., 1991) but will not be considered here.

Concentrations of CO_2 in 1990 of 353 p.p.m.v. are 25% higher than pre-industrial levels at the end of the eighteenth century and are rising at an annual rate of about 0·5% year^{-1}. The vegetation and soils of virgin forests hold 20–100 times more carbon than agricultural systems (Houghton, 1990) and deforestation in temperate and tropical regions between 1850 and 1987 is estimated to have added about 115 Gt to the atmosphere (Houghton et al., 1987). In temperate regions the rate at which virgin forest is brought under cultivation has declined over the last 100 years and forest regrowth following harvesting may act as a sink for CO_2 (Delcourt and Harris, 1980; Armentano and Ralston, 1980; Melillo et al., 1988). A number of other studies, however, show that the conversion of natural forest to even-aged stands or plantations will result in decreased net carbon storage in the system (Harmon et al., 1990; Vitousek, 1991).

Deforestation has continued in the tropics and the rate may be increasing (Houghton, 1990). Setzer and Pereira (1991) suggest that there was a 100% increase in the incidence of fires in the Amazon between 1985 and 1987, and rates of deforestation in several states are rising by 20–30% per annum.

Estimates of the annual carbon release by deforestation, mainly from tropical forests, range from 0·3 to 1·7 Gt year^{-1} (Detweiler and Hall, 1988) to 1–2·6 Gt year^{-1} (Houghton et al., 1987) including 0·2–0·9 Gt C year^{-1} from soils (Bouman, 1990b). The cumulative release of CO_2 from fossil fuels and cement manufacture from 1850 to 1987 is estimated at 200 Gt C (Watson et al., 1990) with current industrial emissions of 5·7 Gt C year^{-1}.

Methane concentration in the atmosphere has more than doubled (from 0·8 to 1·72 p.p.m.v.) since 1860 and is currently increasing at about 1% per year. The present sources of methane, totalling 300–700 Tg year^{-1}, are natural wetlands (20%), aquatic sediments (5%), rice paddies (20%), cattle and other ruminants (16%), termites (4%), biomass burning (10%) and landfill with domestic waste (10%) (Cicerone and Oremland, 1988). Industrial emissions constituting 14% of the total flux according to Cicerone and Oremland (1988) may be a substantial underestimate because of leakages in the gas industry (Crutzen, 1991).

Despite uncertainties over trace gas emissions it can be concluded, because of the magnitude of the fluxes, that future projections of global warming are driven mainly by CO_2 from industrial processes and exploitation of fossil fuels, and biogenic sources of CH_4 related to agricultural land use and resource exploitation by urban populations. However, if the climate changes, carbon pools in plants and soils which have developed under present environmental conditions, may shift to new sink/source relationships with largely unknown consequences for atmospheric composition and future climates.

IV. FUTURE PATTERNS OF CLIMATE CHANGE AND IMPACT ON VEGETATION

A. Scenarios of Future Climates

For the purposes of this discussion the IPCC "Business as Usual" scenario (Houghton et al., 1990) is adopted (assuming no controls on trace gas emissions) in which CO_2 concentrations, including equivalents of other gases, increase to twice pre-industrial levels by the year 2030. If climate forcing occurs there will be delays in the temperature response to the gas composition of the atmosphere because the thermal capacity of the oceans buffers the rate of change in air temperatures. This is known as the realized temperature change and is estimated as about 0·3 °C per decade. There will be considerable geographical variation in the rate of change, which may have selective effects on the physiology and ecology of the biota, but these are difficult to predict. Most general circulation models (GCMs) used to simulate climate change have assumed an equilibrium situation in which the trace gas composition of the atmosphere has stabilized at twice the radiative forcing of pre-industrial conditions. In fact, stabilization of climate will take many decades after any changes in trace gas emissions because of the long lag-time for equilibration of gas and heat sinks. The value of carrying out simulations of equilibrium climate change is that regional climate patterns can be computed. Equilibrium simulations therefore do not predict when a suite of new climate conditions will be manifested but allow assessment of their potential ecological consequences.

A number of recent simulations of the equilibrium climate response to a doubling of CO_2 (Mitchell et al., 1990) have confirmed the earlier estimates of an increase in global surface temperatures over the next century by 3·0 ± 1·5°C (Bolin 1986; Manabe and Stouffer, 1980) with enhanced warming at high latitudes. Above 50–60°C in the northern hemisphere temperatures could increase by 50–100% (4·5–6°C) of the global mean in summer and by a factor of three (8–12°C) in winter. In high latitudes, soil moisture will generally decrease in summer due to higher evapotranspiration and increase in winter as a consequence of increased precipitation, or more precipitation falling as snow, or earlier snow-melt. In mid-latitudes (35–55°) soil moisture may be reduced by 17–20% during the northern summer. In the warmer climate, snow-melt and summer drying begin earlier, reducing soil moisture levels in summer. The insulative effects of snow cover on soil temperatures are of particular significance because decomposition (Bleak, 1970) and methanogenesis (Whalen and Reeburgh, 1988) can continue later in the season if the soil remains unfrozen. Late snow-fall after the soil has frozen can prolong the spring melt while soils can remain unfrozen under snow cover when air temperatures are sub-zero (Mackay and Mackay, 1974). Ping (1987) shows major differences in the mean monthly soil temperature variation between soils in the coastal and interior regions of Alaska. Although the interior site was located at a latitude 3° further north with much lower minimum mean air temperatures ($-22·2$°C in the interior vs. $-11·6$°C at the coast) the amplitude of soil temperatures was lower because of snow cover. Comparisons were also made between the effects of vegetation cover in the two localities. In the interior site removal of the insulative effects of the canopy and moss cover on the forest floor would raise the mean annual soil temperature by 3·4°C and increase the active soil depth. In the warmer coastal site the increase in soil temperature would be 1·6°C.

B. Effects of Climate Change on Vegetation

As a consequence of global warming, caused by a doubling of CO_2, the north temperate and boreal regions are likely to undergo the largest regional changes in temperature and soil moisture regimes. A 2°C rise in mean surface temperatures could cause a 250–300 km displacement of the southern boundary of the permafrost; a 3°C warming could lead to more than a 25% reduction of permafrost in Canada (Barry, 1984). The drier summer conditions could result in a northerly shift in grain-producing areas in North America and Russia bringing historically untilled soils under arable cultivation (Newman, 1980; Smit *et al.*, 1988). For each 1°C rise in mean air temperature the Corn Belt in the United States is predicted to shift 144 km to the north and 100 km to the east.

A number of simulations using forest growth models have predicted that water stress will be a primary cause of forest death in the southern and central regions of North America as climate changes (Emanuel *et al.* 1985 a,b; Solomon, 1986; Shugart *et al.*, 1986, Pastor and Post, 1988).

Emanuel *et al.* (1985a,b) used the Holdridge life zone classification related to predicted changes in global temperatures and precipitation. Under the constraints defined by the climate predictions considered earlier their simulations showed little change in low latitudes but a decrease in the areas covered by boreal forest and tundra by 37% and 32% respectively. The model predicted the boreal forest zone extending northwards over 42% of the present area covered by tundra. At the southern limits of boreal forest, wet forest is replaced by cool temperate forest and boreal moist forest, and boreal moist forest by temperate steppe. There could also be a lag of about 100 years for the replacement of boreal forest by temperate forest if the rate of die-back exceeds the migration rate of species from the south (Shugart *et al.*, 1986). The time required for soil changes to allow seedling establishment, even if the seed is dispersed to a new site, could also delay the period of establishment resulting in a net loss of carbon from forest biomass during the shift in the type of forest cover (Solomon, 1986).

There are many uncertainties involved in these scenarios of changes in vegetation types because of the present limitations in predicting regional climates and the rate at which changes may occur. The rate of change will be critical in determining whether plant communities adapt, whether conifers are successionally replaced by broad-leaf forests or whether die-back of forests, and resulting changes in microclimate, soils and vegetation cover, preceed the establishment of new climax communities for the region.

V. CLIMATE-RELATED PATTERNS OF SOIL CARBON POOLS AND FLUXES

The relative sizes of carbon pools in soils and biota are shown in Fig. 2. Most estimates of carbon in soil organic matter are within the range 1300–1600 Gt ($1 \text{ Gt} = 10^{15}\text{g} = 1 \text{ Pg}$) (Bouman, 1990b) which is about twice the carbon in terrestrial biomass (664 Gt) according to Olson *et al.* (1983). Gross primary production by terrestrial plants fixes 120 Gt year^{-1} or about 17% of the carbon in the atmosphere (725 Gt) per year (Bolin, 1986). Higher plants respire 50% of the fixed carbon and 50% (net primary production) is mainly respired by soil micro-organisms including ~ 0.5–2% returned as CH_4 rather than CO_2 (Lashof, 1989).

Processes in the plant and soil systems are primarily regulated by climate. Given the the magnitude of the carbon fluxes through the biosphere, relatively small changes in temperature and moisture could result in substantial feedbacks to atmospheric composition. Seasonal oscillations of atmospheric CO_2 recorded in the northern hemisphere are mainly the consequence of the seasonal asynchrony of photosynthesis, respiration and decomposition, unlike the humid tropics where these processes occur simultaneously. The oscillations in the boreal region, which dominate the seasonal drawdown for the globe, show increasing amplitude with time which may reflect increased forest production (D'Arriego et al., 1987). Whether soils become a

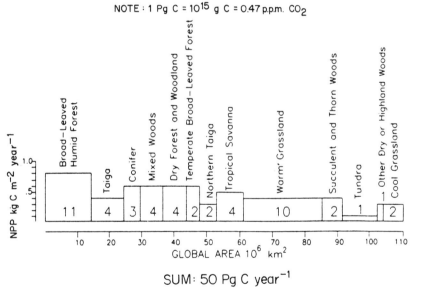

Fig. 2. Pools and fluxes of carbon in major terrestrial ecosystem types. A, distribution of net primary production; B, biomass and C, soil carbon pools. The total area occupied by each ecosystem type is represented by the horizontal axis with flux or density on the vertical axis; the area is therefore proportional to the global production or storage in each ecosystem type. (Source: Lashof (1989) based on data in Olson et al. (1983) and Zinke et al. (1984).)

Fig. 2. Continued

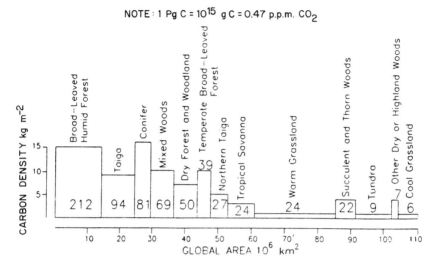

LIVE BIOMASS
(b) CARBON STORAGE IN MAJOR WORLD ECOSYSTEMS (Pg C)

NOTE: 1 Pg C = 10^{15} g C = 0.47 p.p.m. CO_2

SUM: 664 Pg C

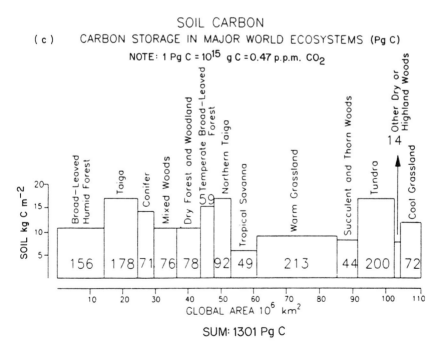

SOIL CARBON
(c) CARBON STORAGE IN MAJOR WORLD ECOSYSTEMS (Pg C)

NOTE: 1 Pg C = 10^{15} g C = 0.47 p.p.m. CO_2

SUM: 1301 Pg C

172   J. M. ANDERSON

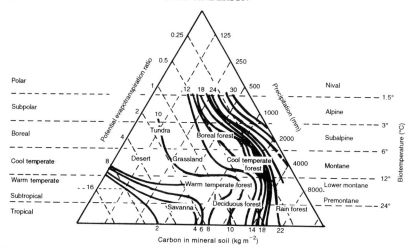

Fig. 3. Isoclines of mean soil carbon density from samples classified according to major ecosystem types. Only the carbon in mineral soil to a depth of 1 m was included in the analysis. (Source: after Post *et al.*, 1982.)

source or sink of CO_2 or CH_4 is determined by the effects of climate on vegetation and geochemical conditions maintaining the present size of soil carbon pools and the magnitude of microbial responses to changes in temperature and moisture.

A. Global Distribution of Soil Carbon Pools

The carbon pools in major ecosystem types shown in Fig. 3 reflect the effect of temperature and moisture on the balance between carbon inputs in plant litter and losses by decomposition and mineralization of soil organic matter (SOM). The quotient of potential evaporation (PET) divided by precipitation (P), the evapotranspiration ratio, is an index of potential water availability in soils in relation to the mean annual temperature. Where rainfall exceeds evapotranspiration this value is <1 whereas in systems subject to moisture deficits the value of the index is >1.

In ecosystems such as steppe grassland or temperate forest with a potential evaporation (PE) ratio near unity the carbon pools are about $10\ kg\ m^{-2}$ but the drier zones of the temperate region, subtropics and subhumid tropics depart from this relationship. Organic matter contents of temperate and tropical soils with soil moisture regimes which are udic (little or no water stress during at least 9 months per year) or ustic (moisture stress for 3–6 months duration) are shown in Table 2. Ustic tropical soils have significantly lower C content at all depths than the udic soils because moisture limits plant production more than microbial activity in soils during the dry season. In

Table 2
Carbon content of soils with udic (moist) or ustic (semi-arid) regimes
in the temperate and tropical regions (Sanchez *et al.*, 1982)

| Depth (cm) | Carbon content (%) | | | |
	Temperate		Tropical	
	Udic	Ustic	Udic	Ustic
0−15	2·18	1·38	1·91	1·23
0–50	1·36	0·95	1·13	0·90

addition high soil temperatures, particularly at the start of the wet season when moisture is not limiting, result in high rates of decomposition which can exceed litter inputs. In the temperate region the udic soils have higher organic matter contents in topsoil but no significant differences at greater depth reflecting less extreme temperature regimes and higher organic matter inputs.

According to Post *et al.* (1982) where the PE ratio exceeds 0·5 very different ecosystem types, ranging from tundra to evergreen rainforest, contain carbon pools of approximately the same order: 21·8 kg m^{-2} in tundra and 19·1 kg m^{-2} in wet tropical forest; 11·6 kg m^{-2} in moist boreal forest and 11·4 kg m^{-2} in moist tropical forest. The dynamics of these pools, however, are very different. In high latitudes precipitation of only a few hundred millimetres exceeds annual PE and soils in poorly drained areas are often waterlogged. Temperature regimes in soils at plant rooting depth are pergelic (mean annual temperature [MAT] < 0°C) or cryic (MAT 0–8°C) (Foth and Schaffer, 1980). In the wet tropics the mean soil temperature shows little variation around 25°C and precipitation of 3000–5000 mm or greater is far in excess of PE (1200 mm). The season tropical forests with a marked dry season have higher MATs (∼27°C) and lower precipitation in the range 1800–2500 mm year^{-1}.

Net production is the tundra (40–100 g C m^{-2} year^{-1}) is low and decomposition is limited by sub-zero temperatures in soils for up to 7 months of the year and waterlogged conditions in meadows during the brief summer. Hence decomposition is inhibited by soil conditions more than production is limited by climate and between 10 and 20% (Schell, 1983) and 40% (Svensson and Rosswall, 1984) of annual net primary production (NPP) may be stored in peats. Overall, wetlands, bogs and peatlands probably constituted a sink for 0·3 Gt C year^{-1} until recent times (Armentano and Menges, 1986). In boreal forests, NPP is about 200–300 g C m^{-2} year^{-1} (Fig. 2) with turnover time on the forest floor of ∼25–40 years (Flanagan and Van Cleve, 1983) compared with average litter fall in humid tropical rainforests of 930 g m^{-2} year^{-1} with a turnover time 0·9 years; the high tropical rates reflecting favourable conditions for decomposition over most of the year or very high rates during the wet season in strongly seasonal environments (Anderson and Swift, 1983; Swift and Anderson, 1989). In tropical forests, therefore, a major proportion of leaf litter decomposition occurs on the soil surface and the small remaining fraction must be more recalcitrant to maintain the large carbon pool in mineral soil.

In contrast to this emphasis on relatively uniform geographical patterns other studies have emphasized the wide regional variation in SOM contents which shows that local factors strongly modify the effects of climate on soil carbon pools (Sanchez et al., 1982; Anderson and Swift, 1983; Bouman, 1990b). Brown and Lugo (1982) adopted a similar approach to Post et al. (1982) in analysing relationships between temperature and precipitation, and carbon pools in tropical forest biomass and SOM. Soil C showed a significant ($P = 0.05$) negative exponential relationship with temperature/precipitation (T/P) but only 23% of the variance was explained by the regression. Much of the unexplained variance was contributed by forests with high rainfall regimes which had higher SOM contents than predicted reflecting soil and vegetation effects overriding climate effects (e.g. Anderson et al., 1983). The rainforests in this group represent about 25% of the tropical forest biome. In cooler climates differences in aspect can modify climate. This is particularly well illustrated, on a landscape scale, by studies in interior Alaska by Van Cleve and colleagues (Van Cleve et al., 1986). This region has a topography of rolling uplands with paper birch and white spruce stands on south-facing aspects and productive, permafrost-free sites, to cold, north-facing aspects occupied by black spruce, most of which are underlain by permafrost (Fig. 4). The climate is strongly continental with minimum temperatures as low as $-50°C$ and summer maxima as high as $+35°C$. The heat sum (degree days [DD] above 0°C) for these forest soils ranges from < 750 DD for black spruce, 800–1100 DD for birch and white spruce stands, to as much as 2200 DD for some aspen stands (Van Cleve et al., 1983). In addition, the thick forest floor and moss-cover of the forest floor in the black spruce insulate the subsoil so that the active layer above the permafrost is only 25–30 cm even in summer compared with a rooting depth > 100 cm in well-drained soils with a high heat sum (Viereck et al., 1983). Cumulative degree days for these Alaskan forest soils was found to explain 71% of the variation in forest floor depth (F cm) according to the function: $F = 33.02 - 0.027\,DD$. Litter production, ranging from as little as 15–142 g m^{-2} year^{-1} in black spruce forests to 268–653 g m^{-2} year^{-1} in balsam poplar stands (Yarrie, 1983) was a positive function of the heat sum. Similar local variation can be detected on an even finer scale in tundra where biomass and SOM can show as much variation over a few metres from the centre of polygons to the stony, frost-heaved margins, as between mean values for different ecosystem types (Heal et al., 1981). Data from three tundra sites (Table 3) show that NPP

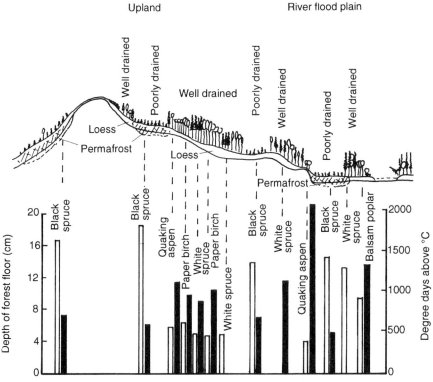

☐ Depth of forest floor (cm)
■ Degree days above 0°C

Fig. 4. Relationships between soils, vegetation, forest floor depth (white bar) and microclimate in a topographic sequence from floodplains to uplands in interior Alaska. The heat sum (degree days, shown as a black bar) is the cumulative product of days with soil temperatures (10 cm depth) above 0°C. (Source: Van Cleve *et al.*, 1983.)

ranged from 114 g C m^{-2} year^{-1} in lichen heath to 351 g C m^{-2} year^{-1} in wet meadow whereas soil carbon ranged between 3·6 and 45 kg C m^{-2} respectively. In the lichen heath, production and decomposition are temperature- and moisture-limited whereas in the tundra meadow decomposition is impeded by waterlogging leading to the accumulation of peaty soils. Small local variations in topography and aspect can therefore have major effects on the balance between production and decomposition and confound comparisons of carbon accumulation rates in chronosequences (Schlesinger, 1990).

In boreal and tundra systems, low temperatures are the main factors maintaining large soil carbon pools in relation to plant production. There is

Table 3
Estimates of organic matter and net primary production in tundra sites (after Heal *et al.*, 1981)

Site	Habitat	Soil carbon $(kg\ m^{-2})$*	Net primary production $(g\ C\ m^{-2}\ year^{-1})$
Point Barrow (Alaska)	Grass–sedge meadow	17	130
Abisko (Sweden)	Stordalen mire	45	172
Hardangervidda (Norway)	Lichen heath	3·6	114
	Dry meadow	14	225
	Wet meadow	30	351
	Birch wood	15	328

*Calculated from organic matter assuming 45% carbon content.

also considerable local variation in waterlogging, permafrost depth and thermal regimes which may result in complex non-linear responses to climate changes. This variation in the variables currently controlling CO_2 and CH_4 fluxes encompasses the range of conditions predicted for future climates and provides the basis for assessing their impact on soil processes.

B. Temperature and Moisture Controls of CO_2 Fluxes

The efflux of CO_2 from soil ("soil respiration") theoretically represents an integrated measure of carbon mineralization from all the different carbon pools in soil and over the year should equal above- and below-ground inputs from plants under steady state conditions. Latitudinal gradients in soil respiration broadly follow trends in litter production (Schlesinger, 1977, Raich and Nadelhoffer, 1989) but an additional component of the flux is derived from root respiration. Various reviews suggest that root respiration contributes about 50% of the flux in forests and 20–30% in grassland (Schlesinger, 1977; Singh and Gupta, 1977; Coleman and Sasson, 1980; Raich and Nadelhoffer, 1989). Raich and Nadelhoffer (1989) show from a very extensive data set that carbon allocation to roots in forest ecosystems is a positive function of litterfall: as litterfall increases from 70 to 500 g m^{-2} year^{-1} total below ground allocation increases from 260 to 1100 g m^{-2} year^{-1}. The major proportional change in below-ground allocation occurred as litterfall increased to 200 g m^{-2} year^{-1} possibly indicating improved soil conditions for tree growth. Because of the difficulties in differentiating root and microbial respiration *in situ*, measurements of CO_2 efflux from soils are an unreliable quantitative measure of carbon mineralization from litter and SOM. None the less, measurement of soil respiration in the field, or by using intact cores in the laboratory, is a sensitive indicator of temperature and moisture controls on microbial processes.

The temperature response of microbial respiration is frequently expressed as a Q_{10} function which expresses the fold increase in respiration for a 10°C rise in temperature. Q_{10} is usually assumed to be around 2 for biological systems but values between 1·6 and 3·7 have been recorded for microbial respiration for many temperate and tropical systems over a temperature range between 10 and 40°C (Schlesinger, 1977; Singh and Gupta, 1977). Generally, Q_{10} values appear to be lower in warm regimes and higher under cold regimes. For example, soil respiration (Anderson, 1973) and microbial respiration in leaf litter (de Boois, 1974) in warm temperate woodlands followed a $Q_{10} = 2$ relationship but in tundra systems the temperature response for litter decomposition was 3·7 between 0 and 10°C (Bunnell et al., 1977) and 4 for tundra meadow soils averaged over 10°C increments from − 10 to 25°C (Flanagan and Bunnell, 1976). It has been suggested (Flanagan and Veum, 1974; Flanagan, 1978) that this inflated temperature response enables the microbial community to maintain comparable decomposition rates to temperate regions near the soil surface during the few summer months over which the soils are thawed. Conversely, Peterson and Billings (1975) showed simple linear relationships between soil respiration and temperature in tundra and, in studies on subarctic woodland soils, Moore (1981) found Q_{10} values of 2·0–2·3 (between 1 and 20°C) and concluded that litter decomposition was simply limited by low temperatures in these systems. In well-drained tundra and taiga soils microbial respiration is temperature-related over the range − 5 to 25°C (Heal et al., 1981).

High soil-water contents, on the other hand, certainly limit the temperature response by micro-organisms. In waterlogged tundra, taiga and bog systems, microbial respiration shows a non-linear response to moisture with optimum rates between 200 and 600% (dry weight organic matter) and declining at higher and lower moisture contents (Bunnell et al., 1977; Heal et al., 1981). In most temperate/boreal forest systems (Anderson, 1973; Schlesinger, 1977; Moore, 1984) and mesic grasslands (Coleman and Sasson, 1980) soil respiration is rarely limited by moisture deficit although litter layers may become desiccated. In arid grasslands − 1·3 MPa moisture tension has been found to reduce microbial respiration to about 10% of values at field capacity (− 0·03 MPa) and is still detectable below − 8 MPa (Wildung et al., 1975; Hunt, 1978).

C. Controls of CH_4 Production in Soils

Details of the physiological controls of methanogenesis and fluxes in agroecosystems have been reviewed elsewhere (Cicerone and Oremland, 1988; Bouman, 1990) and the present brief discussion will be limited to tundra systems which are likely to be affected by warming and changes in the depth of the water table.

The development of anaerobic conditions for methanogenesis depends upon high water content limiting oxygen diffusion in soil and an available carbon source. Available carbon not only promotes the development of low oxygen tensions through the action of facultative anaerobes but fermentation, which is the main process of carbon mineralization in flooded soils, also produces carboxyllic acids and other reduced organic carbon compounds which form the substrates for methanogens. Temperature also has direct and indirect effects on microbial processes and gas solubilities resulting in highly elevated Q_{10} responses of 5·5–8·0 over 2–12°C (Svensson and Rosswall, 1984) and hence most of the annual flux occurs when the soils are thawed (Whalen and Reeburgh, 1988). Sebacher *et al.* (1986), however, found a poor correlation between methane fluxes and soil temperatures in a range of Alaskan sites but significant emissions were recorded at 2–4°C indicating the potential important effects of an earlier spring thaw in these systems. Emissions were logarithmically related to water level with highest rates from tundra meadows flooded with up to 20 cm of water.

Svensson and Rosswall (1984) found that CH_4 production was restricted to the upper layers of peat at a depth of about 10 cm just below where the redox potential reached lowest values. Methane may be consumed in the more superficial oxidized layers to a depth of 5 cm (Whalen and Reeburgh, 1990). Net fluxes measured at the surface show very wide temporal and spatial variability determined by soil conditions, topography and vegetation cover. Seasonal measurements of net CH_4 fluxes in Alaskan tundra by Whalen and Reeburgh (1988) showed that flux rates associated with tussocks of *Carex* and *Eriophorum* were five to eight times higher than those from intertussock areas (Fig. 5). It was suggested that a possible explanation for this pattern was that the root/rhizome systems act as conduits for CH_4 bypassing the surface oxidative layers as has been shown in rice paddies (Cicerone and Shetter, 1981).

Soil respiration rates can occur at moisture tensions below wilting point at which plant production is inhibited (Wildung *et al.*, 1975). Increased soil temperatures in semi-arid regions presently under grassland are likely to result in gradual declines in soil carbon pools after an initial stimulation of plant production caused by enhanced nutrient mineralization (Schimel *et al.*, 1990). In boreal and tundra systems there is conflicting evidence of elevated temperature responses by micro-organisms which could result in large releases of CO_2 and CH_4 in response to increased length of the growing season and the amplitude of seasonal changes. Changes in the active soil depth, melt-water runoff, waterlogging and evapotranspiration will determine the

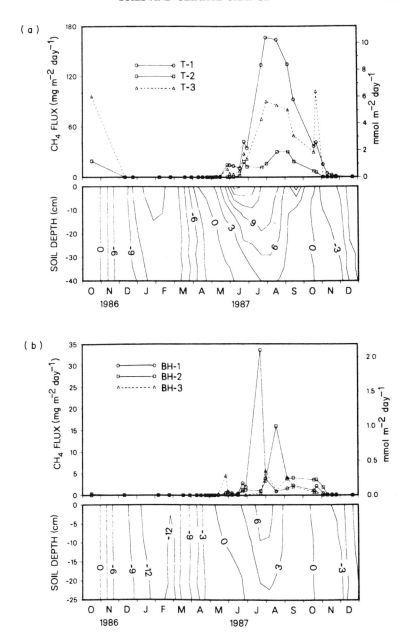

Fig. 5. Net methane flux to the atmosphere and soil temperature isotherms (3°C contour intervals) for *Eriophorus* tussocks (A), and inter-tussock areas (B) in Alaskan tundra meadow. (Source: Whalen and Reeburgh, 1988.)

magnitude of carbon sources in the region and the relative fluxes of CO_2 and CH_4.

VI. EFFECTS OF CLIMATE ON LITTER
DECOMPOSITION

The decomposition rate of litter deposited on the soil surface is dominated by the effects on temperature and moisture conditions on microbial processes. The microclimates on the soil surface, imparted by snow or vegetation cover, and physicochemical conditions within the soil can modify the simple expression of climate on decomposition rates.

Materials which contain carbon and nutrients in forms which are optimum for microbial activity (high-quality resources) will decompose rapidly under favourable conditions but if the availability of carbon and nutrients is markedly suboptimum (low resource quality) micro-orgamisms may not be able to express the potential level of physiological activity set by favourable climatic conditions. These conditions of low resource quality in litter are particularly defined by high concentrations of lignin and phenolics, and low concentrations of nitrogen (Swift et al., 1979). In herbaceous materials with low lignin and high nitrogen concentrations, such as green legume materials, polyphenol concentrations can determine rates of microbial activity (Palm and Sanchez, 1991). Other studies (Fox et al., 1991) using similar green manures have not confirmed this relationship.

For litters such as grass, herbs and crop residues with lignin concentrations up to about 10–15%, decomposition rates can be predicted from nitrogen concentrations or C/N ratios (Hunt, 1977; Singh and Gupta, 1977; Juma and McGill, 1986; Seastedt, 1988; Taylor et al., 1989). At higher lignin concentrations, the initial lignin or the lignin/N ratio often appears to be a major factor determining litter decomposition by micro-organisms (Aber and Melillo, 1980; Berg and Staff, 1981; Melillo et al., 1982; Berg and Agren,1984; Harmon et al., 1989). There have also been a number of exceptions to this pattern for a wide range of litter types and systems (Schlesinger and Hasey, 1981; Anderson et al., 1983; Moore, 1984; Taylor et al., 1989) suggesting that the generality of lignin as the primary decomposition-rate determinant for leaf litter may also be overstated. It may be concluded that while initial resource quality of litters provides a general basis for predicting decomposition rates in a particular environment there are many exception which point to an incomplete understanding of decomposition-rate determinants, particularly in extreme environments.

A. Isolating Climate Effects on Decomposition from Other Factors

The clearest expression of climate controls on litter decomposition will be manifested under conditions where resource quality and microclimate variables are experimentally controlled. Figure 6 shows the results of such an experiment in which ^{14}C-labelled ryegrass enclosed in tubes stoppered with fine mesh was incubated in similar soils in the UK and Nigeria. The shape of the decomposition curve was identical in both cases but the time course for similar mass losses was four times faster in Nigeria (MAT 26·1°C) than in the UK (MAT 8·9°C). Further analysis of data from a large number of studies using similarly labelled materials in a wide range of temperate and tropical environments showed that climate explained most of the variation in decomposition rates (Jenkinson et al., 1991). Notable exceptions to this relationship were two allophanic soils in which decomposition was slower than expected (see below).

A similar approach, using a standard litter, has been adopted to investigate climatic controls over litter decomposition (Berg et al., 1984a; Meentemeyer and Berg, 1986; Dyer et al., 1990). Standard Scots pine needles (0·41% N, 25·7% lignin) in litter bags were placed annually and sampled over 3 years so that mass losses between 10 and 45% could be obtained. These constraints on the data collection were defined because initial mass losses are dominated by leaching of soluble materials, which is predominantly a physical process and not closely related to temperature, and mass loss rates above 45% were increasingly dominated by the decomposition-rate retarding effects of lignin and microbial secondary products (Berg and Agren, 1984). Daily means of temperature and moisture in the soil surface layer (0–4 cm) were derived from climate variables using the SOIL model (Jansson and Berg, 1985). In a study extending over a NW European transect (Fig. 7a) forest sites were selected which were mainly Scots pine stands where ground vegetation was sparse or absent. MATs ranged from about 1 to 9°C and mean annual precipitation from < 500 mm to > 1000 mm (Fig. 7b). The two response functions showing highest coefficients of determination for mean mass loss over the whole transect were the derived, mean soil temperature, accounting for 81% of variation, and mean annual, actual evapotranspiration (AET) accounting for 55% of variation (Fig. 8). Regressions for the whole data set showed a poor fit for percentage mass-loss rates per day against soil temperature, soil moisture and other simulated response functions, indicating considerable local variation in microclimate and soil biological properties which confounded the broad trends. These local effects were clearly illustrated in an earlier experiment using the same methods in 14 Scots pine, four lodgepole pine, eight Norway spruce and two birch sites in Sweden (Berg et al., 1984b). The derived soil variables explained 95–99% of mass losses

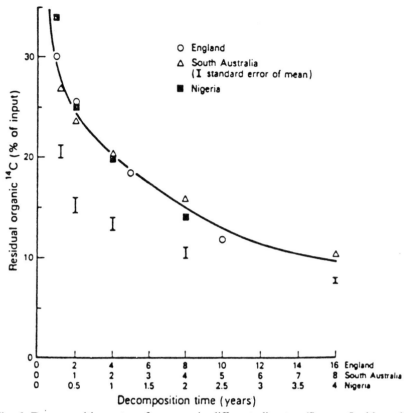

Fig. 6. Decomposition rates of ryegrass in different climates. (Source: Ladd *et al.*, 1985.)

within one Scots pine forest but only 25% of variance for all sites. The fit of the model was improved to explain 70% of variation if the Norway spruce sites were excluded; an effect attributed to stand microclimates on these spruce sites.

B. Climate Resource Quality Interactions in Litter Decomposition

Not surprisingly the interactions between litter resource quality and climate complicate these regional patterns. Dyer *et al.* (1990) used an extensive data set of 92 litter bag studies, mainly in boreal and warm temperate forests, to investigate relationships between decomposition rates, litter resource quality (lignin and nitrogen) and climate variables. AET explained 65% of the variation in first-year mass losses (10 and 90%) from a wide range of plant

Fig. 7a. Site locations and characteristrics in a transect used to investigate the decomposition rates of standard *Pinus sylvestris* needles in relation to climatic factors. (Source: Berg *et al.*, 1991.)

litters from trees, shrubs and some herbaceous species. But the combination of AET with nitrogen and/or lignin concentrations did not substantially change this relationship. When the data for warm sites were analysed separately the variation explained by AET ($R^2 = 0.57$) was improved by the inclusion of lignin concentration ($R^2 = 0.78$) but AET, lignin and N explained little of the differences in mass losses from litter in the cool sites. This result is partially a consequence of Dyer *et al.* (1990) restricting their analysis to first-year mass losses of litters. In taiga and tundra soils it can take several seasons for the resource quality differences between plant litters to be manifested in soils where the temperature rises above 0°C for only 4–5 months per year. Some of the unexplained variation in decomposition rates may also be attributable to leaching of soluble organics which can account for up to 20% of first-year mass losses from litter. Leaching is primarily a physical process and hence is not a function of climate controls on biological processes. Under temperate conditions this leaching phase may be a transient phase of litter decomposition and may exert little control on the value of the decomposition constant (Harmon *et al.*, 1990).

Fig. 7b.

Plot No.	Plot of standard transect	Site name	Latitude/ longitude	Altitude (m)	Annual precipitation (mm)	Annual mean temp. (°C)	Tree species	Understorey type	Soil type
1	+	Kevo	69°45'N; 27°01'E	90	420	2·4	Scots pine	Lichens	Cambic podzol
2	+	Harads	66°08'N; 20°53'E	58	650	1·3	Scots pine	Lichens	Cambic podzol
3:1	+	Manjärv	65°47'N; 20°37'E	135	700	1·0	Scots pine	Lichens	Ferric podzol
3:2		Manjärv	65°47'N; 20°37'E	135	700	1·0	Scots pine	Dwarf shrub	Ferric podzol
3:3		Manjärv	65°47'N; 20°37'E	135	700	1·0	Scots pine	Low herbs	Ferric podzol
4:23		Norrliden	64°21'N; 19°46'E	260	595	1·3	Scots pine	Dwarf shrub	Ferric podzol
6:51	+	Jädraås	60°49'N; 16°30'E	185	720	3·8	Scots pine	Dwarf shrub	Ferric podzol
7	+	Brattforsheden	59°38'N; 14°58'E	178	850	5·2	Scots pine	Dwarf shrub	Ferric podzol
8	+	Nennesmo	58°16'N; 13°35'E	155	930	6·2	Scots pine	Dwarf shrub	Ferric-Org.podzol
9	+	Målilla	57°25'N; 15°40'E	105	670	6·2	Scots pine	Dwarf shrub	Ferric podzol
10:1		Mästocka	56°36'N; 13°15'E	135	1070	6·8	Scots pine	Dwarf shrub	Ferric-Org.podzol
10:2		Mästocka	56°36'N; 13°15'E	135	1070	6·8	Norway spruce	None	Ferric-Org.podzol
11		Böda	57°15'N; 17°01'E	10	650	7·5	Western red cedar	None	Histosol
12	+	Vomb	55°39'N; 13°39'E	46	770	7·0	Scots pine	Grass	Humic Cambisol
13	+	Ehrhorn	53°00'N; 09°57'E	81	730	8·0	Scots pine	Dwarf shrub	Ferric-Org.podzol
14	+	Ede	52°02'N; 05°42'E	45	765	9·3	Scots pine	Grass	Humic Cambisol

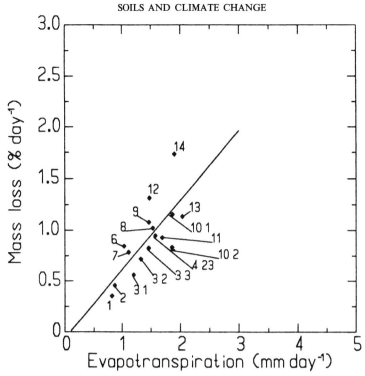

Fig. 8. Relationship between mass loss from standard *Pinus sylvestris* needles and evapotranspiration in a European transect from northern Sweden to Belgium. Numbers refer to the site locations shown in Fig. 7a. (Source: Berg *et al.*, 1991.)

It might be concluded that under cool climatic conditions in high latitudes the decomposition rates of plant materials with a wide spectrum of resource quality characteristics are convergent because of temperature limitations on microbial activity over much of the year. Flanagan and Van Cleve (1983), however, showed that the decomposition rate-constant (k) for F-layer material in black spruce, white spruce, aspen and birch stands was an inverse function of percentage of lignin and a positive function of percentage of N. This is a consequence of feedbacks between litter quality, decomposition rates and nutrient availability to trees which are generally recognized (Adams *et al.*, 1970; Zimka and Stachurski, 1976; Miller, 1981; Vitousek, 1982). Sites with low nutrient availability select for species with low nutrient demands, these produce litter of low quality which decomposes slowly further reducing nutrient availability. In the black spruce sites of interior Alaska on water-logged or northerly slopes this loop is amplified by the development of moss cover which insulates the forest floor, further slowing nutrient returns to the trees (Fig. 9). On the well-drained warm sites more productive species such as birch and aspen produce a high-quality litter which decomposes rapidly,

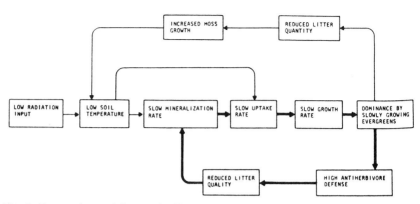

Fig. 9. Interactions of factors leading to low forest productivity in cold, nutrient deficient, boreal forests. (Source: Chapin, 1986.)

maintaining rapid turnover and high nitrogen availability on the forest floor (Flanagan and Van Cleve, 1983). As will be considered below, there is experimental evidence that increasing the heat sum of the cold sites could change the direction of the feedback loops and promote the development of more productive systems with smaller soil carbon pools.

VII. SOIL ORGANIC MATTER DYNAMICS AND CLIMATE CHANGE

There are two key aspects relating to the impact of climate change on soils. The first is the stability of organic carbon in soil and the rate of mineralization in response to changes in plant and soil conditions determining the size of the pool. The second is the effects of changes in the active soil depth which will initiate processes of SOM stabilization which are presently impeded in soils which are frozen for more than half of the year.

A. The Response of SOM Pools to Disturbance

Under undisturbed natural vegetation the organic matter content of soil, determined by climate and vegetation, is relatively constant and carbon inputs as dead organic matter are equal to outputs as CO_2. If future climate change scenarios are realized, with accompanying shifts in vegetation cover and litter inputs, conditions affecting the residence time of organic matter in soils will change to some extent. An insight into the stability of soil organic matter pools to changes in environmental conditions is provided by changes in land use from forests to arable agriculture or grassland and the effects of forest management.

Losses of SOM when forests are converted to agricultural systems are extensively documented (Schlesinger, 1977, 1984; Buringh, 1984; Allen, 1985). When soils are first cultivated organic matter in the topsoil (0–30 cm) declines rapidly and then levels off to about 20–60% of the carbon content under natural vegetation. The time course of carbon losses depends upon the climate and soil type. The conversion of tropical forests results in carbon losses from 7 to 54% of the initial pool over 3 years (Harcombe, 1977). Some tropical soils developed on volcanic material (Andepts) have high organic matter contents but show negligible carbon losses after disturbance (Harcombe, 1977; Werner, 1984). Under productive agriculture with high organic matter inputs total carbon contents may not decline so rapidly but the use of carbon isotope dating showed that after the conversion of forested land to sugar cane plantations half of the original soil carbon remained after 50 years of continuous cultivation.

In temperate regions, the clear felling of forests results in a rapid decline in forest floor mass by 20–50% over 10 years which slowly recovers during successional development of forest cover (Covington, 1981; Federer, 1984; Pastor and Post, 1986). Comparisons of cultivated with native or uncultivated land in temperate regions have shown that the concentration of organic-C may decrease by 13–60% in surface horizons in less than 20 years depending upon initial levels, soil type and management practice (Buringh, 1984; Juma and McGill, 1986). Rates of organic matter losses in chernozems are slow compared with other soils under grassland. Cultivation of steppe results in the mineralization of 30 g C m^{-2} year^{-1} out of 5000 g C m^{-2} in the top 10 cm of soil. Steady state is reached in about 200 years with losses of about 36% of the initial reserve (Titlyanova, 1987).

Radiocarbon dating has shown that organic matter in the soils of northern coniferous forests is contemporary compared with much older carbon distributed to much greater depth in most temperate and tropical soil profiles. In 16 forest profiles developed under Scots pine in northern Sweden (65–67° N) the mean age of carbon in the B-horizon, extending to a depth of less than 1 m, was generally less than 800 years (Table 4). In contrast, the mean age in temperate and tropical soils can increase on average by 46 years cm^{-1} down the profile to a depth of 1–2 m where organic carbon may have a mean age of 10 000 years (Sharpenseel and Schiffman, 1977).

These studies show that most temperate and tropical soils contain organic matter which is turning over quite rapidly and is maintained in dynamic equilibrium with litter inputs, while about half of the carbon is very stable and effectively inactive on an ecological time scale (Jenkinson and Rayner, 1977). In contrast, in podsols and relatively unweathered soils (e.g. Inceptisols) found at high latitudes, much of the organic carbon is contemporary in

Table 4
Organic matter content (ignition loss) and radiocarbon age of podsol profiles developed under *Pinus sylvestris* in northern Sweden (65°N). Mean annual air temperature -2 to $0°C$, mean annual precipitation 450–550 mm year^{-1} (Tamm and Holmen, 1967)

	Site no.					
Measurement	610	972	991	992	1404	1405
Horizon ignition loss (%)						
A_0	81·9	67·7	79·8	79·8	67·9	69·7
$A_{1,2}$	1·6	1·9	1·2	2·8	1·3	2·4
B	4·2	6·2	2·1	4·6	2·9	3·3
Horizon organic matter content (kg $\times$ 10^3)						
A_0	50	45	55	55	35	45
$A_{1,2}$	14	14	13	21	17	15
B	130	220	80	200	140	150
Age of B horizon in years	500	590	750	630	730	480
(SD)	(65)	(75)	(60)	(65)	(80)	(90)

age. The dynamics of these pools and their susceptibility to changing environmental conditions is therefore central to the impact of climate change on soils.

B. Soil Organic Matter Formation and Stabilization

A number of similar schemes have been used to define the functional constituents of soil organic matter (Jenkinson and Rayner, 1977; Paul and van Veen, 1979; Parton *et al.*, 1987). Parton *et al.* (1987) divide organic matter in soils into "active", "slow" and "passive" fractions which turn over in years, decades and centuries respectively (Fig. 10). Inputs of plant material are divided into the low molecular weight ("metabolic") compounds which are rapidly metabolized by micro-organisms, and the "structural" compounds, such as lignin and cellulose, which decompose at a slower rate. The "active" fraction consists of microbial biomass and metabolic products. The "slow" and "passive" fractions are more or less resistant to microbial attack through processes of physical and chemical stabilization which are largely determined by soil properties.

1. Decomposition and Humification

The resource quality controls on the initial phase of decomposition were considered earlier. Litter decomposition rates follow the typical time-course

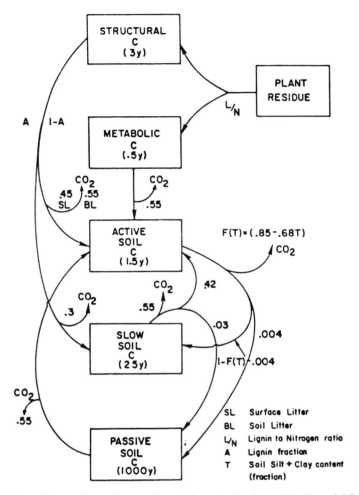

Fig. 10. Flow diagram for carbon pools and fluxes in the CENTURY model for the Great Plains grasslands. (Source: Parton *et al.*, 1987.)

shown in Fig. 10 with the slope of the curve mainly determined by resource quality and physical factors. The initially rapid phase of mass loss reflects losses of materials such as simple sugars, amino acids and other labile constituents. The rate then declines as an increasing proportion of the residual mass comprises not only the more decomposition-resistant plant cell wall compounds. In undisturbed organic forest soils, increasing depth in the forest floor profile approximates to cohorts of ageing litter material. Microbial respiration rates of different forest floor materials from the Alaskan sites considered earlier are shown Table 5 and reflect the general decline in the

Table 5
Mean laboratory respiratory rates (litres O_2 g^{-1} h^{-1} ±SD) for forest floor materials collected in spring from sites in interior Alaska. Material was air dried, moistened to a standard water content (250%) and incubated at 15°C (Flanagan and Van Cleve, 1983)

Vegetation type (Humus)	Layer of forest floor		
	01 (Litter)	021 (Fermentation)	022
Aspen	123·1 ± 53·6	150·9 ± 11·3	68·7 ± 7·8
Birch	159·3 ± 31·4	107·7 ± 0·7	73·6 ± 1·9
White spruce	94·2 ± 9·6	86·4 ± 22·5	59·3 ± 9·5
Black spruce	45·8 ± 11·7	23·4 ± 5·0	10·9 ± 2·3

biodegradability of the material with increasing depth and litter quality differences between the sites.

During the process of decomposition microbial products are also synthesized. Cheshire et al. (1974) showed, for example, that during the decomposition of ^{14}C-labelled hemicellulose, the decrease in xylose and arabinose content and the accumulation of mannose is indicative of microbial polysaccharide production and stabilization. As decomposition progresses, a wide range of new compounds are formed through complex polymerization and substitution reactions; alkyl-C and carboxyl-C groups increase and carbohydrate-C and aromatic-C decrease (Zech et al., 1985). Bracewell et al. (1976) suggest that the degree of carbohydrate depolymerization is related to climate and that relatively unsubstituted aliphatic chains are characteristic of cold, wet soils where decomposition is slow in contrast to highly substituted chains in soils where biological activity is high. Little of the aromatic ring constituents of lignin are utilized as a carbon source by micro-organisms but low molecular weight phenolic degradation products released from the litter undergo oxidative repolymerization to form recalcitrant humic polymers (Stevenson, 1982; Stott et al., 1983). The biochemical composition of microbial secondary products alone can not account for the long residence times of some organic-C fractions and, indeed, are rather similar in biochemical composition across different soil types (Stevenson, 1982). It is, therefore, the stabilization of these compounds by physical and chemical processes which determines their age.

2. Organic Matter Stabilization

Many workers in temperate and tropical regions have shown that clays have an important role in organic matter stabilization (e.g. Jones, 1973; Theng, 1979; Paul, 1984) and soil carbon content generally increases as a positive

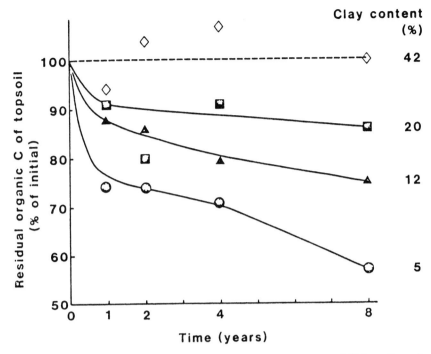

Fig. 11. Losses of organic-C from calcareous soils showing the stabilizing effect of clay content. (Source: Oades, 1988.)

function of clay content (Fig. 11). Oades (1988) points out that a number of variables, including texture, base states, moisture regime and plant production, are covariates with clay content and it is difficult to identify causal factors in different soils. In the case of the data in Fig. 11, however, the soil samples received no plant litter inputs for 8 years after the addition of labelled material and were all located in similar Mediterranean climates in south Australia.

Physical incorporation of organic matter within the clay matrix results in the protection of organic polymers and bacterial cell wall constituents from enzymic attack (McGill and Paul, 1976; Tiessen and Stewart, 1983; Ladd *et al.*, 1985). In field incubation experiments with ^{14}C-labelled legume material carried out by Ladd *et al.* (1985), differences in carbon content between the soils after 8 years, shown in Fig. 11, were due to stabilization of microbial products rather than to differences in litter decomposition rates. Sørensen (1981) also showed that the rate of carbon stabilization was greatest during the rapid initial growth phase of micro-organisms after ^{14}C-labelled cellulose was added to soils of variable clay content.

Examination of thin sections of soils by electron microscope has shown that the stability of carbohydrates is conferred by a coating of clay platelets (Foster, 1985). Theng *et al.* (1986) also provided direct evidence of clay stabilization from ^{13}C-NMR spectra of physical protection in two New Zealand soils where the intercalation of carbon in a smectite clay resulted in a much older radiocarbon age (6800 BP) than would commonly be found in topsoil. Aluminium silicate (allophane) is particularly effective in stabilizing organic matter and accounts for the high organic matter content of Andepts derived from volcanic materials (Tate and Theng, 1980). Montmorillonitic clays also give more protection than illites and kaolinites. Proteins present in the interlamellar spaces of montmorillonite may be almost completely resistant to microbial attack (Jenkinson, 1988) and the incorporation of litter into a montmorillonite clay by earthworms has also been shown in laboratory experiments to stabilize organic matter (Shaw and Pawluk, 1986). In this last example the inhibition of decomposition by occlusion of macroscopic organic matter within the clay as well as physical stabilization processes may have been involved but the mechanisms were not determined.

Cation bridging of colloidal organic particles, including bacteria, to clay microaggregates (< 250 μm) appears to be the principle mechanism of stabilization in clays (Theng, 1979; Oades, 1988). Aggregates above this size range are structurally maintained by the transient binding by roots and fungal hyphae (Tisdall and Oades, 1982) but within the microaggregates the binding mechanisms involve cation bridges between clays and negatively charged humified materials and microbial cell wall debris. This electrostatic binding is very stable and the aggregates below 20 μm are unaffected by tillage and may resist dispersion even by ultrasonics. The major cations involved are Ca^{2+} and Mg^{2+} in neutral and alkaline soils, and Al^{3-} and Fe^{3+} in strongly weathered acid and ferallitic soils. High concentrations of these polyvalent cations not only inhibit the swelling of clays but also organic colloids are more condensed when their functional groups are saturated (Oades, 1988). The concentration of calcium is of particular significance. Low concentrations increase the solubility of organic matter and mineralization of organic carbon in soil; hence liming soil has a long-term effect of stabilizing SOM (Jenkinson, 1988). The complexing of soluble organics by calcium is a particularly significant factor contributing to the depth of SOM distribution in grassland soils (Schoenau and Bettany, 1987). Fulvic acids, for example, are among the most recent of soil carbon fractions but if they move down the profile and are complexed with calcium they appear among the oldest fractions (Goh *et al.*, 1976). Chemical fractionation of carbon in two Canadian soils (Table 6) reveals very different ages of constituents originally formed under grassland and forest (Campbell *et al.*, 1967). Radiocarbon ages of humin, humic acid and fulvic acid fractions in the chernozem were 3–10 times older than in the podsol and mobile humic acids had a mean

Table 6
Carbon dates (mean residence times [MRT] ±SD) of fractions extracted from chernozem and grey podsolic soils in Canada (Campbell et al., 1967)

	Chernozem		Grey-podsolic	
Fraction	% total C	MRT (years)	% total C	MRT (years)
Humin I	58	1135 ± 50	56	335 ± 50
("Calcium" humates*	21	1410 ± 95	(none)	
Humic acids ("mobile")	26	785 ± 50	25	85 ± 45
Fulvic acids	18	555 ± 45	17	50
Unfractionated soil	100	870 ± 50	100	250 ± 60

*Extracted from the Humin I fraction.

radiocarbon age of 785 years compared with 1410 years for "calcium humates".

In conclusion, warm soils with moderate base status tend to be associated with high-quality litters which have rapid initial decomposition rates. Intimate contact with the soil minerals, however, retards the later stages of decomposition. The accumulation and residence time of SOM increases as a function of the content and species of clay, and concentrations of polyvalent cations. The input of organic matter below ground as roots and root exudates may therefore be more important for SOM dynamics than litterfall. As the soil environment becomes colder, organic matter turnover becomes progressively more superficial in the profile and divorced from the underlying mineral horizons (Sims and Nielsen, 1986). Soil organic matter then accumulates as an inverse function of litter resource quality (Agren and Bosatta, 1987) in active and slow pools; there is little stabilization of SOM forming passive fractions and hence the carbon is of contemporary age.

VIII. CONSEQUENCES OF CLIMATE CHANGE FOR PLANT–SOIL INTERACTIONS

It is currently predicted that changes in climate resulting from increased atmospheric concentrations of CO_2, CH_4 and other trace gases will have the greatest impact in the north temperate region and particularly on ecosystems in the boreal and tundra zones. Mobilization of carbon as CO_2 and CH_4 from soils in response to climate change could feed back on climate forcing, modified by the degree to which changes in plants and soils act as sinks for carbon under the combined effects of climate warming, carbon enrichment and changes in nutrient availability. The relationships between climate, soil temperature and moisture, vegetation cover, plant production and demography, fire regime and soil processes are extremely complex and there have been two approaches to understanding the interactions between these variables. On one hand simulation models provide a means of testing the sensitivity of these systems to changes and predictions can be evaluated against patterns of ecosystem structure and functioning within present climate gradients and soil chronosequences. On the other hand, manipulative experiments on whole systems, or compartments, provide a means of identifying response functions which can be used to parameterize the model or to test particular predictions.

Pastor and Post (1988) used a forest growth model (LINKAGES) to run simulations for several sites including mixed spruce/broad-leaf forests within the boreal forest zone. Forest growth was simulated from plots with no vegetation or forest floor present but with the complement of tree species present as seeds and similar nitrogen capital to the sites. Two contrasting soils, sand and silty-clay loam, were selected to represent the range of soil water availability at each site. The basic structure of the model (Fig. 12) is a set of three subroutines (*Tempe, Moist, Decomp*) which determine site conditions (degree days, available soil water and available nitrogen) and a set of three demographic subroutines (*Birth, Grow and Kill*) which calculate the annual establishment, growth and death of individual trees and cohorts of litter production. The two sets of subroutines are linked by *Gmult* which calculates degree-day, water and nitrogen multipliers and the output of litter to the soil *Decomp* subroutine. Further details on the parameterization of the model in Pastor and Post (1986) show that decomposition was related to lignin : nitrogen ratios and AET (as discussed earlier) though for the boreal forest site where AET was low it was assumed that this function would not modify temperature effects. Nitrogen mineralization was expressed as N immobilization by lignin during the initial phase of decomposition and the N concentration at which N is mobilized. No effects of increased CO_2 on tree water relations, carbon fixation or litter quality, or changes in fire regime, were assumed.

The model was run for the equivalent of 200 years under current climate conditions. The climate conditions were then altered linearly to reach the "2 × CO_2" climate over the next 100 years and then run under these conditions for a further 200 years. The results showed that for the northern-most sites (55–59°N) the warming was insufficient to cause the simulated boreal forest to be replaced by hardwood stands (Solomon, 1986). With higher temperatures, productivity and biomass of spruce increased and there were only small differences between soil types. Organic matter in the forest floor increased as a consequence of higher litter production with lower N (and implicitly higher lignin) content.

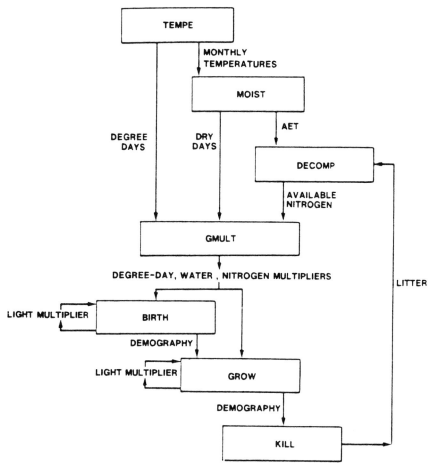

Fig. 12. Flow chart for the LINKAGES forest growth model used to simulate the effects of climate change on North American forests. (Source: Pastor and Post, 1986.)

In the southern boreal and northern hardwood sites (47–54°N) major shifts in the tree species composition occurred. On the silty-clay loams with no decrease in soil water availability the current mixed spruce/fir/hardwood forest was replaced with a hardwood forest (Fig. 13). This forest was more productive for two reasons. Firstly, in the model the hardwoods have faster intrinsic growth rate and can attain higher biomass than either spruce or fir. Secondly, the warmer climate, as well as higher nitrogen and lower lignin concentrations in the hardwood litter, increased soil nitrogen availability and this amplified the effect of warming on productivity. On the sandy soils,

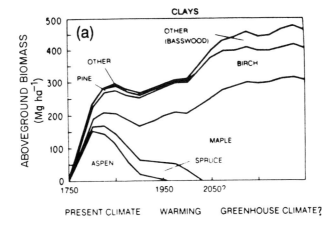

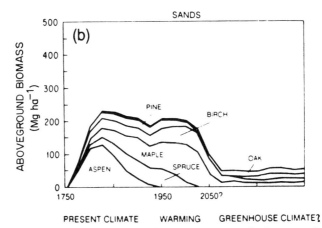

Fig. 13. Predictions of species composition and biomass of Minnesota forests using the LINKAGES forest growth model (Fig. 11) to simulate the effects of climate change produced by doubling atmospheric CO_2. Runs using the same climate parameters were carried out for a clay-loam soil with high water-holding capacity (a) and (b) for a sandy soil with low water-holding capacity. (Source: Pastor and Post, 1986.)

moisture stress resulted in a decrease in forest biomass to 25% of the initial value and soil organic matter also decreased in most cases.

Feedback effects on decomposition and N-mineralization caused by the insulative effects of moss cover and litter accumulation were not included in LINKAGES and, as considered earlier, this can be a critical factor affecting biological processes in soils. Bonan *et al.* (1990) used a similar population based model JABOWA to simulate the effects of the 1, 3 and 5°C warming on the boreal forest complex in interior Alaska (Van Cleve et al., 1983) considered earlier. This model does not explicity simulate nutrient availability but this is implicit in the moss-organic matter accumulation which is correlated with forest temperatures and nutrient availability. Simulated and observed values for biomass, forest floor mass and site conditions (including permafrost depth) showed good agreement for the north-slope black spruce forest. Fire intensity was a function of fuel buildup and fire severity (the depth of forest floor burnt) as a linear function of forest floor moisture content. Runs simulated forest succession for 250 years after a fire and 500 years with recurrent fires. Details of the output will not be considered here. The particular features relevant to this discussion are that on the poorly drained north slopes, dominated by black spruce, a drier climate in the absence of fire resulted in a decrease in the active soil depth. This was a consequence of a drier cover of moss and forest floor material. This reduced the thermal conductivity of these surface layers and thereby the depth of seasonal thawing. With increased precipitation to offset evapotranspiration and 1°C warming the seasonal thaw increased substantially. With recurrent forest fires and 5°C warming the active depth of soil increased (with implicit increases in C and N mineralization), biomasses of birch and aspen increased by 6–10 times and the biomass of black spruce declined. On the south-facing slopes none of the tree species currently found in the uplands of interior Alaska was able to grow under the 3 and 5°C warming and these systems were converted to steppe-like vegetation. If such extensive dieback occurs (cf. Emanuel et al., 1985a,b; Solomon, 1986) carbon will be released from large woody biomass pools as well as soils. In the dry above-ground environment decomposition of standing dead wood can be very slow with a half-life in the order of a century (Harmon et al., 1986).

Various estimates have been made of carbon mobilization from boreal and tundra soils through climate warming. Linkins et al. (1984) calculated that 60 Gt, about 15% of the carbon stores in this region, could be mobilized over 50 years assuming a Q_{10} of 2·5 for microbial responses to predicted warming. Similarly, the Rothamsted model (Jenkinson et al., 1991) has been used to estimate the release of 61 Gt C from the soils of the high northern latitudes over the next 60 years assuming a mean annual temperature increase of 0·06°C in this region and no temperature changes in warm-temperate and tropical zones. Lashof (1989) concluded that the carbon flux from the boreal zone associated with a 4°C global temperature change is probably less than 6 Gt C year^{-1}, with a likely range of 0·5–2 Gt C year^{-1}, and about 1 Gt C year^{-1} from tundra due to changes in temperature and depth of the water table. Woodwell (1986, in Lashof, 1989) estimated an even larger mobilization of 3–10 Pg C per annum (totalling 10–50% of the carbon capital) assuming a 10–15°C temperature rise and a microbial Q_{10} of 1·5–3·0. This estimate may be a substantial overestimate because labile and more

Table 7

Total CO_2 incorporation for intact, vegetated cores from two Alaskan tundra meadows (IBP-2 and Central Marsh) incubated in phytotrons at 4°C and 8°C with water-table levels maintained at the surface or at a depth of -5 cm. The decrease in CO_2 incorporation (net CO_2 production) at 8°C is expressed as a percentage of CO_2 incorporation at 4°C (Billings *et al.*, 1982)

Site	Temperature (°C)	Water-table depth (cm)	Net CO_2 incorporation (g CO_2 m^{-2} season^{-1})	% decrease CO_2 incorporation
IBP-2	4	0	320·8	
	8	0	274·3	15
	4	−5	286·8	
	8	−5	77·8	73
Central Marsh	4	0	214·4	
	8	0	183.8	15
	4	−5	340·4	
	8	−5	58·7	83

recalcitrant carbon pools are not distinguished and the temperature regime used by Woodwell (1986) was weighted by high summer temperatures (Lashof, 1989).

It is impossible to test these predictions but some assessment of the responses of systems to future climates can be derived from manipulative experiments. Billings *et al.* (1982, 1983) used intact cores from two tundra meadows to investigate the effects of elevated temperatures, lowering of the water table and CO_2 enrichment on the capacity of these soils to act as carbon sinks. The cores were removed from the field in a frozen state during winter and maintained for 56 days in an environment chamber. The site-related light and temperature regime was established with July air temperature maxima of 4 or 8°C, and the water table maintained near the surface or at a depth of 5 cm (Billings *et al.*, 1982). The results summarized in Table 7 show that maintaining the water table near the soil surface resulted in similar net CO_2 fluxes at 4 and 8°C with only 15% less storage at the higher temperature. The lowered water table resulted in little change in CO_2 storage at 4°C but at 8°C there was a 73% decrease in CO_2 incorporation into plants and soils and the cores became net sources of CO_2. Closely similar experiments involved incubating cores from one site, Point Barrow, for 88 days with an atmosphere containing 400 or 800 μl l^{-1} CO_2 (Billings *et al.*, 1983). The effects of doubling the atmospheric CO_2 were transitory and less significant than the temperature/water-level treatments.

Analogous field experiments were carried out in Alaska where greenhouses were erected over undisturbed tussock tundra to maintain conditions with current, intermediate or double CO_2 concentrations and temperatures 4°C above ambient (Tissue and Oechel, 1987). Within 3 weeks plants maintained at elevated CO_2 concentrations underwent physiological adjustment and showed no differences in photosynthetic rate or changes in carbohydrate contents. Tiller production was increased under elevated CO_2 and with higher temperatures net carbon storage increased for the 3 years of observations (Oechel and Riechers, 1987, reported by Melillo et al., 1990).

Heal et al. (1981) report that artificially heated tundra meadow soils at Point Barrow evolved 46–65% of the gaseous carbon as methane; though details of the experiments are not given. Methane production was not measured by Billings et al. (1982) and hence the net release of carbon as "greenhouse" gases may have been larger than the measured flux. However, if the soil is aerobic at the soil surface CH_4 is absorbed and oxidized (Whalen and Reeburgh, 1990). An overall balance could therefore be maintained between CH_4 sinks and sources within the mosaic of hummocks and hollows in the tundra meadow.

In the boreal forest, about 400 m² of black spruce forest floor was heated to approximately 9°C above ambient temperature for three summers (Van Cleve et al., 1983). This resulted in a 20% reduction in forest floor mass (8360 to 6488 g m⁻¹) and the humus layer also decreased by 742 g m⁻². This reduction in the soil organic matter was accompanied by an increase in mineral-N and available-P and spruce foliage showed increased rates of photosynthesis, higher concentrations of N (1·05% heated vs. 0·76 control), P (0·14 heated vs. 0·10% control) and K (0·80% heated vs. 0·56% control) in response to the more favourable nutrient and temperature regime. Measurements of litter decomposition were not made but it is reasonable to assume that a more rapid cycling of nutrients through smaller soil pools and larger, more productive tree biomass would result from the continued warming. It would be scientifically dubious and technically difficult to scale up these results to predict the responses of whole stands to a warmer climate; though the results broadly conform to those predicted by Bonan et al. (1990) considered earlier. However, soil drainage is extensively carried out in forest management and affects both tree production and soil carbon mineralization. Dang and Lieffers (1989) showed from analysis of tree rings that over 13–19 years after drainage of peatlands in Alberta, growth of black spruce increased by 76–766% over undrained areas. Lowering the water table to 30–60 cm below the surface in Finnish peat bogs resulted in CO_2-C fluxes shifting from net sinks of 25 g C m⁻² year⁻¹ to net sources of 250 g C m⁻² year⁻¹ (Sivola, 1986). Although mass balances for carbon in biomass and soil are not available for the same stands in boreal forest, these effects are consistent with those of a warmer climate and with the increased seasonal amplitude of atmospheric CO_2 concentrations recorded in the boreal region (D'Arriego et al., 1987).

Ultimately, increased production depends on the availability of nutrients, particularly nitrogen, required to incorporate fixed carbon in biomass. The growth response of northern coniferous forests to nitrogen fertilization is generally recognized (Tamm, 1979; Ballard, 1984). Similarly, in the tundra microcosm experiments considered earlier, for example, no changes in carbon storage were observed in relation to elevated atmospheric CO_2 concentrations but the addition of ammonium nitrate (Billings et al., 1984) increased plant biomass indicating that CO_2 was not limiting production in these systems. Increased atmospheric deposition of nitrogen from agricultural and industrial sources may result in enhanced forest production and carbon storage (Melillo and Gosz, 1983) but this probably amounts to less than 3% of the carbon sources (Peterson and Melillo, 1985) because of the wide excess of CO_2 mobilized from fossil fuels in relation to increased N deposition from anthropogenic sources. Increased nitrogen mobilization associated with carbon mineralization in organic soils is likely to have more widespread effects than N deposition on plant production in response to warmer climates. Schimel et al. (1990) used the CENTURY model to simulate the effects of climate change on grasslands and showed that increases in temperature-associated changes in precipitation caused increases in decomposition and long-term emission of carbon from soil organic matter. Nutrient release associated with the loss of soil organic matter caused increases in net primary production in the short term indicating nutrient interactions are a major control over vegetation response to climate change.

IX. CONCLUSIONS

The magnitude of carbon sources and sinks in the biosphere is a major area of uncertainty in global carbon budgets and projections of future climate change. Vaghjiani and Ravishankara (1991) calculate that the residence time and flux of methane in the atmosphere have been substantially overestimated, implying the existence of larger methane sinks elsewhere; while Mosier et al. (1991) conclude that nitrogen deposition and fertilizer applications reduce CH_4 uptake by soils. The oligotrophic soils of high latitudes may be important sinks under drier conditions as proposed by Whalen and Reeburgh (1990). The question of net balance between CO_2 sources and sinks is also unresolved at a systems level. Tans et al. (1990) suggest that terrestrial vegetation in the northern hemisphere consitutes a sink for $2 \cdot 0$–$3 \cdot 4$ Gt C year^{-1} and even larger net sinks for CO_2 are proposed by Prentice and Fung (1990). On the other hand, Watson et al. (1991) suggest that spatial variability of plankton in the North Sea consitutes a major source of error in attempting to compute carbon budgets. None the less, net increases in CO_2 and CH_4 are manifest and, whatever the sink strengths, it is evident that compensatory processes are not operating to reduce the likelihood of climate forcing by greenhouse gases. This chapter has therefore focused on the controls over organic matter dynamics in soils under changing conditions of climate and vegetation cover.

The impact of climate change is predicted to be greatest in the boreal and tundra zones where there are large pools of organic matter in soils. There is some evidence of inflated physiological responses by micro-organisms which would result in amplified rates of carbon mobilization as CO_2 and possibly CH_4, in response to warming. However, feedbacks between litter resource quality, soil microclimates and nutrient availability to plants modify the direct expression of climate warming on soil processes. Over the next century these soils could constitute sources of atmospheric CO_2 of comparable magnitude to deforestation in the tropics if future climate scenarios are manifested because of the relatively labile characteristics of the SOM.

Soils development in the frigid regions of Canada and Alaska has been through physical rather than chemical weathering processes and soils are mainly shallow Inceptisols and Histosols. In northern Europe podsols (Spodosols) predominate on sandy soils of Pleistocene Age (Foth and Schaffer, 1980) where low molecular weight humus compounds are active in weathering and podsolization processes (Schoenau and Bettany, 1987; Anderson, 1988). In podsols there is little stabilization of soil carbon (as shown by contemporary radiocarbon dates) even in temperate regions because of the low base-status, sandy soils on which these soils, and associated vegetation types, tend to develop. But as soil temperatures and available moisture increase in cool temperate podsols on sandy soils, plant production and the depth distribution of carbon also increase (Shetron, 1974).

On base-rich soils, low soil temperatures limit the interaction of surface litter decomposition with the underlying mineral soil horizons and there is little physical and chemical stabilization of soil organic matter (Sims and Nielsen, 1986). Radiocarbon dating of soils in the cool temperate region shows that carbon storage in passive soil pools is a long-term process reaching equilibria over centuries following the warming of cryic soils. Increased chemical stabilization could occur quite rapidly as the active depth of the soil increases and soluble organics are complexed by cations (Anderson, 1979; Schoenau and Bettany, 1987) particularly in soils developing on calcium-rich glacial tills. The increased rooting depth as soils warm will also promote physical stabilization of organic matter in more developed soils with higher clay content. Schlesinger (1990) suggests that rates of soil carbon storage from tundra to temperate and tropical average $2·4 \pm 0·7 \, g \, C \, m^{-2}$ year^{-1} chronosequences spanning 3000 to 10 000 years. Hence the formation of stabilized carbon accounts for about $0·4 \, Gt \, C$ year^{-1} or $0·7\%$ of terrestrial net primary production and therefore the increased active depth of soils in high latitudes is unlikely to constitute a carbon sink on an ecological time-scale. On a time-scale of decades to centuries, the initial response of plant–soil interactions to warming may be analogous to carbon dynamics shown for a grassland catena on a chernozem (25% clay) in Saskatchewan (Martel and Paul, 1974). Mean radiocarbon age decreased from 545 years BP in shallow top soils on the summit to modern in saturated zones at the slope foot but carbon contents were similar. This suggested that soil carbon pools at the slope foot were maintained by rapid turnover of a more productive system than on the summit, and independent of carbon in the B-horizon at > 50 cm depth containing old (5000–8000 years BP) buried horizons dating from the last ice age.

In comparison with these comparatively slow climate-driven pedogenic processes, shifts in land-use are already affecting major changes in soil carbon pools over much of the temperate and tropical regions. It may be concluded that up to AD 2020 these directs of agriculture and forestry practices are likely to continue to dominate carbon fluxes between atmosphere and soil. Thereafter, our capacity to predict the effects of climate on land-use and natural systems are limited by the regional sensitivity of GCMs and our understanding of the interaction of climate change with CO_2-enrichment on plant/soil relationships. Research into these fields is currently an international priority.

REFERENCES

Aber, J.D. and Melillo, J.M. (1980). Litter decomposition: measuring relative contributions of organic matter and nitrogen to forest soils. *Can. J. Bot.* **58**, 416–421.

Adams, S.N., Jack, W.H. and Dickson, D.A. (1970). The growth of Sitka spruce on pooly drained sites in northern Ireland. *Forestry* **43**, 125–133.

Agren, G.I. and Bosatta, E. (1987). Theoretical analysis of the long-term dynamics of carbon and nitrogen in soils. *Ecology* **68**, 1181–1189.

Allen, J.C. (1985). Soil responses to forest clearing in the United States and the tropics: geological and biological factors. *Biotropica* **17**, 15–27.

Anderson, D.W. (1979). Processes of humus formation and transformation in soils of the Canadian Great Plains. *J. Soil Sci.* **30**, 77–84.

Anderson, D.W. (1988). The effect of parent material and soil development on nutrient cycling in temperate ecosystems. *Biogeochemistry* **5**, 71–97.

Anderson, J.M. (1973). Carbon dioxide evolution from two deciduous woodland soils. *J. appl. Ecol.* **10**, 361–378.

Anderson, J.M. and Swift, M.J. (1983). Decomposition in tropical forests. In: *Tropical Rainforests: Ecology and Management* (Ed. by S.L. Sutton, T.C. Whitmore and A.C. Chadwick), pp. 287–309. Blackwell Scientific, Oxford.

Anderson, J.M., Proctor, J. and Vallack, H.W. (1983). Ecological studies in four contrasting lowland rain forests in Gunung Mulu National Park, Sarawak. III. Decomposition processes and nutrient losses from leaf litter. *J. Ecol.* **7**, 503–527.

Armentano, T.V. and Menges, E.S. (1986). Patterns of change in the carbon balance of organic-soil wetlands of the temperate zone. *J. Ecol.* **74**, 755–774.

Armentano, T.V. and Ralston, C.W. (1980). The role of temperate forests in the global carbon cycle. *Can. J. For. Res.* **10**, 53–60.

Ballard, R. (1984). Fertilization of plantations. In: *Nutrition of Plantation Forests* (Ed. by G.D. Bowen and E.K.S. Nambiar), pp.327–360. Academic Press, London.

Barry, R.G. (1984). Possible CO_2-induced warming effects on the cryosphere. In: *Climatic Changes on a Yearly to Millennial Basis* (Ed. by N.-A. Mörner and W. Karlén), pp.571–604. Reidel, Dordrecht.

Berg, B. and Agren, G. (1984). Decomposition of needle litter and its organic-chemical components—theory and field experiments. Long-term decomposition in a Scots pine forest III. *Can. J. Bot.* **62**, 2880–2888.

Berg, B. and Staff, H. (1981). Decomposition rate and chemical changes of Scots pine needle litter. II Influence of chemical composition. In: *Structure and Function of Northern Coniferous Forests* (Ed. by T. Persson). *Ecol. Bull.* (Stockholm) **32**, 373–390.

Berg, B., Ekbohm, G. and McClaugherty, C. (1984a). Lignin and holocellulose relations during long-term decomposition of some forest litters. Long-term decomposition in a Scots pine forest. IV. *Can. J. Bot.* **62**, 2540–2550.

Berg, B., Jansson, P.-E. and Meentemeyer, V. (1984b). Litter decomposition and climate—regional and local models. In: *State and Change of Forest Ecosystems—Indicators in Current Research* (Ed. by G.I. Agren), pp.389–404. Report No. 13. Swedish University of Agricultural Research, Uppsala. Department of Ecology and Environmental Research, Uppsala.

Berg, B., Jansson, P-E., Oloffson, J. and Reurslag, A. (1991). Decomposition Rates of Scots Pine Needle Litter in a Climatic Transect in N.W. Europe. Report No. 46. Swedish University of Agricultural Sciences, Department of Ecology and Environmental Research, Uppsala.

Billings, W.D., Luken, J.O., Mortensen, D.A. and Petersen, K.M. (1982). Arctic tundra: a sink or source for atmospheric carbon dioxide in a changing environment? *Oecologia* (Berl.) **53**, 7–11.

Billings, W.D., Luken, J.O., Mortensen, D.A. and Petersen, K.M. (1983). Increasing atmospheric carbon dioxide: possible effects on arctic tundra. *Oecologia (Berl.)* **58**, 286–289.

Billings, W.D., Petersen, K.M., Luken, J.O., and Mortensen, D.A. (1984). Interaction of increasing atmospheric carbon dioxide and soil nitrogen on the carbon balance of tundra microcosms. *Oecologia (Berl.)* **65**, 26–29.

Bleak, A.T. (1970). Disappearence of plant material under a winter snow cover. *Ecology* **51**, 915–917.

Bolin, B. (1986). The carbon cycle and predictions for the future. In: *The Greenhouse Effect, Climate Change and Ecosystems* (Ed. by B. Bolin, B.R. Döös, J. Jäger, and R.A. Warwick). *SCOPE* **29**, 93–156. Wiley, Chichester.

Bonan, G.B., Shugart, H.H. and Urban, D.L. (1990). The sensitivity of some high-latitude boreal forests to climatic parameters. *Clim. Change* **16**, 9–29.

Boois, H.M. de (1974). Measurement of seasonal variations in the oxygen uptake of various litter layers of an oak forest. *Plant Soil* **40**, 545–555.

Bouman, A.F. (1990a). Landuse related sources of greenhouse gasses. *Landuse Pol.* **20**, 154–164.

Bouman, A.F. (1990b). Exchange of greenhouse gasses between terrestrial ecosystems and the atmosphere. In: *Soils and the Greenhouse Effect* (Ed. by A.F. Bouman), pp. 62–191. Wiley, Chichester.

Bracewell, J.M., Robertson, G.W. and Tate, K.R. (1976). Pyrolysis-gas chromatography studies on a climosequence of soils in tussock grasslands, New Zealand. *Geoderma* **15**, 209–215.

Brown, R.J.E. (1983). Effects of fire on the permafrost ground thermal regime. In: *The Role of Fire in Northern Circumpolar Ecosystems* (Ed. by R.W. Wein and D.A. Maclean), pp.97–110. Wiley, Chichester.

Brown, S. and Lugo, A.E. (1982). The storage and production of organic matter in tropical forests and their role in the global carbon cycle. *Biotropica* **14**, 161–187.

Bunnell, F. and Scoullar, K.A. (1975). ABISKO II. A computer simulation model of carbon flux in tundra ecosystems. In: *Structure and Function of Tundra Ecosystems* (Ed. by T. Rosswall and O.W. Heal). *Ecol. Bull.* **20**: 425–428.

Bunnell, F., Tait, D.E.N., Flanagan, P.W. and Van Cleve, K. (1977). Microbial respiration and substrate weight loss I. A general model of the influences of abiotic variables. *Soil Biol. Biochem.* **9**, 33–40.

Buringh, P. (1984). Organic carbon in soils of the world. In: *The Role of Terrestrial Vegetation in the Global Carbon Cycle* (Ed. by G.W Woodwell). *SCOPE* **23**, 91–109. Wiley, Chichester.

Campbell, C.A., Paul, E.A., Rennie, D.A. and McCallum, K.J. (1967). Applicability of the carbon dating method to soil humus studies. *Soil Sci.* **104**, 217–224.

Chapin, F.S. (1986). Controls over growth and nutrient use by taiga forest trees. In: *Forest Ecosystems in the Alaskan Taiga* (Ed. by K. Van Cleve, F.S. Chapin, P.W. Flanagan, L.A. Vierech and C.T. Dyrness), pp. 96–111. Springer-Verlag, New York.

Cheshire, M.V., Mundie, C.M. and Shepherd, H. (1974) Transformations of sugars when rye hemicellulose labelled with ^{14}C decomposes in soil. *J. Soil Sci.* **25**, 90–95.

Cicerone, R.J. and Oremland, R.S. (1988). Biogeochemical aspects of atmospheric methane. *Global Biogeochem. Cycl.* **2**, 299–327.

Cicerone, R.J. and Shetter, J.D. (1981). Sources of atmospheric methane: measurement in rice paddies and a discussion. *J. Geophys. Res.* **86**, 7203–7209.

Coleman, D.C. and Sasson, A. (1980). Decomposer sub-system. In: *Grasslands, Systems Analysis and Man* (Ed. by A.I. Breymeyer and G.M. Van Dyne), pp. 609–655. Cambridge University Press, Cambridge.

Covington, W.W. (1981). Changes in the forest floor organic matter and nutrient content following clear-cutting in northern hardwoods. *Ecology* **62**, 41–48.

Crutzen, P.J. (1991). Methane's sinks and sources. *Nature* **350**, 380–381.

Dang, Q.L. and Lieffers, V.J. (1989). Assessment of patterns of response of tree ring growth of black spruce following peatland drainage. *Can. J. For. Res.* **19**, 924–929.

D'Arriego, R., Jacoby, G.C. and Fung, I.Y. (1987). Boreal forests and atmosphere–biosphere exchange of carbon dioxide. *Nature* **329**, 321–323.

Delcourt, H.R. and Harris, W.F. (1980). Carbon budget of the southeastern US biota: analysis of historical changes in trend from source to sink. *Science* **210**, 321–323.

Detweiler, R.P. and Hall, C.A.S. (1988). Tropical forests and the global carbon cycle. *Science* **239**, 42–47.

Dyer, M.L., Meentemeyer, V. and Berg, B. (1990). Apparent controls of mass loss rate of leaf litter on a regional scale. *Scand. J. For. Res.* **5**, 1–13.

Emanuel, W.R., Shugart, H.A. and Stevenson, M.P. (1985a). Climatic change and the broad-scale distribution in terrestrial ecosystem complexes. *Clim. Change* **7**, 29–43.

Emanuel, W.R., Shugart, H.A. and Stevenson, M.P. (1985b). Response to comment: climatic change and the broad-scale distribution in terrestrial ecosystem complexes. *Clim. Change* **7**, 457–460.

Federer, C.A. (1984). Organic matter and nitrogen content of the forest floor of even-aged northern hardwoods. *Can. J. For. Res.* **14**, 763–767.

Flanagan, P.W. (1978). Microbial ecology and decomposition in Arctic tundra and Sub-Arctic taiga ecosystems. In: *Life Sciences, Microbiology and Ecology* (Ed. by M. Loutit), pp. 161–168. Springer-Verlag, Berlin.

Flanagan, P. and Bunnell, F. (1976). Decomposition models based on climatic variables, substrate variables, microbial respiration and production. In: *The Role of Terrestrial and Aquatic Organisms in Decomposition Processes* (Ed. by J.M. Anderson and A. Macfadyen), pp. 437–457. Blackwell Scientific, Oxford.

Flanagan, P.W. and Van Cleve, K. (1983). Nutrient cycling in relation to decomposition and organic matter quality in taiga ecosystems. *Can. J. For. Res.* 13, 795–817.

Flanagan, P.W. and Veum, A.K. (1974). Relationship between respiration, weight loss, temperature and moisture in organic residues in tundra. In: *Soil Organisms and Decomposition in Tundra* (Ed. by A.J. Holding, O.W. Heal, S.F. MacLean and P.W. Flanagan), pp. 249–278. Tundra Biome Steering Committee, Stockholm.

Foster, R.C. (1985). *In situ* localization of organic matter in soils. *Quaest. Entomol.* 21, 609–633.

Foth, H.D. and Schaffer, J.W. (1980). *Soil Geography and Land Use.* Wiley, Chichester.

Fox, R.H., Myers, R.J.K. and Vallis, I (1991). The nitrogen mineralization rate of legume residues in soil as influences by their polyphenol, lignin and nitrogen contents. *Plant Soil* 129, 251–259.

Goh, K.M. Stout, J.D. and Rafter, T.A. (1976). Radiocarbon enrichment of soil organic matter fractions in New Zealand soils. *Soil Sci.* 123, 385–391.

Harcombe, P.A. (1977). Nutrient accumulation by vegetation during the first year of recovery of a tropical ecosystem. In: *Recovery and Restoration of Damaged Ecosystems* (Ed. by J. Cairns, K.L. Dickinson and E.E. Herricks), pp. 347–378. University of Virginia Press, Charlottesville.

Harmon, M.E., Franklin, J.F., Swanson, F.J., Sollins, P., Gregory, S.V., Lattin, J.D. *et al.* (1986). Ecology of coarse woody detritus in temperate ecosystems. *Adv. Ecol. Res.* 15, 133–302.

Harmon, M.E., Baker, G.A., Spycher, G. and Greene, S. (1989). Leaf-litter decomposition in *Picea–Tsuga* forests of Olympic National Park, Washington, USA. *For. Ecol. Manag.* 29.

Harmon, M.E., Ferrell, W.K. and Franklin, J.F. (1990). Effects on carbon storage of conversion of old-growth forests to young forests. *Science* 247, 699–702.

Heal, O.W., Flanagan, P.W., French, D.D. and Maclean, S.F. (1981). Decomposition and accumulation of organic matter. In: *Tundra Ecosystems: a Comparative Analysis* (Ed. by L.C. Bliss, O.W. Heal, and J.J. Moore), pp. 587–633. Cambridge University Press, Cambridge.

Houghton, R.A. (1990). The global effects of deforestation. *Env. Sci. Technol.* 24, 414–422.

Houghton, R.A., Boone, R.D., Fruci, J.R., Hobbie, J.E., Melillo, J.M., Palm, C. A. *et al.* (1987). The flux of carbon from terrestrial ecosystems to the atmosphere in 1980 due to changes in land use: geographic distribution of the global flux. *Tellus* 39B 122–139.

Houghton, R.A., Jenkins, G.J. and Ephraums, J.J. (1990). *Climate Change: The IPCC Scientific Assessment.* Cambridge University Press, Cambridge.

Hunt, H.W. (1977). A simulation model for decomposition in grasslands. *Ecology* 58, 469–484.

Hunt, H.W. (1978). A simulation model for decomposition in grasslands. In: *Grassland Simulation Model* (Ed. by G.S. Innis), pp. 155–184. Springer-Verlag, Berlin.

Jansson, P-E. and Berg, B. (1985). Temporal variation of litter decomposition in relation to simulated soil climate. Long-term decomposition in a Scots pine forest V. *Can. J. Bot.* **63**, 1008–1016.

Jenkinson, D.J. (1988). Soil organic matter and its dynamics. In: *Russell' s Soil Conditions and Plant Growth* (Ed. by A. Wild), pp. 564–607. Longmans Scientific, Harlow.

Jenkinson, D.J. and Rayner, J.H. (1977). The turnover of soil organic matter in some of the Rothamsted classical experiments. *Soil Sci.* **123**, 298–305.

Jenkinson, D.S., Adams, D.E. and Wild, A. (1991). Model estimates of CO_2 emissions from soil in response to global warming. *Nature* **351**, 304–306.

Jones, M.J. (1973). The organic matter content of the savanna soils of West Africa. *J. Soil Sci.* **24**, 42–53.

Juma, N.G. and Mc Gill, W.B. (1986). Decomposition and nutrient cycling in agro-ecosystems. In: *Microfloral and Faunal Interactions in Natural and Agro-ecosystems* (Ed. by M.J. Mitchell and J.P. Nakas), pp. 74–136. Martinus Nijhoff/Dr W. Junk, Hague.

Ladd, J.N., Amato, M. and Oades, J.M. (1985). Decomposition of plant material in Australian soils. III. Residual organic and microbial biomass C and N from isotope-labelled legume material and soil organic matter, decomposing under field conditions. *Aust. J. Soil Res.* **23**, 603–611.

Lashof, D.A. (1989). The dynamic greenhouse: feedback processes that may influence future concentrations of atmospheric trace gases and climate change. *Clim. Change.* **14**, 213–242.

Lashof, D.A. and Ahuja, D.R. (1990). Relative contribution of greenhouse gas emissions to global warming. *Nature* **344**, 529–531.

Linkins, A.E., Melillo, J.M. and Sinsabaugh, R.L. (1984). Factors affecting cellulase activity in terrestrial and aquatic systems. In: *Current Perspectives in Microbial Ecology* (Ed. by M.J. Klug and C.A. Reddy), pp. 572–579. American Society for Microbiology, Washington.

Mackay, J.R. and Mackay, D.K. (1974). Snow cover and ground temperatures, Garry Island, NWT. *Arctic* **27**, 287–296.

Manabe, S. and Stouffer, R.J. (1980). Sensitivity of a global climate model to an increase of CO_2 concentration in the atmosphere. *J. Geophys. Res.* **85**, 5529–5554.

Martel, Y.A. and Paul, E.A. (1974). The use of radiocarbon dating of organic matter in the study of soil genesis. *Soil Sci. Soc. Amer. Proc.* **38**, 501–506.

McGill, W.B. and Paul, E.A. (1976). Fractionation of soil and ^{15}N nitrogen to separate the organic and clay interactions of immobilized N. *Can. J. Sci.* **56**, 203–212.

Meentemeyer, V. and Berg, B. (1986). Regional variation in rate of mass loss of *Pinus sylvestris* litter in Swedish pine forests as influenced by climate and litter quality. *Scand. J. For. Res.* 1, 167–180.

Melillo, J.M. and Gosz, J.R. (1983). Interactions of biogeochemical cycles in forest ecosystems. In: *The Major Biogeochemical Cycles and Their Interactions* (Ed. by B. Bolin and R.B. Cook), pp. 177–222. Wiley, New York.

Melillo, J.M., Aber, J.D. and Muratore, J.F. (1982). Nitrogen and lignin control of hardwood leaf litter decomposition dynamics. *Ecology* **63**, 621–626.

Melillo, J.M., Fruci, J.R., Houghton, R.A., Moore B. III and Skole, D.L. (1988). Land-use change in the Soviet Union between 1850 and 1980: causes of a net release of CO_2 to the atmosphere. *Tellus* **40B**, 116–128.

Melillo, J.M., Callaghan, T.V., Woodward, F.I., Salati, E. and Sinha, S.K. (1990). Effects on ecosystems. In: *Climate Change: The IPCC Scientific Assessment* (Ed. by R.A. Houghton, G.J. Jenkins and J.J. Ephraums), pp. 283–310. Cambridge University Press, Cambridge.

Miller, H.G. (1981). Forest fertilization: some guiding concepts. *Forestry* **54**, 157–167.

Mitchell, J.F.B., Manabe, S., Tokioka, T. and Meleshko, V. (1990). Equilibrium climate change. In: *Climate Change: The IPCC Scientific Assessment* (Ed. by R.A. Houghton, G.J. Jenkins and J.J. Ephraums), pp. 131–172. Cambridge University Press, Cambridge.

Moore, T.R. (1981). Controls on the decomposition of organic matter in subarctic spruce–lichen woodland soils. *Soil Sci.* **131**, 107–113.

Moore, T.R. (1984). Litter decomposition in a sub-arctic, spruce–lichen woodland in eastern Canada. *Ecology* **65**, 299–308.

Mosier, A., Schimel, D., Valentine, D., Bronson, K. and Parton, W. (1991). Methane and nitrous oxide fluxes in native, fertilized and cultivated grasslands. *Nature* **350**, 330–332.

Newman, J.E. (1980). Climate change impacts on the growing season of the North American cornbelt. *Biometeorol.* **7**, 128–142.

Oades, J.M. (1988). The retention of organic matter in soil. *Biogeochemistry* **5**, 35–70.

Olson, J.S., Watts, J.A. and Allison, L.J. (1983). *Carbon in Live Vegetation of the Major World Ecosystems*. National Technical Information Service, Springfield, Virginia.

Palm, C.A. and Sanchez, P.A. (1991). Nitrogen release from the leaves of some tropical legumes as affected by their lignin and polyphenol contents. *Soil Biol. Biochem.* **23**, 83–88.

Palm, C.A., Houghton, R.A. and Melillo, J.M. (1986). Atmospheric carbon dioxide from deforestation in Southeast Asia. *Biotropica* **18**, 177–188.

Parton, W.J., Schimel, D.S., Cole, C.V. and Ojima, D.S. (1987). Analysis of factors controlling soil organic matter levels in Great Plains grasslands. *Soil Sci. Soc. Amer. J.* **51**, 1173–1179.

Pastor, J. and Post, W.M. (1986). Influence of climate, soil moisture, and succession on forest soil carbon and nutrient cycles. *Biogeochemistry* **2**, 3–27.

Pastor, J. and Post, W.M. (1988). Responses of northern forests to CO_2-induced climate change. *Nature* **334**, 55–58.

Paul, E.A. (1984). Dynamics of organic matter in soils. *Plant Soil* **76**, 275–285.

Paul, E.A. and van Veen, J.A. (1979). The use of tracers to determine the dynamic nature of organic matter. In: *Modelling Nitrogen from Farm Wastes* (Ed. by J.K.R. Gasser), pp. 75–132. Applied Science Publishers, London.

Peterson, B.J. and Melillo, J.M. (1985). The potential storage of carbon caused by eutrophication of the biosphere. *Tellus* **37B**, 117–127.

Peterson, K.M. and Billings, W.D. (1975). Carbon dioxide flux from tundra soils and vegetation as related to temperature at Barrow, Alaska. *Amer. Mid. Nat.* **94**, 88–98.

Ping, C.L. (1987). Soil temperature profiles of two Alaskan soils. *Soil Sci. Soc. Amer. J.* **151**, 1010–1018.

Post, W.M., Emanuel, W.R., Zinke, P.J. and Stagenberger, A.G. (1982). Soil carbon pools and world life zones. *Nature* **298**, 156–159.

Prentice, K.C. and Fung, I.Y. (1990). The sensitivity of terrestrial carbon storage to climate change. *Nature* **346**, 48–51.

Raich, J.W. and Nadelhoffer, K.J. (1989). Below ground carbon allocation in forest ecosystems: global trends. *Ecology* **70**, 1346–1354.

Ramanathan, V., Cess, R.D. Harrison, E.F., Minis, P., Barkstrom, B.R., Ahmad, E. and Hartmann, D. (1989). Cloud-radiative forcing and climate: results from the Earth Radiation Budget Experiment. *Science* **243**, 57–63.

Rind, D., Chiou, E.W., Chu, W., Larsen, J., Oltmans, S., Lerner, J. (1991). Positive water vapour feedback in climate models confirmed by satellite data. *Nature* **349**, 500–503.

Sanchez, P.A., Gichuru, M.P. and Katz, L.B. (1982). Organic matter in major soils of the tropical and temperate regions. *Trans. 12th Int. Congr. Soil. Sci. (New Delhi)* **1**, 99–114.

Scharpenseel, H.W. and Schiffman, H. (1977). Radiocarbon dating of soils: a review. *Z. Pflanzenernaehr. Bodenk.* **140**, 159–174.

Scharpenseel, H.W., Schomacker, M. and Ayoub, A. (1990). *Soils on a Warmer Earth*. Elsevier.

Schell, D.M. (1983). Carbon-13 and Carbon-14 abundances in Alaskan aquatic organisms: delayed production from peat in arctic food webs. *Science* **219**, 1068–1071.

Schimel, D.S., Parton, W.J., Kittel, T.G.F., Ojima, D.S. and Cole, C.V. (1990). Grassland biogeochemistry links to atmospheric processes. *Clim. Change* **17**, 13–26.

Schlesinger, W.H. (1977). Carbon balance in terrestrial detritus. *Ann. Rev. Ecol. Syst.* **8**, 51–81.

Schlesinger, W.H. (1984). Soil organic matter: a source of atmospheric CO_2. In: *The Role of Terrestrial Vegetation in the Global Carbon Cycle: Measurement by Remote Sensing* (Ed. by G.M. Woodwell). *SCOPE* **23**, 111–131. Wiley, Chichester.

Schlesinger, W.H. (1990). Evidence from chronosequence studies for a low carbon-storage potential of soils. *Nature* **348**, 232–234.

Schlesinger, W.H. and Hasey, M.M. (1981). Decomposition of chaparral shrub foliage: losses of organic and inorganic constituents from deciduous and evergreen leaves. *Ecology* **62**, 762–774.

Schoenau, J.J. and Bettany, J.R. (1987). Organic matter leaching as a component of carbon, nitrogen, phosphorus and sulfur cycles in a forest, grassland and gleyed soil. *Soil Sci. Soc. Amer. J.* **51**, 646–651.

Seastedt, T.R. (1988). Mass, nitrogen and phosphorus dynamics in foliage and root detritus in tallgrass prairie. *Ecology* **69**, 59–65.

Sebacher, D.I., Harriss, R.C., Bartlett, K.B., Sebacher S.M. and Grice, S.S. (1986). Atmospheric methane sources: Alaskan tundra bogs, an alpine fen and a sub-artic boreal marsh. *Tellus* **38B**, 1–10.

Setzer, A.W. and Pereira, M.C. (1991). Amazonia biomass burnings in 1987 and an estimate of their tropospheric emissions. *Ambio* **20**, 19–22.

Shaw, C. and Pawluk, S. (1986). Faecal microbiology of *Octolasion tyrtaeum, Aporrectodea turgida* and *Lumbricus terrestris* and its relation to the carbon budgets of three artificial soils. *Pedobiologia* **29**, 377–389.

Shetron, S.G. (1974). Distribution of free iron and carbon as related to available water in some forested sandy soils. *Soil Soc. Sci. Amer. Proc.* **38**, 359–362.

Shugart, H.H., Antanovsky, M. Ya, Jarvis, P.G. and Sandford, A.P. (1986). Assessing the responses of global forests to the direct effects of increasing CO_2 and climatic change. In: *The Greenhouse Effect, Climate Change and Ecosystems* (Ed. by B. Bolin, B.R. Döös, J. Jäger and R.A. Warwick). *SCOPE* **29**, 93–56. Wiley, Chichester.

Sims, Z.R. and Nielsen, G.A. (1986). Organic carbon in Montana soils as related to clay content and climate. *Soil Sci. Soc. Amer. J.* **50**, 1269–1271.

Singh, J.S. and Gupta, S.R. (1977). Plant decomposition and soil respiration in terrestrial ecosystems. *Bot. Rev.* **43**, 449–528.

Sivola, J. (1986). Carbon dioxide dynamics in mires reclaimed for forestry in eastern Finland. *Ann. Bot. Fenn.* **23**, 59–67.

Smit, B., Ludlow, L. and Brklacich, M. (1988) Implications of a global climatic warming for agriculture: a review and appraisal. *J. Env. Qual.* **17**, 519–527.

Solomon, A.M. (1986). Transient response of forests to CO_2-induced climate change: simulation modelling experiments in eastern North America. *Oecologia (Berl.)* **68**, 567–579.

Sørensen, L.H. (1981). Carbon–nitrogen relationships during the humification of cellulose in soils containing different amounts of clay. *Soil Biol. Biochem.* **13**, 313–321.

Stevenson, F.J. (1982). *Humus Chemistry: Genesis, Composition, Reactions.* Wiley, New York.

Stott, D.E., Kassim, G., Jarrell, W.M., Martin, J.P. and Haider, K. (1983). Stabilization and incorporation into biomass of specific plant carbons during biodegradation in soil. *Plant Soil* **71**, 15–26.

Svensson, B.H. and Rosswall, T. (1984). *In situ* methane production from acid peat in plant communities with different moisture regimes in a subarctic mire. *Oikos* **43**, 341–350.

Swift, M.J. and Anderson, J.M. (1989) Decomposition. In: *Ecosystems of the World 14B. Tropical Rain Forest Systems* (Ed. by H. Leith and M.J.A. Werjer), pp. 547–569. Elsevier Science, Amsterdam.

Swift, M.J., Heal, O.W. and Anderson, J.M. (1979). *Decomposition in Terrestrial Ecosystems.* Blackwell Scientific, Oxford.

Tamm, C.O. (1979). Nutrient cycling and productivity of forest ecosystems. In: *Impact of Intensive Harvesting on Forest Nutrient Cycling* (Ed. by A.L. Leaf), pp. 2–22. State University of New York, College of Environmental Science and Forestry, Syracuse.

Tamm, C.O. and Holmen, H. (1967). Some remarks on soil organic matter turnover in Swedish podzol profiles. *Meddel. Nor. Skogs.* **85**, 69–88.

Tans, P.P., Fung, I.Y. and Takahashi, T. (1990). Observational constraints on the global atmospheric CO_2 budget. *Science* **247**, 1431–1438.

Tate, K.R. and Theng, B.K.G. (1980). Organic matter and its interactions with inorganic soil constituents. In: *Soils with Variable Charge* (Ed. by B.K.G. Theng), pp. 225–249. New Zealand Society of Soil Science, Lower Hutt.

Taylor, B.R., Parkinson, D. and Parsons, W.F.J. (1989). Nitrogen and lignin content as predictors of litter decay rates: a microcosm test. *Ecology* **70**, 97–104.

Theng, B.K.G. (1979). *Formation and Properties of Clay–Polymer Complexes.* Developments in Soil Science. Elsevier, Amsterdam.

Theng, B.K.G., Churchman, G.J. and Newman, R.H. (1986). The occurrence of interlayer clay–organic complexes in two New Zealand soils. *Soil Sci.* **142**, 262–266.

Tiessen, H. and Stewart, J.W.B. (1983). Light and electron microscopy of stained microaggregates: the role of organic matter and microbes in soil aggregation. *Biogeochemistry* **5**, 312–322.

Tisdall, J.M. and Oades, J.M. (1982). Organic matter and water-stable aggregates in soils. *J. Soil Sci.* **33**, 141–163.

Tissue, D.T. and Oechel, W.C. (1987). Response of *Eriophorum vaginatum* to elevated CO_2 and temperature in the Alaskan Arctic tundra. *Ecology* **68**, 401–410.

Titlyanova, A.A. (1987). Carbon cycle in the soil compartment of agro-ecosystems. *Trans. 13th Int. Cong. Soil Sci. Hamburg* **6**, 925–933.

Vaghjiani, G.L. and Ravishankara, A.R. (1991) New measurements of the rate coefficient for the reaction of OH with methane. *Nature* **350**, 406–409.

Van Cleve, K., Dyrness, C.T., Viereck, L.A. Fox, J., Chapin, F.S. and Oechel, W. (1983). Taiga ecosystems in interior Alaska. *Bioscience* **33**, 39–44.

Van Cleve, K., Chapin, F.S., Flanagan, P.W., Vireck, L.A. and Dyrness, C.T. (1986). *Forest Ecosystems in the Alaska Taiga. Ecological Studies* **57**, 160–189. Springer-Verlag, Heidelberg.

Viereck, L.A., Dyrness, C.T., Van Cleve, K., and Foote, M.J. (1983). Vegetation, soils and forest productivity in selected forest types in interior Alaska. *Can. J. For. Res.* **13**, 703–720.

Vitousek, P.M. (1982). Nutrient cycling and nutrient use efficiency. *Amer. Nat.* **119**, 553–572.

Vitousek, P.M. (1991). Can planted forests counteract increasing atmospheric carbon dioxide? *J. Env. Qual.* **20**, 348–354.

Wang, W-C., Dudeck, M.P., Liang, X-Z. and Kiehl, J.T. (1991). Inadequacy of effective CO_2 as a proxy in simulating the greenhouse effect of other radiatively active gases. *Nature* **350**, 573–577.

Watson, A.J., Robinson, C., Robertson, J.E., Williams, P.J. le B. and Fasham, M.J.R. (1991). Spatial variability in the sink for atmospheric CO_2 in the North Atlantic. *Nature* **350**, 50–53.

Watson, R.T., Rodhe, H., Oeschger, H. and Siegenthaler, U. (1990). Greenhouse gasses and aerosols. In: *Climate Change: The IPPC Scientific Assessment* (Ed. by R.A. Houghton, G.J. Jenkins and J.J. Ephraums), pp. 1–40. Cambridge University Press, Cambridge.

Werner, P. (1984). Changes in soil properties during tropical wet forest succession. *Biotropica*, **16**, 43–50.

Whalen, S.C. and Reeburgh, W.S. (1988). A methane flux time series for tundra environments. *Global Biogeochem. Cycles* **2**, 399–409.

Whalen, S.C. and Reeburgh, W.S. (1989) A methane flux transect along the trans-Alaska pipeline haul road. *Tellus* **42**, 237–249.

Whalen, S.C. and Reeburgh, W.S. (1990). Consumption of atmospheric methane to sub-ambient concentrations by Tundra soils. *Nature* **346**, 160.

Wildung, R.E., Garland, T.R. and Buschbom, R.L. (1975). The interdependent effects of soil temperature and water content on soil respiration rate and plant root respiration in arid grassland soils. *Soil Biol. Biochem.* **7**, 373–378.

Yarrie, J. (1983). Environmental and successional relationships of the forest communities of the Porcupine River drainage, interior Alaska. *Can. J. For. Res.* **13**, 721–728.

Zech, W., Kogel, I., Zucker, A. and Alt, H. (1985). CP-MAS-[13]-NMR-Spektren organischer Lagen einer Tangelrendzina. *Z. Pflanzen. Bodenk.* **148**, 481–488.

Zimka, J.R. and Stachurski, A. (1976). Regulation of C and N transfer to the soil of forest ecosystems and the rate of litter decomposition. *Bull. Acad. Pol. Sci.* **24**, 127–132.

Zinke, P., Stangenberger, A., Post, W., Emanuel, W. and Olson, J. (1984). Worldwide Organic Carbon and Nitrogen Data. Report ORNL/TM-8857, Oak Ridge National Laboratory, Oak Ridge.

Predicting the Responses of the Coastal Zone to Global Change

P.M. HOLLIGAN and W.A. REINERS

ADVANCES IN ECOLOGICAL RESEARCH VOL. 22
ISBN 0–12–013922–7

I. INTRODUCTION

The margins of continents and oceans—where land and sea meet—are regions of high physical and biological diversity that are heavily utilized by man for residential, agricultural, commercial (including transportation), waste disposal, recreational and military purposes, as well as fishing, mariculture, and the extraction of energy and mineral resources. It is estimated that > 50% of the world's human population lives on coastal plains (Ray, 1989) and, with this proportion increasing, it is here that much of the global investment into agricultural, urban and industrial development is taking place. The dynamic nature of the interface between land and sea (Oppenheimer, 1989), makes the coastal zone particularly vulnerable to global change as a consequence of the direct (physical disturbance, pollution, etc.) and indirect (climatic) effects of man on coastal environments. The economic costs dealing with the resultant environmental problems over the next few decades are potentially enormous (Hekstra, 1989).

A. Definitions of the Coastal Zone

The coastal zone is functionally defined as "that space in which terrestrial environments influence marine (or lacustrine) environments and vice versa" (Carter, 1988). A legal definition has been accepted by international agreement as the 200 nautical mile limit from land over which coastal nations exert sovereignty (Economic Exclusion Zone). From an oceanographic perspective the coastal or shelf seas extend from the land margin to the edge of the continental shelf at 100–200 m depth (Fig.1).

For scientific purposes, the functional definitions of the coastal zone and coastal ocean depend on the nature and scales of the processes that characterize the land–ocean boundary (see Lasserre and Martin, 1986; Skreslet, 1986; Jansson, 1988; Walsh, 1988). The landward boundary, for example, can be defined as the mean tidal high water mark, the limit of saltwater wedges in underlying aquifers, or the region of salt aerosol deposition. The ocean boundary may be considered in terms of bathymetry, the influence of fresh water (including terrestrially derived dissolved and suspended matter), or various physical processes that distinguish shallow shelf waters from deep ocean waters (Eisma, 1988b). The lack of a rigorous set of definitions leads to considerable ambiguity in terminology, but the generally accepted terms illustrated in Fig. 2 will be followed here.

Determination of the length of the world's coastline is a classic problem in fractal geometry; estimates lie between 500 000 and 1 000 000 km (Hekstra, 1989). Together the coastal plains and seas occupy only about 8% of the

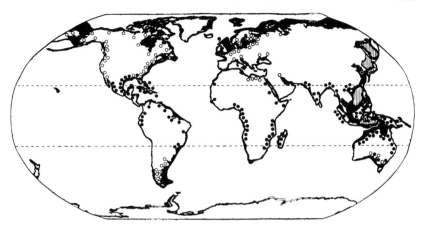

Fig. 1. Map of the globe showing the continental shelf (0–150 m) in black (adapted from Postma and Zijlstra, 1988) and the main distributions of coastal mangrove (filled circles) and saltmarsh (empty circles) (adapted from Chapman, 1977). The areas of vertical shading indicate water > 150 m deep surrounded by shallower water.

global surface area, but their combined biological productivity is approximately equivalent to that of upland regions (27% of surface area) and of the oceans (65% of surface area) (Ray, 1989). Coastal ecosystems, including mangrove salt marsh, coral reef and benthic algal communities, cover only ~0·4% of surface area (Walsh, 1988) but are some of the most productive on earth and are also active transformation sites for materials transported from land to ocean. In these terms the coastal zone represents a complex boundary zone, or reticulum, of varying width between land and sea that is continuous with the river systems and is critical to the earth system as a region of intense geomorphological and geochemical activity.

B. Main Issues of Environmental Change in the Coastal Zone

The whole coastal zone has now been altered to some degree by man as a result of the modification of riverine discharges (Meybeck and Helmer, 1989), the exploitation of marine resources (Bardach, 1989), and various shoreline development and engineering activities (Bruun, 1989). Coastal ecosystems are also affected by natural variations in climate (e.g. Aebischer *et al.*, 1990), and will be further influenced by anthropogenic climate changes. Such factors act synergistically on the coastal environment so that attempts to predict modification of the coastal zone by climate change must take account of the direct effects of man, and vice versa. It is necessary, therefore, to understand both how the coastal zone is responding now to human interference and, against this background, how it has responded in the past and might respond in the future to any change in global climate or sea level. The influence of coastal ecosystems on the trace gas composition of the atmosphere may also be significant in terms both of the high rates of biological productivity contributing to the global budgets of these gases, and of variations in gas fluxes with the rise and fall of sea level during glacial cycles acting as a feedback mechanism on the climate system. This issue has received surprisingly little attention in studies of global change.

Changes in the coastal zone fall into four categories:

(1) Those related to the human use and management of land and fresh water (see Petts, 1984; Newman and Fairbridge, 1986), which affect the transport of materials to the coastal zone via rivers (Meade, 1988; Meybeck and Helmer, 1989), groundwater (Zektzer *et al.*, 1973) and atmosphere (UNESCO, 1989; Duce, 1991), and alter fluxes to the sea of suspended matter (Milliman, 1990), nutrients (Billen *et al.*, 1991) and pollutants (Kullenberg, 1986; UNEP, 1990).

(2) Those related to the uses and management of coastal seas, including exploitation of living resources and mariculture (Bardach, 1989), the various adverse effects of transport, dredging, dumping and mineral extraction (see UNEP, 1990), coastal engineering and protection (Bird, 1985; Bruun, 1989), and the introduction of exotic species (e.g. Carlton *et al.*, 1990).

(3) Those related to climate change resulting from natural and anthropogenic fluctuations of the trace gas composition of the atmosphere (Mitchell, 1989; Warrick and Farmer, 1990). Regional and global shifts in temperature, cloud cover, precipitation, and wind stress will affect all coastal environments, as well as the growth and reproduction of marine organisms. Significant feedback effects are possible (Lashof, 1989) as a result of alterations to biogeochemical and physicochemical (e.g. stability of methane clathrates in coastal sediments [MacDonald, 1990]) properties.

(4) Those related to sea-level rise, itself a consequence of climate warming (Peltier and Tushingham, 1989; UNESCO, 1990; IPCC, 1990). Recent "best" estimates are about $+20$ cm and $+60$ cm over the present level by about the years 2030 and 2100 respectively which represent a two- to three-fold increase in the present rate of sea-level rise. Impacts include

---

Fig. 2. Diagrammatic section of the coastal zone and coastal ocean to show the main hydrographic boundaries.

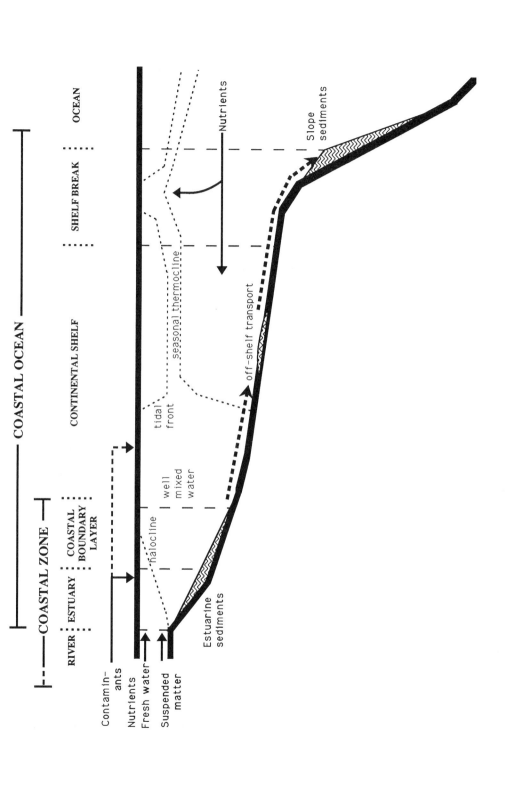

COASTAL OCEAN

COASTAL ZONE

RIVER ESTUARY

COASTAL BOUNDARY LAYER

CONTINENTAL SHELF

SHELF BREAK

OCEAN

Contaminants
Nutrients
Fresh water
Suspended matter

halocline

well mixed water

tidal front

seasonal thermocline

off-shelf transport

Nutrients

Estuarine sediments

Slope sediments

permanent and temporary inundation, coastal erosion and salt penetration into fresh waters, as well as changes in tide and wave characteristics
(Warrick and Farmer, 1990; Vellinga and Leatherman, 1989; Hekstra,
1989). The effects of increases both in mean sea level and in the
frequency of extreme conditions associated with storms need to be
considered.

C. Implications of Global Change for the Coastal Zone

Although many of the causes and consequences of change in the coastal zone
are local or regional by nature, they are of global significance in terms of
large-scale ecological and geomorphological processes which are likely to be
altered irreversibly over decadal time-scales relevant to the future needs and
impacts of human society. Environmental disturbance introduces disorder to
ecosystems with immediate, and generally harmful, consequences for the
exploitation and management of living resources (Glasby, 1988), but methods to quantify and evaluate such disorder in the coastal zone are not
available. Furthermore a lack of knowledge about interactions between
physical, chemical and biological processes leads to great uncertainty in
predictions of the form and impact of change and of possible feedback effects
on climate. Material transport and transformation processes in the coastal
zone have a central role in the global cycling of matter between the biosphere
and atmosphere (Fig. 3). At times of high sea level the biological turnover
times for key elements such as C and N in the sea may be significantly
reduced, as organic matter from land is held and oxidized more efficiently on
the continental shelves.

Efforts to develop ecologically sensible economic models (Ray, 1989) for
continuing use of the coastal zone are likely to depend on better "environmental accounting" (Glasby, 1988) of human activities. Making the best
choice between no response to change, adaptive response to change, and
limitation response to the cause of change is extremely difficult (for discussions of this problem in relation to sea-level rise see Broadus, 1989 and
Hekstra, 1989), and will require much better information on coastal zone
dynamics than is presently available.

In this chapter we summarize our knowledge of how the coastal zone is
likely to be affected by global change. Although much has already been
written on this particular problem, we attempt here to develop a global
perspective of how coastal ecosystems function (IGBP, 1990) which depends
on the integration of terrestrial and marine sciences.

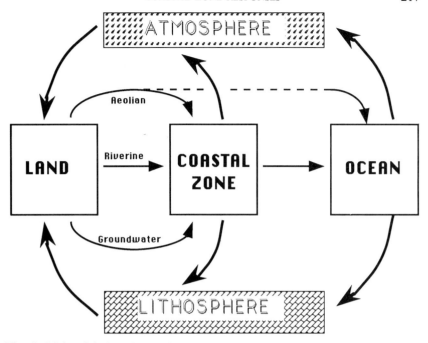

Fig. 3. Main global pathways for the transport and transformation of matter, showing the central importance of the coastal zone.

II. NATURE OF THE COASTAL ZONE

A. Classification of the Coastal Zone

By any definition, the coastal zone is a highly variable environment (see Wolfe and Kjerfve, 1986) for which quantitative methods of comparing different regions are not available. Maps exist for particular properties of the coastal zone (e.g. Davies, 1972), but are usually based on too little information for such properties to be quantified on a regional or global basis as can be done for the continents or oceans. Equally, the heterogeneity of coastal environments makes it difficult to extrapolate from the results of detailed local or regional studies.

In spite of this variability, coastal zones can be classified in terms of certain well-defined physiographic, physico-chemical and biological features (Carter, 1988). The processes governing the development of different types of coastlines and continental shelves have been summarized by Eisma (1988b). Global data on the fluxes of fresh water and suspended matter from the continents to the oceans are reviewed by Milliman and Meade (1983), and a subsequent analysis by Milliman (1990) demonstrates the wide regional differences for both land sources and ocean sinks (Fig. 4). The geomorphology of coastal environments reflects not only the geological setting and sediment inputs, but also the physical conditions affecting sediment transport (climate, tidal mixing, etc.), important chemical transformations at the boundary between salt and fresh water and on the surfaces of suspended particles (Stumm, 1987), and various biological processes concerned with the stability and colonization of sediment and rock surfaces and the formation of carbonate banks and reefs (Spencer, 1988). The properties of offshore waters are strongly influenced by the global and regional patterns of tidal energy dissipation (Miller, 1966), and of upwelling (Barber and Smith, 1981).

There have been few attempts to combine both physical and biological parameters in a global classification scheme for coastal environments. In the context of the development of conservation policies, Hayden *et al.* (1984) used biogeographic and physical oceanographic information to define a broad set of coastal realms that were globally consistent but, by necessity, based on a very limited subset of criteria. The use of classification systems to help predict future changes in coastal ecosystems will require information about fundamental geomorphological, physical and ecological properties. Suitable computer-based methods for analysing such data sets are now being developed for geographical applications in coastal regions (Fricker and Forbes, 1988).

B. Exchanges of Materials

The riverine transport of terrigenous materials into coastal seas has been relatively well studied so that global estimates for certain elements, organic carbon, nutrients, as well as various pollutants, are available (GESAMP, 1987; Meybeck and Helmer, 1989; Degens *et al.*, 1990). Considerably less is known about atmospheric transport in respect of the coastal zone (GESAMP, 1989; Duce, 1991) although processes determining the transfer of continental dust to the oceans have received considerable attention (Pye, 1987), and recent regional investigations have shown that, for certain materials, atmospheric inputs to the sea can exceed riverine ones (Martin *et al.*, 1989; Duce, 1991). The contribution of groundwater exchange (Zektzer *et al.*, 1973, 1983) is poorly understood. Recent reviews by Eisma (1988a) and Degens *et al.* (1990) indicate that rivers are the dominant mechanism globally

---

Fig. 4. Land sources and marine sinks for fluvial sediment and water discharged into the oceans. (Source: Milliman, 1990, by permission.)

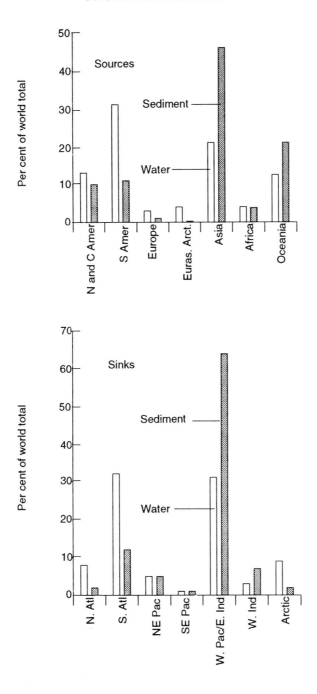

by which land influences the coastal environment, although atmospheric inputs of certain contaminants from urban and industrial areas are locally more important.

The hydrological cycle (Berner and Berner, 1987) is of fundamental importance in determining the nature of land–ocean interactions at the global scale. Inputs of fresh water and associated suspended and dissolved matter into the coastal ocean are a function of catchment area, geology, vegetation and climate and, on account of the relative buoyancy of fresh water, also affect mixing and circulation processes in estuaries and the coastal boundary layer as well as their chemical and biological properties (see Skreslet, 1986; Jansson, 1988). The spatial scales of these effects depend on the magnitude and seasonality of river flow. The influence of the Amazon water can be detected in mid-Atlantic (Muller-Karger *et al.*, 1988) and the sedimentary fans of large deltas extend right across the continental shelf and slope. Even for coasts without large rivers, episodic events such as the spring snow melt or a storm flood can carry suspended matter directly from land to the deep ocean (Fig. 5).

Riverine delivery of inorganic nutrients (Peterson *et al.*, 1988; Meybeck and Helmer, 1989) enhances the biological productivity of inshore coastal waters but for coastal oceans in general it is a relatively minor source of nutrients compared to ocean inputs across the shelf break. The nature of the catchment basin and associated freshwater ecosystems affects the concentrations and proportions of nutrients in river water reaching the sea (Billen *et al.*, 1991).

Organic matter carried from land to sea via rivers and the atmosphere corresponds approximately to 1% of terrestrial productivity (Ittekkot, 1988; Spitzy and Ittekkot, 1991) and is a significant component of the energy budget for coastal ecosystems. Part of this material is refractory and buried in nearshore sediments (Berner, 1989).

Materials carried from sea to land other than water vapour include salt in marine aerosols (Franzen, 1990) and in groundwater, and sulphur compounds derived largely from volatile dimethylsulphide (DMS) released by certain marine plants (Andreae, 1986)

C. Physico-chemical Properties

The enormous heat capacity of the sea and the interchange of energy through evaporation and condensation confer a large moderating influence on air temperature and humidity in coastal zones. Such climatic effects are felt most strongly where cold upwelled water or warm ocean currents extend onto the

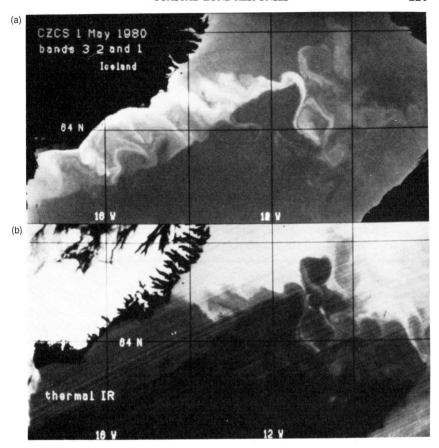

Fig. 5. Coastal Zone Color Scanner (Nimbus-7 satellite) images of the N. Atlantic Ocean southeast of Iceland, 1 May 1980. (a) Colour composite image showing the riverine discharge of suspended matter (light shades) onto the continental shelf with entrainment into an oceanic frontal region. (b) Infra-red thermal image showing the frontal boundary between relatively warm N. Atlantic water (dark shades) and cold Norwegian Sea water.

continental shelf, and are of great ecological significance to terrestrial ecosystems (Walter, 1983).

The coastal zone is a region of high energy dissipation on account of interactive oceanic and atmospheric forcing associated with topographic discontinuities (continental slope, coastal mountains) and with density gradients caused by freshwater inflows and by the seasonal heating/cooling cycle (see Brink, 1987; Postma, 1988; and Blanton, 1991 for reviews). These processes define high- and low-energy shelf systems in which contrasting conditions for the transport and transformation of materials derived from the land, oceans and atmosphere are encountered. It is this variable physical and chemical environment that underlies the great diversity of animal, plant and microbial life within the coastal zone.

Deltas, estuaries and coastal lagoons are major sites for the transformation and accumulation of organic matter and sediment. All are inherently variable over a wide range of temporal and spatial scales (Ketchum, 1983; Wolfe and Kjerfve, 1986) so that mean boundary conditions are not a good indicator of net fluxes from rivers to the coastal ocean. In particular the factors that determine exchanges between estuaries and the open sea, involving the dynamics of freshwater plumes and the coastal boundary layer (Fig. 2), are still not well understood (e.g. Dronkers, 1988; Griffin and LeBlond, 1990).

There is a considerable body of information about sediment transport and accumulation within estuaries (Dyer, 1986; Nichols, 1986; Stevenson et al., 1988) and deltas (e.g. Milliman et al., 1987; Wells and Coleman, 1987; Day and Templet, 1989; Wright et al., 1990). Estuaries may or may not be sites of active sediment accumulation depending on local factors (topography, sediment supply rate, hydrodynamic properties, etc.) and trends in sea level (Allen, 1990). Although conditions of continuous sediment accumulation can extend outwards across the continental shelf especially off large deltas, many shelves are characterized by relict or temporary deposits with the modern sedimentary material that escapes the estuaries being transported outwards to the continental slope (Ittekkot et al., 1986; Salge and Wong, 1988; Jouanneau and Latouche, 1989; Palanques et al., 1990). These processes of sediment accumulation and transport play an important part in determining chemical changes both to the particulate material itself and to dissolved material through particle–water reactions.

Estuaries are sites of complex chemical interactions related to salinity gradients from fresh to salt water, to phase transformations involving particle–water reactions, and to biological processes that cause biogeochemical transformations of both particulate and dissolved matter. Church (1987), Mantoura (1987), Kempe (1988) and Burton (1988) review and discuss various aspects of estuarine chemistry particularly with respect to processes that determine non-conservative gradients in chemical properties between fresh and salt water. Particularly important in the context of environmental change are losses of matter to the sediments and atmosphere, including trace gases such as N_2O and CH_4, that result from anthropogenic additions of nutrients, organic matter and pollutants to coastal waters.

D. Biological Properties

Favourable conditions of light and nutrient availability, determined by turbulence and by inputs of nutrients from ocean and land, maintain relatively high rates of primary production in coastal environments (Paasche, 1988). Values per unit area for shelf waters are typically three times higher than for the oceans (Walsh, 1988), and more than 10 times higher in the case of certain coastal upwelling systems (Barber and Smith, 1981; Payne et al., 1987). Coastal ecosystems, including salt marshes, mangrove swamps, mudflats, seagrass and kelp beds and coral reefs, are even more productive (Chapman, 1977; Lewis, 1981; McRoy and Lloyd, 1981; Stevenson, 1988; Twilley, 1988; Charpy-Roubaud and Sournia, 1990). As a result the coastal ocean supports > 90% of the world's fishery yield (Walsh, 1988), representing 87% of the finfish yield and almost all the shellfish catch (Postma and Zijlstra, 1988). The global energy flow patterns and their constraints on continental shelves are reviewed by Cushing (1988).

A global evaluation of the primary productivity of coastal ecosystems has been carried out only for benthic algae and phytoplankton (Charpy-Roubaud and Sournia, 1990) and even for these groups there are many uncertainties. Algal growth may be limited by light as in turbid estuaries (e.g. Cloern, 1987; Turner et al., 1990) and deep well-mixed water columns, or by nutrients (Smith, 1984; Howarth, 1988). The supply of nutrients is controlled both by physical mixing processes and by biogeochemical recycling in water and sediments (Postma, 1988; Walsh, 1988), and gives rise to considerable variability in space and time of algal biomass particularly in regions of physical discontinuities or fronts.

Comparative studies of marine and freshwater plant communities suggest that the former are generally more productive (Lugo et al., 1988) and support a greater biomass of heterotrophic organisms (Nixon, 1988). This difference is attributable to tidal energy which tends to maintain a well-oxygenated environment in which nutrients are recycled more efficiently.

The fate of primary organic carbon within the coastal zone and ocean affects both ecological and geochemical processes. Transport pathways and degradability are critical factors in determining patterns of secondary production and of carbon storage in marine sediments (see Skreslet, 1986; Jansson, 1988). Certain marginal ecosystems, such as mangrove (Twilley, 1988) and saltmarsh (Hopkinson, 1988) communities, export organic matter whereas others such as mudflats are significant sinks. Most algal material with a low fibre content is digestible by animals in contrast to higher plant detritus which is only slowly broken down by microbial action (Mann, 1988). These differences are reflected in the communities of consumer organisms. Further offshore phytoplankton may be grazed in the water column or sink to the bottom, and the exact nature of the coupling between pelagic and benthic ecosystems has a significant impact on fish recruitment (Townsend and Cammen, 1988).

E. Biogeochemical Processes

High redox conditions dominate surface soils and sediments and most of the water column of coastal ecosystems, so that organic matter is readily reoxidized (Henrichs and Reeburgh, 1987). This process leads to the regeneration of nutrients otherwise bound in organic or inorganic complexes, and to the release of CO_2.

However, anaerobic conditions prevail in the soils of poorly drained coastal plains and in some coastal sediments and bottom waters, particularly within poorly ventilated basins and estuaries, in situations of high organic production, and more generally in subsurface sediments. Organic matter may be preserved, or mineralized by various reactions that yield various trace gases including oxides of N, sulphides and methane (e.g. Marty et al., 1990) and also affect the bioavailability of metals, phosphorus and other substances.

The coastal oceans are known to play an important part in the carbon economy of the global ocean (Deuser, 1979; Henrichs and Reeburgh, 1987; Berner, 1989). In the context of uncertainties about global budgets for anthropogenic CO_2 (Tans et al., 1990) it is surprising that more attention has not been given to the fate of terrestrial organic carbon transferred to the sea in particulate and dissolved forms which together exceeds the quantity of organic carbon buried in marine sediments. In order to maintain the global cycle of carbon the ocean must be a net source of CO_2 (Smith and Mackenzie, 1987) and, as indicated in Fig. 6, the main efflux to the atmosphere is almost certainly from inshore coastal waters.

Recent observations suggest that much of the organic carbon buried in marine sediments might be of terrestrial origin (Ittekkot and Haake, 1990) which, in the light of evidence that export from land has increased with changes in land use (Berner, 1989), raises important questions about what factors determine the fate of this material in the marine environment. They also imply that organic matter derived from marine organisms is more efficiently recycled than terrestrial organic matter.

The few studies made of the CO_2 system in coastal waters indicate rapid changes at times of high biological activity, with surface waters acting as a source or sink for atmospheric CO_2 depending on the balance between photosynthesis and respiration and on the influence of river waters (Pegler and Kempe, 1988). Studies on the fate of plant carbon on the continental

ORGANIC CARBON AND CO₂ FLUXES

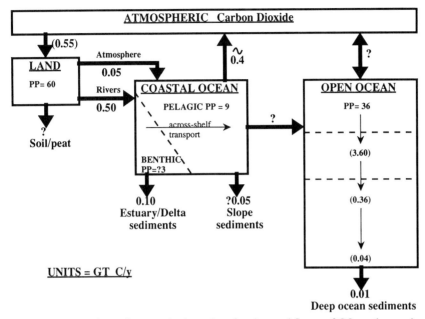

UNITS = GT C/y

Fig. 6. Estimates of net photosynthetic carbon fixation and fluxes of CO_2 and organic carbon between the land, coastal ocean and open ocean. Sources of information include Smith and Mackenzie (1987), Berner (1989), Walsh (1989), Charpy-Roubaud and Sournia (1990), Spitzy and Ittekkot (1990), and Duce (1991). The assumed flux of 0.55 GT C/y from the atmosphere to land balances the flux from land to ocean.

shelf (Rowe, 1987) also suggest efficient utilization by pelagic and benthic organisms. By contrast, Walsh (1989) has hypothesized that a large proportion of the organic matter produced during the spring phytoplankton outburst is exported from the shelf to slope sediments and that this pathway may be an important potential sink for anthropogenic CO_2.

The significance of shelf carbonate formation for the carbon budget of the global ocean is not well understood. The quantities of carbon fixed in this form are large at times of high sea level when the shelves are flooded, and represent a strong perturbation of the marine carbon cycle between glacial and interglacial periods.

Various studies have demonstrated that denitrification in anaerobic coastal sediments (Christensen et al., 1987; Seitzinger, 1988; Law and Owens, 1990) is a significant source of the greenhouse gas nitrous oxide, but much less is know about nitrification in aerobic coastal environments (Henriksen and Kemp, 1988) as a potential source of N_2O. Denitrification may also represent an important mechanism for the removal of anthropogenic nitrate, with close coupling to the oxidation of (and therefore control by) organic carbon (Smith et al., 1989).

Another important biogeochemical process occurring in the coastal zone and open ocean is the emission of dimethylsulphide (DMS) derived from macroalgae and phytoplankton (Andreae, 1986). DMS is oxidized in the atmosphere to sulphate, and contributes both to the sea-to-land transport of SO_4 (Turner et al., 1988) and to the acidity of rainfall. It is also thought to play an important role in the formation of cloud condensation nuclei in the marine atmosphere (Charlson et al., 1987) with the potential for direct feedback effects on climate. The biological factors that lead to the formation of free DMS from precursor compounds are only partly understood (Kiene, 1990).

III. EFFECTS OF HUMAN ACTIVITIES

A. Delivery of Fresh Water

The drainage of fresh water from land to sea as part of the global hydrological cycle (Berner and Berner, 1987) has a profound influence on the coastal zone through the geomorphological (Carter, 1988) and geochemical impacts of suspended and dissolved materials that are transported mainly by rivers (Milliman, 1990; also see Fig. 4) and through the effects of the fresh water on the hydrodynamics of the coastal boundary layer (Blanton, 1991). Such processes have been strongly modified by the exponential increase over the last few decades in the numbers of dams and other structures that impound fresh water on land (Petts, 1984) for human use.

It has been calculated that the total discharge of fresh water from the continents has been reduced by about 15% since the 1950s (Newman and Fairbridge 1986), a quantity that is equivalent to a change in sea level of about -0.7 mm year^{-1} over this period which is substantial compared to the estimated present rate of sea level rise of about $+1.5$ mm year^{-1}. Such a reduction is expected to affect vegetation at the land–sea boundary in terms both of salt incursion and of evapotranspiration by wetland ecosystems, and is also likely to lead to greater salt penetration into groundwater reservoirs. These types of change are difficult to attribute specifically to freshwater management as many other factors are involved.

Another consequence of river control is that seasonal differences in flow rates become relatively small. In the absence of periods of high flow not only does the carrying capacity for suspended and dissolved matter fall, but estuarine circulation patterns and salinity distributions are modified. Changes in the estuarine residence times for water, as well as for associated particulate and dissolved matter, are likely to have far-reaching effects on chemical processes (Morris, 1990). Again the broader ecological conse-

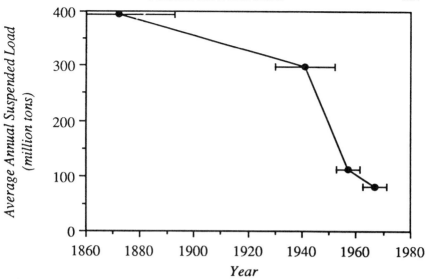

Fig. 7. Declining annual suspended load of the Mississippi River. Horizontal bars show periods for which average data was obtained (data from Kesel, 1988, as presented in At The Land–Sea Interface: A Call For Basic Research, published by Joint Oceanographic Institutions Inc., Washington DC, 1990).

quences are not well understood, although the direct effects of fresh water on the coastal ecosystems has received attention (see Skreslet, 1986).

B. Transport and Fate of Suspended Matter

Long-term records of sediment discharge for major rivers show clearly the reductions associated with freshwater management (e.g. Fig. 7 for the Mississippi). In the case of the Nile it is estimated that a combination of sediment starvation due to the Aswan High Dam, continuing consolidation of modern sediments, and sea-level rise will lead to submergence of much of the delta region within 30 km of the present coastline during the next century (Milliman *et al.*, 1989; Stanley, 1990); accelerated erosion along parts of the coast has been observed (Smith and Abdel-Kader, 1988), and changes in hydrography and nutrient levels of coastal waters have caused lower biological productivity (Halim, 1991). However, within any large delta, natural changes in the patterns of drainage lead to variations in sediment supply on a subregional basis, so that not all subsidence is attributable to the activities of man (Wells and Coleman, 1987).

   By contrast, forest clearance and agricultural practices of man tend to enhance soil erosion and sediment delivery to the coastal zone (Degens *et al.*, 1990) particularly in regions where mountains extend to the coast. Severely affected areas today are Madagascar and Indonesia where large-scale changes in the turbidity of coastal waters are reported (but poorly documented).

The effects of variations in sediment supply by rivers on the dynamics of estuarine and coastal marsh ecosystems are not well known. Since sea level has been relatively constant for the last 3000 years (e.g. Ters, 1987), most estuaries in regions not affected by isostatic changes in coastline elevation are thought to have reached some equilibrium condition with respect to inputs and outputs of suspended matter (e.g. Allen, 1990). The long-term effects of imbalances in sediment supply on the development of coastal marshes have been discussed by Stevenson et al. (1988). Other activities of man that alter coastal erosion and accretion processes include the drainage of marshes, coastal engineering and dredging.

Variations in concentrations of suspended matter in estuarine and coastal waters will alter chemical and biological processes as a result of the effects of particle–water interactions (Stumm, 1987), of sedimentation rate on diagenetic processes, and of changes in turbidity on light penetration and rates of primary production. Few attempts to monitor such interactions over appropriate time periods have been made although the type and scale of probable changes can be inferred from contemporary studies of heavily impacted estuaries (Kempe, 1988).

At larger scales man also influences sedimentation processes on the continental slope. For example, studies of the Ebro River margin in the northwestern Mediterranean (Palanques et al., 1990) have shown that both the rate of sedimentation and the nature of the sediment particles have changed in response to management of the river. Pollutants can be traced from coastal sources to sites of deposition on the slope. Other shelf regions are presumably being affected in a similar way.

C. Chemical Modification

There is now a very extensive literature on the causes and implications of changes in the delivery of nutrients and of a wide range of contaminants to the coastal zone. Recent accounts are provided by GESAMP (1987) and UNEP (1990), and here only a brief summary of some of the major effects is given.

The eutrophication of coastal waters due to the addition of excess nutrients by human practices on land causes various problems related to the adverse effects of plankton blooms and of the deoxygenation of bottom waters (Smetacek et al., 1991). The trend of increasing concentrations of nitrate in the waters of the German Bight (Fig. 8) is typical of changes in many estuaries (e.g. Meybeck et al., 1988) and coastal seas such as the Baltic

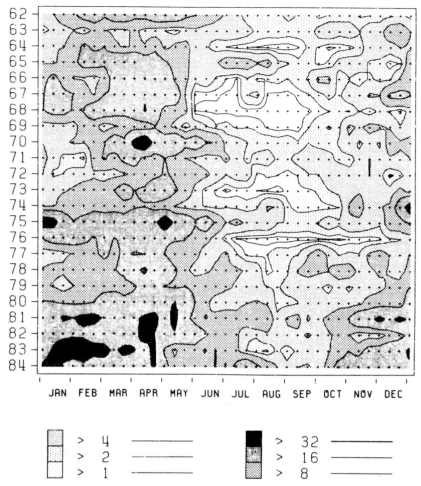

Fig. 8. Seasonal changes in nitrate concentrations (mmol N m^{-3}) from 1962 to 1984 in the North Sea off Helgoland. (Source: Radach *et al.*, 1990, by permission.)

(Cederwall and Elmgren, 1990; Wulff *et al.*, 1990) and the Adriatic (Degobbis, 1989). The effect of higher nutrient levels as one possible cause of more frequent and extensive phytoplankton blooms is relatively well documented (Paerl, 1988; Smetacek *et al.*, 1991), although not all such events can be attributed directly to eutrophication (Richardson, 1989). By contrast, the responses of benthic algae, especially in estuaries, to nutrient loading are much less well known.

The adverse impacts of phytoplankton blooms result from toxic effects on animals (Underdal *et al.*, 1989), from low palatability for gelatinous species such as *Phaeocystis* (Lancelot *et al.*, 1987), or from bottom anoxia caused by sinking and decomposition of organic matter (Justic *et al.*, 1987; Rosenberg, 1990; Wassmann, 1990). The occurrence of such blooms is often linked to increased availability of inorganic nitrogen, which is widely considered to limit primary production in marine waters, but changes in nutrient ratios are likely to be a significant factor in certain situations, particularly where depletion of silicate causes diatoms to be replaced by other types of algae (Officer and Ryther, 1980).

Processes by which nutrients might be removed from the water column are important for predicting the long-term effects of eutrophication, especially for semi-enclosed bodies of water with long residence periods. For example, there is increasing evidence both from field observations and from controlled experiments (Ronner, 1985; Smith *et al.*, 1989; Nowicki and Oviatt, 1990) for the loss of significant proportions of nitrate-nitrogen as a result of denitrification, which must be accounted for in nitrogen budgets.

As summarized in UNEP (1990), the contaminants of most concern in the marine environment include heavy metals, synthetic organic compounds, radionuclides and hydrocarbons. Plastic litter is also a considerable problem in some areas. Many detailed studies of the fate and effects of such compounds have been carried out in rivers, estuaries and coastal seas (e.g. Kullenberg, 1986; Pentreath, 1987; Salomons *et al.*, 1988; Allan *et al.*, 1990), and regional summaries of the status of marine environments published (e.g. Wright and Phillips, 1988; Phillips and Tanabe, 1989). Attention is now turning to the development of physiological and ecological techniques for the early detection of marine pollution (Bayne *et al.*, 1988), and recent work suggests that these will become of considerable practical value for environmental monitoring (Gray *et al.*, 1990). One important conclusion of the UNEP report (p. 54) is that "Early views, that there is no demonstrable causal link between human disease and bathing in contaminated sea water, can no longer be supported". Remarkably little information on this aspect of marine pollution has been published.

How much additional organic matter is transferred from land to sea as a result of human activities is not known, although Berner (1989) suggests that burial rate of organic carbon in nearshore environments may have doubled in the last 200 years mainly as a result of losses from soils. If such terrestrially derived material is largely refractory then the transfer from soil to marine sediments will have no overall affect on atmospheric CO_2. However, it seems more likely that some significant fraction is oxidized with the tendency to increase the net efflux of CO_2 to the atmosphere. The possible effects on carbon burial of higher sedimentation rates due to increased riverine sediment discharges with forest clearance and of enhanced productivity due to eutrophication have also been discussed (Deuser, 1979; Walsh, 1989), but direct evidence that these processes represent a significant sink for anthropogenic CO_2 is still lacking.

D. Ecosystem Modification

A combination of over-exploitation of living resources, coastal development and pollution has led to the modification and loss of marine habitats in many coastal regions. The worst affected areas are tropical coasts with high population densities such as Southeast Asia (Pauly and Thia-Eng, 1988; Gomez, 1988), and temperate coasts close to strong sources of nutrients and pollutants (Kempe, 1988; Elmgren, 1989). Few baseline ecological surveys have been carried out in the past so that objective assessments of the extent and types of change are usually not possible, although quantitative analyses of energy flows within ecosystems do enable general conclusions to be reached about the overall impact of man on marine food chains (Jansson and Jansson, 1988; Elmgren, 1989).

More is known about how man influences particular types of plant and animal communities. Thus coral reefs are damaged by various environmental changes including mining, eutrophication and various activities that affect turbidity of the water and sedimentation (Kuhlmann, 1988; Rogers, 1990; Brown et al., 1990). Both salt marshes and mangroves (see Fig. 1) have been exploited by man for a long time (Queen, 1977; Walsh, 1977), but only recently has some understanding been reached of how disturbance of these important margin ecosystems affects various aspects of coastal ecology including fish survival and recruitment (Twilley, 1988; Fortes, 1988).

As on land the introduction of species to new coastal regions can have marked ecological effects. Recently documented examples include the spread of Asian clams in San Francisco Bay (Carlton et al., 1990) and of the brown seaweed *Sargassum muticum* from Japan along European coasts (Rueness, 1989).

Studies on the effects of ultra-violet irradiation on marine organisms have indicated that ozone depletion in the upper atmosphere could deleteriously affect coastal marine ecosystems in a number of ways (Worrest and Hader, 1989), although the overall implications remain largely speculative.

IV. EFFECTS OF CLIMATE CHANGE

A. Natural Variations in Climate

Most studies of the effects of climate change on coastal ecosystems have concerned variations in the distributions either of particular species or of community types. There are many documented examples of marine organisms becoming more or less abundant at sites close to their limits of distribution, and these can often be correlated to trends in a certain climatic parameter such as temperature. However, the causes of shifts in community structure are generally much harder to identify precisely (Southward, 1980), partly because it appears that rather subtle changes in climatic conditions can induce large ecological changes that reflect the sensitive dynamical nature of marine food chains to climate and to climate-dependent factors such as nutrient levels and salinity, and partly because the direct effects of climate are difficult to distinguish from those of the activities of man.

Much of the information on the effects of climate change comes from long-term studies of plankton (Colebrook, 1986) and of fisheries (Bardach, 1989; Corten, 1990). Although unambiguous interpretation of the observations remains difficult, detection of consistent patterns of relationship between changes in marine biota and in climate are beginning to allow the development of empirical predictive models (McGowan, 1990). Records of change over periods of 100 years or longer, based either on historical information (Southward *et al.*, 1988) or on sediment records (Soutar and Isaacs, 1974; Lange *et al.*, 1990) will become increasingly important for testing hypotheses on the impacts of climate change on coastal ecosystems.

Two important conclusions can be drawn from recent studies of climate-induced changes on coastal ecosystems. Firstly, whole food chains can be affected as shown by Aebischer *et al.* (1990) who described parallel trends in data for plankton, fish and seabirds from the North Sea (Fig. 9). Secondly, climate changes at the scale of ocean basins influence coastal seas over comparable scales; thus El Nino events, apart from changes to the Peruvian fishery, are thought to influence fisheries off Australia (Harris *et al.*, 1988) and plankton off California (Lange *et al.*, 1990). The first attempts have now been made to predict the impacts of climate change on fish and invertebrate populations in coastal waters (Bakun, 1990; Frank *et al.*, 1990). Despite many uncertainties in the models, concerning both the mechanisms by which physical parameters influence biological processes and the non-linear nature of biological interactions within perturbed systems, significant changes in the distribution and abundance of commercially important species are expected. Examples of such changes have already been observed (Bardach, 1989), although the nature of links to climate change has not always been established.

B. Temperature

Temperature influences the growth and distribution of marine organisms both directly in relation to physiological processes (i.e. characteristic temperature-dependent growth and reproductive behaviour) and indirectly through temperature (and salinity) effects on water-column stratification and seasonal ice cover. Climate warming due to the greenhouse gases is expected to be most marked at high latitudes, especially in the northern hemisphere where

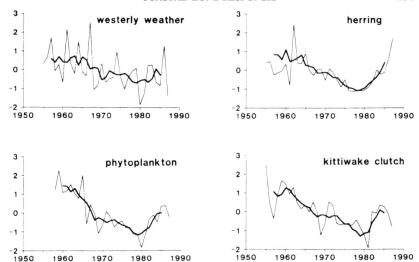

Fig. 9. Data for the NE Atlantic region showing relative changes in frequency of westerly weather, in abundances of phytoplankton and herring, and in kittiwake chick production from 1955 to 1987. (Source: adapted from Aebischer *et al.*, 1990, by permission.)

large changes in the extent of the arctic pack ice and permafrost have been forecast, and is likely to affect most strongly the extensive coastal and shelf sea areas of northern Canada and USSR. Other expected responses include poleward movements of frost-sensitive communities such as mangroves, and of warmer water species into environments presently characterized by anomalously cold water. A good example of the latter is the Gulf of Maine where relatively large fluctuations in bottom water temperature and in the distribution of certain temperature-sensitive species at the edge of their geographical distributions have occurred this century (Frank *et al.*, 1990).

Exposure of marine ecosystems to damaging high temperatures is also likely in connection with a warming trend or with an increase in the frequency of anomalously high water or air temperatures. Coral damage has been reported in the tropics during El Nino events (Warwick *et al.*, 1990). Comparable effects could occur at any latitude, and intertidal organisms are likely to be particularly sensitive.

Some reported changes in fisheries have been related to variations in the length of summer stratified conditions and in summer maximum temperatures (e.g. Harris *et al.*, 1988). Temperature changes may affect reproductive performance, the availability of food, predation, and the timings of migrations to and from spawning and nursery areas (Frank *et al.*, 1990), so that even in relation to just a single climatic parameter biological predictions will continue to be speculative.

C. Wind

Wind has a strong influence on both upwelling and seasonal stratification in coastal seas, affecting pelagic productivity through nutrient and light availability for phytoplankton growth. The correlation between decreases in the frequency of westerly winds and in plankton biomass over the last 30 years for the eastern North Atlantic (Aebischer et al., 1990) has been attributed to later stratification in the spring, and therefore delayed plankton growth, due to increased mixing by northerly winds (Dickson et al., 1988). The same change in the atmospheric pressure field is also linked to intensification of coastal upwelling off the Iberian peninsula further to the south and to variations in fish stocks (Dickson et al., 1988).

The relationship between climate warming, wind and coastal upwelling has been examined by Bakun (1990) who postulates that greater thermal and pressure contrasts between the land and ocean at mid-latitudes will enhance upwelling and affect coastal fisheries. However, the nature of the relationship between wind and fish abundance is complex and not yet fully amenable to numerical modelling studies.

Such effects are essentially regional in scale although parallel changes may occur in all the main ocean basins. Also it is the spatial patterns of marine productivity that are varying and not necessarily total productivity. When the other responses to climate change are taken into account the difficulties of predicting global marine fish production are only too apparent.

D. Extreme Events

Quantitative observations on the effects of extreme climatic events on the coastal zone and its resources are difficult to make for various technological and logistic reasons, especially for the low-frequency events. The range of time-scales of variability in estuarine systems is illustrated in Fig. 10 for a number of environmental parameters. By way of example, our understanding of the impact of major floods or storms occurring once every few years on estuarine flushing and sediment dynamics is extremely poor (see discussion by Allen, 1990).

The importance of extreme events for coastal dynamics is well illustrated by the study of Dolan et al., (1990) who show that a single storm on the Atlantic coast of North America lasting < 5 days had the sand transport potential equivalent to 66% of the total for an average year. Comparable effects of gales on sea salt deposition at inland sites have been described for southern Sweden (Franzen, 1990). Studies of coastal sediment deposits have demonstrated large-scale alterations to coastal geomorphology and ecosystems in the past due to climate or climate-related (e.g. sea-level) changes.

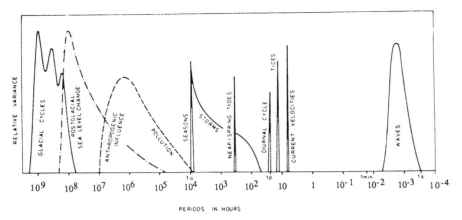

Fig. 10. Time-scales of variability in estuaries due to environmental factors. (Source: Kempe, 1988, by permission.)

Perhaps the most spectacular are those caused by tsunamis (Smith and Dawson, 1990).

A more familiar consequence of climate variability is temperature extremes which, in the coastal zone, can severely affect nearshore ecosystems that are normally protected by the ameliorating influence of the sea. The effects of high temperatures on corals associated with "El Nino" events have already been noted (Warwick *et al.*, 1990), and it is likely that other types of coastal communities will be similarly affected especially within shallow water and intertidal zones under conditions of anomalous warmth.

V. EFFECTS OF SEA-LEVEL CHANGE

Recent reviews have described the causes and consequences, both physical and economic, of sea-level rise (Hekstra, 1989, Vellinga and Leatherman, 1989; Broadus, 1989). Despite large regional differences in the change of relative sea level (in some areas it is falling rather than rising due to continuing postglacial isostatic adjustment), there is general agreement that sea level is rising at a rate significantly faster than the mean for the late Holocene due to a combination of thermal expansion of sea water and melting of ice as the climate warms, and that the rate of rise will increase in the future. It should be noted that the effects of a rise in mean sea level are rather different from those of a rise in extreme high water levels (i.e. surge height) under particular climatic conditions. Another important factor especially in estuaries may be increases in tidal amplitude related to natural or anthropogenically-caused changes in bathymetry (Kempe, 1988).

A. Deltas, Estuaries

Severe impacts of sea-level rise are already being felt in delta and estuarine regions (e.g. Milliman *et al.*, 1989) partly because they are low lying and heavily populated, and partly because they are strongly perturbed by human activities. These lead to contamination of water and sediments by various pollutants, to reduced rates of sediment accumulation, and to enhanced subsidence and erosion of existing sediment deposits. The direct effects of sea-level rise vary from one locality to another but are now severe in most major delta systems including the Mississippi (Day and Templet, 1989), Nile and northern Bay of Bengal (Milliman *et al.*, 1989) resulting in the loss of agricultural land as well as wetlands with valuable natural resources, and in the salinization of fresh water. Similar changes are occurring in estuaries (Kempe, 1988).

Reliable prediction of future impacts will depend on a much better understanding of the processes controlling the sediment delivery and ecology of deltas and estuaries (e.g. Wells and Coleman, 1987; Madden *et al.*, 1988; Allen, 1990), and on an integrated approach to environmental management of coastal wetlands (Ruddle, 1987; Day and Templet, 1989). The latter must assume adaptive exploitation by humans rather than attempt to meet fixed needs as ecosystems are modified by environmental change.

B. Coasts, Islands

Inundation of low-lying islands (Wells and Edwards, 1989) and coastline regression (Hekstra, 1989) are immediate effects of sea-level rise. Rates of submergence depend both on local coastal processes and on the speed and duration of the increase in sea level.

For coral islands the growth of the reefs can either keep pace or catch up with limited sea-level rise. Buddemeier and Smith (1988) estimate that corals in general have a maximum upward growth rate of about 10 mm year^{-1} which is somewhat greater than the most recent estimates (~ 5 mm year^{-1}) of sea level rise over the next few decades. However, other restraints such as reef mining and increasing turbidity of the surrounding waters will only serve to reduce the capacity of reefs to prevent island erosion.

For barrier islands and beaches the balance between sediment supply and removal is critical, particularly in cases where the geomorphology of the coastline is being adapted by man (Bruun, 1989). Reduced sediment supply to coastal waters due to lower estuarine inputs or to protection of beaches and eroding cliffs can have serious impacts on coastline dynamics over distances of 100 km or more in the down-drift direction (Clayton, 1989). The best policies for dealing with beach erosion depend on local conditions, ranging from no action to protective structures or nourishment procedures.

C. Coastal Ecosystems

Natural communities of plants and animals play a crucial role in determining the response of the coastal zone to changes in sea level. Coastal conservation (e.g. Hatcher *et al.*, 1989; Kenchington and Agardy, 1990) probably represents the most effective method of minimizing the adverse impacts of sea-level rise. Biogeomorphological processes (Spencer, 1988) include sediment supply through biomineralization (particularly by organisms that form calcium carbonate) and biological erosion (boring and grazing organisms on rock surfaces), and sediment stabilization and binding by benthic animals and plants (algae, seagrasses) and by mangrove, saltmarsh and sand dune communities.

Saltmarshes (Stevenson *et al.*, 1988) and mangrove forests (Fortes, 1988; Twilley, 1988; Hekstra, 1989) are sites of active sediment accumulation and are able to accommodate sea-level rise, at least within certain limits, provided their growth is not inhibited by drainage or cutting. As sea level rises there will be a tendency for salt to penetrate coastal soils and groundwater so that the maintenance of such plant communities may depend on the availability of space for less salt-tolerant species to spread inland. In situations where agriculture or other types of land use create fixed boundary conditions such adaptation is likely to be inhibited, threatening the survival of such plant communities.

Benthic algae (Admiraal, 1984; Charpy-Roubaud and Sournia, 1990) have an important stabilizing influence on intertidal and subtidal mud and sand surfaces but much less is known about how their growth is affected by, and responds to, changes in environmental conditions, including the rate and type of sediment supply and estuarine eutrophication and pollution. These organisms are also a major source of organic matter for benthic animals which burrow and sort sediments so that changes in their productivity are likely to have far-reaching implications for predicting the effects of sea-level rise on coastal sedimentary environments.

VI. RESPONSES OF THE COASTAL ZONE TO GLOBAL CHANGE

A. Geological Perspective

From a geological perspective, the changes in the coastal zone that may result from future fluctuations in climate and sea-level fluctuation are not without precedent. Rates of upland and shoreline erosion and rates of sediment deposition in deltas and estuaries have always been sensitive to changes in the hydrological cycle (Starkel, 1989) and in sea level, and mass extinctions of invertebrates are associated with marine transgressions (Hallam, 1989). Global change is expected to lead to new quasi-equilibria in the geomorphological, ecological and biogeochemical properties of the coastal zone to which humankind will adapt. One option is the building of engineering structures to protect river terraces, shorelines, bays, deltas, marshes, etc., which may provide short-term solutions to the impacts of change (Hekstra, 1989), but introduce the risk of environmental disaster when they fail.

The immediate future holds two possible departures from the past concerning the rate and type of change. The predicted rates of temperature increase and sea-level rise are rapid even compared to events at the end of the last glaciation and certainly lie beyond human experience during the late Holocene historical period. For example, reconstructions of sea level (e.g. Ters, 1987) suggest that a consistent global rise > 5 mm year $^{-1}$ has not occurred during the last 8000 years. Other human-caused effects such as the release of toxic wastes into coastal waters are unparalleled in geological history. However, the transport of large quantities of suspended matter and inorganic nutrients by rivers as a result of deforestation and agricultural development may be somewhat similar to conditions during the onset of glaciation, at least at higher latitudes, when coastal sediments would have have been exposed to erosion as sea level fell.

B. Scale and Pervasiveness of Components of Change

The various aspects of global change cover a very wide range of spatial and temporal scales making evaluation of impact and risk an extremely complex issue. To illustrate this problem the range of time-scales of environmental forcing in estuarine systems is shown in Fig. 10. All the processes indicated in this diagram are to some extent now influenced by man, either indirectly through climate and sea-level effects or directly as a result of coastal engineering, freshwater management, pollution, and so on.

The spatial scales of anthropogenic impacts in the coastal zone are less easy to specify. The increase in greenhouse gases is a truly global phenomenon because of rapid mixing in the atmosphere, and the secondary climatic effects as well as sea-level rise (itself a derivative of climate warming) are generally considered as global issues. However, for different reasons they both show strong regional variations so that some parts of the globe remain relatively unaffected (Dickinson, 1987). Even greater uncertainties stem from a consideration of the importance of episodic events as opposed to changes in mean conditions. Thus changes in storm frequency and intensity, which are known to have great impacts on coastal ecosystems, may occur globally but show strong regional differences.

In contrast to the effects of climate and sea level, other human influences on the coastal zone have generally been considered as localized on account of the uneven distributions of population density and types of human activity. On the other hand, with the rapid increase over the last few decades in the number of people living in the coastal zone and greater awareness of the widespread degradation in coastal environments (UNEP, 1990), the effects of changes in land use, eutrophication, overfishing, discharge of wastes and perturbation of biogeochemical cycles are now rightly considered as global problems in terms both of their overall social and economic impacts and of their potential importance in relation to the control the of the global environment (Fig. 3). An indicator of the global significance of man's activities is the possible two to three-fold increase during historical times in the mass of material moved by rivers (see Glasby, 1988). Only when there is a comprehensive global basis for assessing the causes and consequences of change in the coastal zone, as is beginning to be developed for the continental land masses and ocean basins, will it be possible to make sound decisions on critical management issues.

C. Ecological Responses

Offshore pelagic ecosystems are constantly being affected by variations in climate and, more recently, by over-exploitation of fish stocks. The changes in biomass of plankton and fish can be very marked, with particular fisheries appearing and disappearing in time. Although such events are spectacular, as illustrated by El Nino events in the Pacific, they are regional in nature and reflect the variability of the ocean–atmosphere system. Any warming of the climate will tend to affect the positions of boundaries between different coastal water masses especially in relation to cold currents, with corresponding shifts in biogeographic boundaries and fishery distributions (Frank *et al.*, 1990).

Inshore waters and coastal habitats are directly influenced by climate, and also by sea-level changes, by climate-dependent processes on land and by a wide range of human activities. Few if any coastal ecosystems are now unaffected by global change, and all recent observations suggest that ecological modifications are increasing in extent and severity in parallel with population pressure. Coastal wetlands and estuaries are most seriously threatened due to habitat loss associated with coastal development (including agriculture and mariculture practices—see Ruddle, 1987), to changes in freshwater flows (Skreslet, 1986), and to pollution (UNEP, 1990). The main global concerns are loss of biodiversity, reduced value of living resources, and the practical difficulties of ecosystem restoration. The latter problem stems from the realization that coastal ecosystems such as mangrove forests and seagrass beds play a vital role both in the life cycles of many marine fish and in preventing coastal erosion.

Stress and eutrophication of marine communities lead to the disappearance of some species and increased dominance by others (e.g. Gray et al., 1990). Energy transfer within ecosystems becomes less efficient (Glasby, 1988; Jansson and Jansson, 1988) so that biomass yields at higher trophic levels are much reduced. Solutions to these problems through ecological management will only be acquired through a better understanding of how coastal marine ecosystems function.

D. Biogeochemical Responses and Feedback

Variations in the atmospheric concentrations of greenhouse gases, particularly CO_2, ultimately underlie many of the global changes discussed in this chapter. Feedback effects on the climate system are possible if the impacts of global change on the coastal zone have a significant influence on global carbon fluxes (Fig. 6). Over time-scales of decades to centuries it appears that global change will modify the carbon cycle more strongly in coastal environments than in the oceans but whether this will lead to an increase or decrease in the capacity for carbon storage (i.e. negative or positive feedback with respect to the atmosphere) is not at all certain.

There is agreement that the coastal ocean is a net source of atmospheric CO_2 but precise estimates of the total flux or of changes in the flux are not available. The main unknown factors concern the fate of terrestrial organic matter in the sea (Ittekkot & Haake, 1990), the rates of accumulation of organic matter (terrestrial + marine) in estuarine and slope sediments (Berner, 1989; Walsh, 1989), and rates of calcium carbonate formation and dissolution on shelves especially at higher latitudes. All have been altered by variations in land discharges of organic carbon, dissolved inorganic carbon, inorganic nutrients and sediments, and by modification of biogeochemical processes in coastal sediments in response to changes in sedimentation rates and to ecotoxicological effects on benthic organisms, but there is insufficient information to quantify these effects globally. In general it appears that coastal zones are likely to become stronger sources of CO_2 as land use changes and marine erosion tend to recycle organic carbon presently stored in soils and coastal sediments.

Other trace gases such as N_2O, NO and methane produced in nitrogen-rich, or low-redox zones of coastal ecosystems, including agriculturally developed coastal plains, contribute to the greenhouse effect. Although evidence that the coastal zone is a major source of these gases is lacking, reconfiguration of coastal ecosystems through sea-level rise, the mobilization of organic matter, and the penetration of salt into freshwater systems could change emission rates. In the long term melting of permafrost in arctic coastal zones may result in the degassing of presently trapped clathrate methane (MacDonald, 1990). Again positive feedback on the levels of greenhouse gases in the atmosphere seems most likely.

Coastal eutrophication is probably causing increased emissions of dimethylsulphide (DMS), especially as certain bloom-forming phytoplankton species such as *Phaeocystis* are known to be major sources. In terms of potential effects on the terrestrial biosphere related to acid deposition or to climatic phenomena associated with the formation of cloud condensation nuclei (Charlson *et al.*, 1987), variations in coastal, as opposed to ocean, sources are likely to be particularly important since the residence time of sulphur gases in the lower atmosphere is relatively short.

Knowledge of the role and importance of the coastal zone in global biogeochemical cycles is still very limited. However, the high rates of biological productivity and the active transformations of terrestrial organic matter in this region indicate that coastal ecosystems contribute significantly to the total fluxes of key elements such as C, N and S.

VII. MANAGEMENT OF THE COASTAL ZONE

Coastal environments and living resources have a limited capacity for exploitation by man (Ray, 1989) and are being irreversibly damaged in many parts of the world. Effective management policies need to be based on an interdisciplinary approach to understanding how coastal ecosystems work and on minimal disturbance to the natural physical and biogeochemical pathways of energy flow (e.g. Jansson and Jansson, 1988; McManus, 1988; Day and Templet, 1989). Detailed studies of particular marine ecosystems that have been strongly perturbed by man, such as the Baltic Sea, have demonstrated the complexity of ecological problems caused by eutrophication and other factors (Jansson and Jansson, 1988). Understanding how sediment transport and coastal geomorphology in estuaries and delta systems is affected by land use is equally challenging. In such cases a combination of observational studies, simulation modelling and monitoring will be required to establish optimal procedures for regulating man's activities and allow sustainable exploitation of resources.

An extensive literature has developed recently on various aspects of coastal-zone management, including the land margin (Cendrero, 1989; Charlier, 1989), mariculture (Fernandez-Pato, 1989) and fish stocks (Pope, 1989, Wilson *et al.*, 1990), and problems concerning the resolution of conflicting interests (Johnson and Pollnac, 1989; Charlier, 1989). Also statements on national policies for coastal management have been published—e.g. for Norway (Andresen and Floistad, 1988). Two common themes emerge; coastal pollution has to be controlled within strict limits, and coastal development curtailed especially in relation to tourism. If these steps are not taken coastal resources will not be sustainable and coastal ecosystems will have little capacity to respond to global changes in climate and sea level. Considerable advances have been made in assessing the impacts of sea-level rise (Broadus, 1989) although great uncertainties remain as to how to deal with it.

Various types of coastal models are being developed, including ones for catchment basin studies of pollutants (Krysanova et al., 1989), coastal landscape dynamics (Costanza et al., 1990), ecosystem functioning (Lindeboom et al., 1989), and coupled physical–biological interactions in coastal oceans (Hofmann, 1991). In each case realistic simulations are being achieved, giving encouragement that the models will be useful for predictive purposes and deciding between different management options.

Environmental monitoring is an integral part of assessing the results of predictive environmental models and management programmes. Satellite remote sensing is used increasingly for large-scale ecological and environmental studies of the coastal zone (Holligan et al., 1989; Dutrieux et al., 1990; Jensen et al., 1990), whereas various types of biological monitoring (Stebbing and Harris, 1988; Gray et al., 1990) are now employed for pollution impact studies. Other new methods include optical sensors to determine the biological and physical properties of marine sediments (O'Connor et al., 1989), and a range of in situ instrumentation for continuous measurements of hydrographic variables. Such technological and analytical advances will greatly facilitate the evaluation of coastal management procedures.

VIII. RESEARCH NEEDS

A general objective for studies of global change in coastal zones might be to develop a predictive understanding of the global impacts of changes in land use, climate and sea level on coastal ecosystems, with special emphasis on the interactions between changing conditions on land and at sea, and on possible feedback effects to the physical environment.

Research priorities to meet such an objective should include studies of:

• the effects of changes in delivery of fresh water, nutrients and sediments from land to sea on coastal geomorphology and ecology,
• the effects of change in climate and freshwater runoff on the physical dynamics of the coastal ocean (circulation, mixing, location of fronts, etc.),

- the ecological and biogeochemical impacts of changes in the inputs and fate of sediments, organic matter and contaminants in the coastal zone, and
- the effects of rising sea level and changing storm frequency and intensity on the land margin.

The coastal zone is the arena in which land and sea interact and, in order to address changes on the coastal zone, strong links with terrestrial and oceanic (and also atmospheric) global research programmes are vital. Additional specific requirements of a coastal programme are likely to include:

1. A Geographic Data Base and Classification System

Geographically referenced data for world coastal zones will be required for developing an understanding of the dynamic properties of coastal ecosystems and for making predictions about the impacts of global change. Important relevant parameters fall into the following categories:

- geomorphological data from the landward margin of the coastal plain to the continental slope,
- hydrographic data including residual currents, tides, locations of frontal boundaries, and distributions of temperature, salinity, nutrients (N, P, Si in their several forms), inorganic particulates in terms of size class distributions and total bulk, and organic matter,
- meteorological data including solar radiation, air temperatures, humidity, wind and precipitation,
- biological data including type, biomass, and productivity of the main plant, microbial and animal communities.

Although varying degrees of spatial/temporal coverage and accuracy are inevitable, such data are needed for identifying gaps in information, for the support of intensive process studies (see below), and as a basis for model-based predictions of the impacts of global change on coastal zones.

The world's coastal ecosystems will never be fully described through direct observations and measurements. In order to extrapolate and interpolate from well-understood areas, it will be necessary to apply results from designated classes of conditions to other geographic areas belonging to those same classes. Such procedures necessitate a broadly accepted classification system for the variables that characterize coastal zones and oceans. Development of a multivariate classification system must be directly linked with a global data base.

2. Process Studies

Process studies are designed to determine by quantitative experimentation the cause and effect of particular phenomena. Most research work in the coastal zone already involves site-specific process studies. Within the context of a co-ordinated coastal zone project they would be directed towards critical analysis of spatial and temporal variability, the effects of changing environmental conditions, and the development of predictive models.

Examples of process studies that might be included in such a programme are the elucidation of net sediment deposition and erosion patterns with altered river flows, the impacts of eutrophication on primary production in the coastal zone, the effects of contaminants on decomposition and burial rates of organic carbon, or the relationship between beach erosion and wave characteristics. In each case new observational and modelling techniques are leading to significant advances in understanding the dynamic properties of coastal zones.

3. Integrated, Large System Studies

At some stage, the various phenomena that drive coastal zone systems have to be addressed in an integrated way. One approach is to undertake large-scale, interdisciplinary studies based on catchment basin–coastal zone units. A limited number of such projects in contrasting coastal regions would have two advantages. First, they would provide a new level of understanding of land–ocean interactions for a range of geomorphological and ecological conditions. Second, and more important from a scientific point of view, they would enable interpolative rather than extrapolative approaches to quantitative assessments of the whole coastal zone.

4. Model Development and Testing

Development of predictive capacity is an underlying philosophy of global change science. Observations alone, while they eventually yield understanding, do not provide quick answers to social needs and are not the most effective means of doing environmental science.

Given the need for prediction, the enormous variability of coastal zones globally, and the broad range of possible scenarios for change, the only recourse is through numerical modelling. Model development must be fundamental to any co-ordinated research effort, from the level of single process studies to predictions of the integrative effects of global change. The following kinds of problems are of particular importance:

- modelling of interactive processes across appropriate time and space scales,
- scaling up from rate measurements to models of regional or global fluxes,
- modelling the direction and strength of feedback processes,
- developing methods and acquiring field data for model evaluation,
- providing linkages between different types of models (hydrodynamic, ecosystem, water quality, etc.)

5. International Co-ordination

A number of coastal zone research programmes are being undertaken by individual countries and by United Nations agencies, and more are being planned. Co-ordination between these programmes is limited, given the global nature of many coastal problems and the general applicability of simulation and predictive models for the coastal zone.

Better collaboration is also needed between terrestrial, marine, and atmospheric scientists studying global environmental problems in coastal zones. The International Geosphere–Biosphere Programme (IGBP, 1990) provides the opportunity to plan and execute new interdisciplinary studies of this key component of the earth's biosphere.

IX. CONCLUSIONS

As stated by Oppenheimer (1989), the coastal zone "will bear the brunt of the combined effects" of environmental pollution, climate and sea-level change, and enhanced ultraviolet radiation. We must recognize that such problems cannot be eliminated by environmental management, only ameliorated (Glasby, 1988), and that in developing new management policies "our abilities to control the ecological processes of the coastal zone are still negligible" (Ray, 1989). Strategies for the use of natural resources of the coastal zone must be based on a proper understanding of the geomorphological, ecological and biogeochemical processes that characterize the dynamic boundary between land and sea, and supported by appropriate predictive models of the impacts of global change on coastal ecosystems (Costanza *et al.*, 1990).

Even with the best scientific understanding and models, progress on these issues will continue to be hampered by social and political barriers in many countries that prevent long-term planning for the protection of coastal environments, the proper means for curbing coastal pollution (including allocation of responsibility), and the resolution of conflicts over use of the coastal-zone resources. Management plans require better information than is presently available about the rate at which the coastal zone is being altered, and about how it is likely to be further modified over the next few decades. Such predictions should include an objective assessment of the socio-economic implications of change.

Any management strategy for the coastal zone must include conservation of coastal wetlands (mangroves, saltmarsh, lagoons) and nearshore benthic communities (mudflats, reefs, etc.). Against a background of predicted changes in the hydrology, sea level and storm frequency, these ecosystems play a crucial role in maintaining the physical stability of the coastal zone as well as in the global cycling of elements such as C, N and S.

ACKNOWLEDGEMENTS

The second author acknowledges the support of the International Geosphere–Biosphere Programme and the Wissenschaftskolleg zu Berlin for
financial support that enabled the preparation of this chapter.

REFERENCES

Admiraal, W. (1988). The ecology of estuarine-inhabiting diatoms. In: *Progress in
Phycological Research, Vol. 3* (Ed. by F.E. Round and D.J. Chapman), pp. 269–
322. Biopress, London.
Aebischer, N.J., Coulson, J.C. and Colebrook, J.M. (1990). Parallel long term trends
across four marine trophic levels and weather. *Nature* **347**, 753–755.
Allan, R.J., Campbell, P.G.C. Forstner, U. and Lum, K. (Eds) (1990). *Fate and
Effects of Toxic Chemicals in Large Rivers and Their Estuaries. Sci. Total Environ.*
97/98. Elsevier, Amsterdam.
Allen, J.R.L. (1990). The Severn Estuary in Southwest Britain: its retreat under
marine transgression, and fine sediment regime. *Sediment. Geol.* **66**, 13–28.
Andreae, M.O. (1986). The ocean as a source of atmospheric sulfur compounds. In:
The Role of Air–Sea Exchange in Geochemical Cycling (Ed. by P. Buat–Menard),
pp. 331–362. *NATO ASI Series C*, **185**. Reidel, Dordrecht.
Andresen, S. and Floistad, B. (1988). Sea use planning in Norwegian waters: National
and international dimensions. *Coastal Mgmt* **16**, 183–200.
Bakun, A. (1990). Global climate change and intensification of coastal upwelling.
Science **247**, 198–201.
Barber, R.T and Smith, R.L. (1981). Coastal upwelling ecosystems. In: *Analysis of
Marine Ecosystems* (Ed. by A.R. Longhurst), pp. 31–69. Academic Press, London.
Bardach, J.E. (1989). Global warming and the coastal zone. *Clim. Change* **15**, 117–
150.
Bayne, B.L. Clarke, K.R. and Gray, J.S. (Eds) (1988). Biological effects of pollutants.
Results of a workshop. *Mar. Ecol. Prog. Ser.* **46**, 1–278.
Berner, R.A. (1989). Biogeochemical cycles of carbon and sulfur and their effect on
atmospheric oxygen over phanerozoic time. *Palaeogeogr. Palaeoclimatol. Palaeoe-
col.* **75**, 97–122.
Berner, E.K. and Berner, R.A. (1987). *The Global Water Cycle*. Prentice-Hall,
Englewood Cliffs, New Jersey.
Billen, G., Lancelot, C. and Meybeck, M. (1991). N, P and Si retention along the
aquatic continuum from land to ocean. In: *Ocean Margin Processes in Global
Change* (Ed. by R.F.C. Mantoura, J.-M. Martin and R. Wollast), Wiley, Chiches-
ter.
Bird, E.C.F. (1985). *Coastline Changes—A Global Review*. Wiley Interscience, Chi-
chester.
Blanton, J.O. (1991). Circulation processes along oceanic margins in relation to
material fluxes. In: *Oceanic Margin Processes in Global Change*. (Ed by R.F.C.
Mantoura, J.-M. Martin and R. Wollast), pp. 145–164. Wiley, Chichester.
Broadus, J.M. (1989). Impacts of future sea level rise. In: *Global Change and Our*

Common Future. Papers from a Forum (Ed by R.S. DeFries and T.F. Malone), pp. 125–138. National Academy Press, Washington, DC.

Brink, K.H. (1987). Coastal ocean physical processes. *Rev. Geophys. Space Phys.* **25**, 204–216.

Brown, B.E., Le Tissier, M.D.A. Scoffin, T.P. and Tudhope, A.W. (1990). Evaluation of the environmental impact of dredging on intertidal coral reefs at Ko Phuket, Thailand, using ecological and physiological parameters. *Mar. Ecol. Prog. Ser.* **65**, 273–281.

Bruun, P. (1989). Coastal engineering and use of the littoral zone. *Ocean & Shoreline Mgmt* **12**, 495–516.

Buddemeier, R.W. and Smith, S.V. (1988). Coral reef growth in an era of rapidly rising sea level: Predictions and suggestions for long-term research. *Coral Reefs* **7**, 51–56.

Burton, J.D. (1988). Riverborne materials and the continent–ocean interface. In: *Physical and Chemical Weathering in Geochemical Cycles* (Ed. by A. Lerman and M. Meybeck) pp. 299–320. Kluwer, Dordrecht.

Carlton J.T., Thompson, J.K. Schemel, L.E. and Nichols, F.H. (1990). Remarkable invasion of San Francisco Bay (California, USA) by the Asian clam *Potamocorbula amurensis*. I. Introduction and dispersal. *Mar. Ecol. Prog. Ser.* **66**, 81–94.

Carter, R.W.G. (1988). *Coastal Environments. An Introduction to the Physical, Ecological and Cultural Systems of Coastlines.* Academic Press, Orlando, Florida.

Cederwall, H. and Elmgren, R. (1990). Biological effects of eutrophication in the Baltic Sea, particularly the coastal zone. *Ambio* **19**, 109–112.

Cendrero, A. (1989). Land-use problems, planning and management in the coastal zone: An introduction. *Ocean & Shoreline Mgmt* **12**, 367–381.

Chapman, V.J. (Ed.) (1977). *Ecosystems of the World. 1. Wet Coastal Ecosystems.* Elsevier, Amsterdam.

Charlier, R.H. (1989). Coastal zone: Occupance, management and economic competitiveness. *Ocean & Shoreline Mgmt,* **12**, 383–402.

Charlson, R.J., Lovelock, J.R. Andreae, M.O. and Warren, S.G. (1987). Oceanic phytoplankton, atmospheric sulphur, cloud albedo and climate. *Nature* **326**, 655–661.

Charpy-Roubaud, C. and Sournia, A. (1990). The comparative estimation of phytoplanktonic, microphytobenthic and macrophytobenthic primary production in the oceans. *Mar. Micro. Food Webs* **4**, 31–57.

Christensen, J.P., Murray, J.W., Devol, A.H. and Codispoti, L.A. (1987). Denitrification in continental shelf sediments has major impact on the oceanic nitrogen budget. *Global Biogeochem. Cycles* **1**, 97–116.

Church, T.M. (1987). Advances in the chemistry of rivers, estuaries, microcosms and salt marshes. *Rev. Geophys.* **25**, 1431–1438.

Clayton, K.M. (1989). Sediment input from the Norfolk Cliffs, Eastern England—a century of coast protection and its effect. *J. Coast. Res.* **5**, 433–442.

Cloern, J.E. (1987). Turbidity as a control on phytoplankton biomass and productivity in estuaries. *Contin. Shelf Res.* **7**, 1367–1381.

Colebrook, J.M. (1986). Environmental influences on long-term variability in marine plankton. *Hydrobiology* **142**, 309–325.

Corten, A. (1990). Long-term trends in pelagic fish stocks of the North Sea and adjacent waters and their possible connection to hydrographic changes. *Neth. J. Sea Res.* **25**, 227–235.

Costanza, R., Sklar, F.H. and White, M.L. (1990). Modeling coastal landscape dynamics. *Bioscience* **40**, 91–107.

Cushing, D.H. (1988). The flow of energy in marine ecosystems, with special reference to the continental shelf. In: *Continental Shelves. Ecosystems of the World. 27* (Ed. by H. Postma and J.J. Zijlstra), pp. 203–230. Elsevier, Amsterdam.

Davies, J.L. (1972). *Geographic Variation in Coastal Development*. Oliver & Boyd, Edinburgh.

Day, J.W. and Templet, P.H. (1989). Consequences of sea level rise: Implications for the Mississippi Delta. *Coastal Mgmt* **17**, 241–257.

Degens, E.T., Kempe, S. and Richey, J.E. (1990). Summary: Biogeochemistry of major world rivers. In: *Biogeochemistry of Major World Rivers* (Ed. by E.T. Degens, S. Kempe and J.E. Richey). *SCOPE* **42**, 18–39. Wiley, Chichester.

Degobbis, D. (1989). Increased eutrophication of the northern Adriatic Sea. *Mar. Poll. Bull.* **20**, 452–457.

Deuser, W.G. (1979). Marine biota, nearshore sediments, and the global carbon balance. *Org. Geochem.* **1**, 243–247.

Dickinson, R.E. (1987). How will climate change? The climate system and modelling of future climate. In: *The Greenhouse Effect, Climate Change and Ecosystems* (Ed. by B. Bolin, B.R. Doos, I. Jager and R.A. Warrick). *SCOPE* **29**, 207–270. Wiley, NY.

Dickson, R.R., Kelly, P.M., Colebrook, J.M., Wooster, W.S. and Cushing, D.H. (1988). North winds and production in the eastern North Atlantic. *J. Plankt. Res.* **10**, 151–169.

Dolan, R., Inman, D.L. and Hayden, B. (1990). The Atlantic coast storm of March 1989. *J. Coastal Res.* **6**, 721–725.

Dronkers, J. (1988). Inshore/offshore water exchange in shallow coastal systems. In: *Coastal–Offshore Ecosystem Interactions. Lecture Notes on Coastal and Estuarine Studies. Vol. 22* (Ed. by B.-O. Jansson), pp. 3–39. Springer–Verlag, Berlin.

Duce, R.A. (1991). Chemical exchange at the air–coastal sea interface. In: *Ocean Margin Processes in Global Change* (Ed. by R.F.C. Mantoura, J.-M Martin and R. Wollast), pp. 91–110. Wiley, Chichester.

Dutrieux, E., Denis, J. and Populus, J. (1990). Application of SPOT data to a baseline ecological study of the Mahakam delta mangroves (East Kalimantan, Indonesia). *Oceanol. Acta* **13**, 317–326.

Dyer, K.R. (1986). *Coastal and Estuarine Sediment Dynamics*. Wiley, Chichester.

Eisma, D. (1988a). The terrestrial influence on tropical coastal seas. *Mitt. Geol.-Palaont. Inst. Hamburg* **66**, 289–317.

Eisma, D. (1988b). An introduction to the geology of continental shelves. In: *Continental Shelves. Ecosystems of the World. 27* (Ed. by H. Postma and J.J. Zijlstra), pp. 39–81. Elsevier, Amsterdam.

Elmgren, R. (1989). Man's impact on the ecosystems of the Baltic Sea: energy flows today and at the turn of the century. *Ambio* **18**, 326–332.

Fernandez–Pato, C. (1989). Mariculture developments; environmental effects and planning. *Ocean & Shoreline Mgmt* **12**, 487–494.

Fortes, M.D. (1988). Mangrove and seagrass beds of East Asia: Habitats under stress. *Ambio* **17**, 207–213.

Frank, K.T., Perry, R.I. and Drinkwater, K.F. (1990). The predicted response of northwest Atlantic invertebrate and fish stocks to CO_2-induced climate change. *Trans. Am. Fisheries Soc.* **119**, 353–365.

Franzen, L.G. (1990). Transport, deposition and distribution of marine aerosols over southern Sweden during dry westerly storms. *Ambio* **19**, 180–188.

Fricker, A. and Forbes, D.L. (1988). A system for coastal description and classification. *Coastal Mgmt* **16**, 111–137.

GESAMP (1987). Land–Sea Boundary Flux of Contaminants: Contributions from Rivers. Reports & Studies No. 32. UNESCO, Paris.

GESAMP (1989). Atmospheric Input of Trace Species to the World Ocean. Report and Studies No. 38. World Meteorological Organization, Geneva.

Glasby, G.P. (1988). Entropy, pollution and environmental degradation. *Ambio* 17, 330–335.

Gomez, E.D. (1988). Overview of environmental problems in the East Asian seas region. *Ambio* 17, 166–169.

Gray, J.S., Clarke, K.R., Warwick, R.M. and Hobbs, G. (1990). Detection of initial effects of pollution on marine benthos: an example from the Ekofisk and Eldfisk oilfields, North Sea. *Mar. Ecol. Prog. Ser.* 66, 285–299.

Griffin, D.A. and LeBlond, P.H. (1990). Estuary/ocean exchange controlled by spring–neap tidal mixing. *Estuar. Cstl. Shelf Sci.* 30, 275–297.

Halim, Y. (1991). The impact of man's alterations of the hydrological cycle on ocean margins. In: *Ocean Margin Processes in Global Change* (Ed. by R.F.C. Mantoura, J.-M. Martin and R. Wollast), pp. 301–328. Wiley, Chichester.

Hallam, A. (1989). The case for sea-level change as a dominant causal factor in mass extinction of marine invertebrates. *Phil. Trans. R. Soc. Lond. B* 325, 437–455.

Harris, G.P., Davies, P., Nunez, M. and Meyers G. (1988). Interannual variability in climate and fisheries in Tasmania. *Nature* 333, 754–757.

Hatcher, B.G., Johannes, R.E. and Robertson, A.I. (1989). Review of research relevant to the conservation of shallow tropical marine ecosystems. *Oceanogr. Mar. Biol. Ann. Rev.* 27, 337–414.

Hayden, B.P., Ray G.C. and Dolan, R. (1984) Classification of coastal and marine environments. *Environ. Conserv.* 11, 199–207.

Hekstra, G.P. (1989) Global warming and rising sea levels: the policy implications. *The Ecologist* 19, 4–13.

Henrichs, S.M. and Reeburgh, W.S. (1987). Anaerobic mineralization of marine sediment organic matter: rates and role of aerobic processes in the oceanic carbon economy. *Geomicrobiol. J.* 5, 191–237.

Henriksen, K. and Kemp, W.M. (1988). Nitrification in estuarine and coastal marine sediments. In: *Nitrogen Cycling in Coastal Marine Environments* (Ed. by T.H. Blackburn and J. Sorensen), *SCOPE* 33, 207–249. Wiley, Chichester.

Hofmann, E.E. (1991). How do we generalize coastal models to global scale? In: *Ocean Margin Processes in Global Change*. (Ed. by R.F.C. Mantoura, J.-M. Martin and R. Wollast), pp. 401–418. Wiley, Chichester.

Holligan, P.M., Aarup, T. and Groom, S.B. (1989). The North Sea: Satellite colour atlas. *Contin. Shelf Res.* 9, 667–765.

Hopkinson, C.S. (1988). Patterns of organic carbon exchange between coastal ecosystems. The mass balance approach in salt marsh ecosystems. In: *Coastal–Offshore Ecosystem Interactions. Lecture Notes on Coastal and Estuarine Studies. Vol. 22* (Ed. by B.-O. Jansson), pp. 122–154. Springer-Verlag, Berlin.

Howarth, R.W. (1988). Nutrient limitation of net primary production in marine ecosystems. *Ann. Rev. Ecol.* 19, 89–110.

IGBP (1990). *The International Geosphere–Biosphere Programme: A Study of Global Change. The Initial Core Projects*. Ch. 4. How changes in land use affect the resources of the coastal zone, and how changes in sea level and climate alter coastal ecosystems. Report No. 12. Stockholm.

IPCC (Intergovernmental Panel on Climate Change) (1990). *Climate Change. The IPCC Scientific Assessment 9. Sea Level Rise.* WMO/UNEP, pp. 257–281. Cambridge University Press, Cambridge.

250 P. M. HOLLIGAN AND W. A. REINERS

Ittekkot, V. (1988). Global trends in the nature of organic matter in river suspensions. *Nature* 332, 436–438.

Ittekkot, V. and Haake, B. (1990). The terrestrial link in the removal of organic carbon in the sea. In: *Facets of Modern Biogeochemistry* (Ed. by V. Ittekot, S. Kempe, W. Michaelis and A. Spitzy), pp. 318–325. Springer-Verlag, Berlin.

Ittekkot, V., Safiullah S. and Arain, R., (1986). Nature of organic matter in rivers with deep sea connections: The Ganges–Brahmaputra and Indus. *Sci. Total Environ.* 58, 93–107.

Jansson, B.O. (Ed.) (1988). *Coastal–Offshore Ecosystem Interactions. Lecture Notes on Coastal and Estuarine Studies. Vol.22.* Springer-Verlag, Berlin.

Jansson, A.M. and Jansson, B.-O. (1988). Energy analysis approach to ecosystem redevelopment in the Baltic Sea and Great Lakes. *Ambio* 17, 131–136.

Jensen, J.R., Ramsey, E.W., Holmes, J.M., Michel, J.E., Savitsky, B. and Davis, B.A. (1990). Environmental sensitivity index (ESI) mapping for oil spills using remote sensing and geographic information system technology. *Int. J. Geogr. Inf. Syst.* 4, 181–201.

Johnson, J.C. and Pollnac, R.B. (1989). Introduction to managing marine conflicts. *Ocean & Shoreline Mgmt* 12, 191–198.

Jouanneau, J.M. and Latouche, C. (1989). Continental fluxes to the Bay of Biscay: Processes and behaviour. *Ocean & Shoreline Mgmt* 12, 477–485.

Justic, D., Legovic, T. and Rottini-Sandrini, L. (1987). Trends in oxygen content 1911–1984 and occurrence of benthic mortality in the northern Adriatic Sea. *Estuar. Cstl. Shelf Sci.* 25, 435–445.

Kempe, S. (1988). Estuaries – their natural and anthropogenic changes. In: *Scales and Global Change* (Ed. by T. Rosswall, R.G. Woodmansee and P.G. Risser), pp. 251–285. Wiley, Chichester.

Kenchington, R.A. and Agardy, M.T. (1990). Achieving marine conservation through biosphere reserve planning and management. *Environ. Conserv.* 17, 39–44.

Kesel, R.H. (1988). The decline in suspended load of the Lower Mississippi River and its influence on adjacent wetlands. *Env. Geol. Water Sci.* 11, 271–281.

Ketchum, B.H. (Ed.) (1983). *Ecosystems of the World. 26. Estuaries and Enclosed Seas.* Elsevier, Oxford.

Kiene, R.P. (1990). Dimethyl sulfide production from dimethlylsulfoniopropionate in coastal seawater samples and bacterial cultures. *Appl. Environ. Microbiol.* 56, 3292–3297.

Krysanova, V., Meiner, A. Roosaare, J. and Vasilyev, A. (1989). Simulation modelling of the coastal waters pollution from agricultural watershed. *Ecol. Modelling* 49, 7–29.

Kuhlmann, D.H.H. (1988). The sensitivity of coral reefs to pollution. *Ambio* 17, 13–21.

Kullenberg, G. (Ed.) (1986). *Contaminant Fluxes Through the Coastal Zone. Rapp. P.-v. Reun. Cons. Int. Explor. Mer.* 186.

Lancelot, C., Billen, G., Sournia, A., Weisse, T., Colijn, F., Veldhuis, M.J.W. Davis, A. and Wassmann, P. (1987). *Phaeocystis* blooms and nutrient enrichment in the continental coastal zones of the North Sea. *Ambio* 16, 38–46.

Lange, C.B., Burke, S.K. and Berger, W.H. (1990). Biological production off southern California is linked to climatic change. *Clim. Change* 16, 319–329.

Lashof, D.A. (1989). The dynamic greenhouse: Feedback processes that may influence future concentrations of atmospheric trace gases and climatic change. *Clim. Change* 14, 213–242.

Lasserre P. and Martin, J-M. (1986). *Biogeochemical Processes at the Land–Sea Boundary*. Elsevier Oceanography Series **443**.

Law, C.S. and Owens, N.J.P. (1990). Denitrification and nitrous oxide in the North Sea. *Neth. J. Sea Res*. **25**, 65–74.

Lewis, J.B. (1981). Coral reef ecosystems. In: *Analysis of Marine Ecosystems* (Ed. by A.R. Longhurst), pp. 127–158. Academic Press, London.

Lindeboom, H.J., van Raaphorst, W., Ridderinkhof, H. and van der Veer, H.W. (1989). Ecosystem model of the western Wadden Sea: a bridge between science and management. *Helgol. Meeresunt.* **43**, 549–564.

Lugo, A.E., Brown, S. and Brinson, M.M. (1988). Forested wetlands in freshwater and salt-water environments. *Limnol. Oceanogr*. **33**, 894–909.

MacDonald, G.J. (1990). Role of methane clathrates in past and future climates. *Clim. Change* **16**, 247–281.

Madden, C.J., Day, J.W. and Randall, J.M. (1988). Freshwater and marine coupling in estuaries of the Mississippi River deltaic plain. *Limnol. Oceanogr*. **33**, 982–1004.

Mann, K.H. (1988). Production and use of detritus in various freshwater, estuarine, and coastal marine ecosystems. *Limnol. Oceanogr*. **33**, 910–930.

Mantoura, R.F.C. (1987). Organic films at the halocline. *Nature* **328**, 579–580.

Martin J.-H., Elbaz-Poulichet, F., Guieu, C., Loye-Pilot, M.-D. and Han. G. (1989). River versus atmospheric input of material to the Mediterranean Sea: An overview. *Mar. Chem.* **28**, 159–182.

Marty, D., Esnault, G., Caumette, P., Ranavaison-Rambeloarisoa E. and Bertrand J.-C. (1990). Dénitrification, sulfato-réduction et méthanogenèse dans les sédiments superficiels d'un étang saumâtre méditerranéen. *Oceanol. Acta* **13**, 199–210.

McGowan, J.A. (1990). Climate and change in oceanic ecosystems: The value of time-series data. *Trends Ecol. Evol.* **5**, 293–299.

McManus, J.W. (1988). Coral reefs of the ASEAN region: Status and management. *Ambio* **17**, 189–193.

McRoy, C.P. and Lloyd, D.S. (1981). Comparative function and stability of macrophyte-based ecosystems. In: *Analysis of Marine Ecosystems* (Ed. by A.R. Longhurst), pp. 473–489. Academic Press, London.

Meade, R.H. (1988). Movement and storage of sediment in river systems. In *Physical and Chemical Weathering in Geochemical Cycles* (Ed. by A. Lerman and M. Meybeck), pp. 165–179. Kluwer, Dordrecht.

Meybeck, M. and Helmer, R. (1989). The quality of rivers: From pristine stage to global pollution. *Palaeogeogr. Palaeoclim. Palaeoecol.* **75**, 283–309.

Meybeck, M., Cauwet, G., Dessery, S., Somville, M., Gouleau, D. and Billen, G. (1988). Nutrients (organic C, P, N, Si) in the eutrophic River Loire (France) and its estuary. *Estuar. Cstl. Shelf Sci*. **27**, 595–624.

Miller, J.R. (1966). The flux of tidal energy out of the deep oceans. *J. Geophys. Res.* **71**, 2485–2489.

Milliman, J.D. (1990). River discharge of water and sediment to the oceans: Variations in time and space. In: *Facets of Modern Biogeochemistry* (Ed. by V. Ittekkot, S. Kempe, W. Michaelis and A. Spitzy), pp. 83–90. Springer-Verlag, Berlin.

Milliman, J.D., Broadus, J.M. and Gable, F. (1989). Environmental and economic implications of rising sea level and subsiding deltas: the Nile and Bengal examples. *Ambio* **18**, 340–345.

Mitchell, J.F.B. (1989). The greenhouse effect and climate change. *Rev. Geophys.*, **27**, 115–139.

Morris, A.W. (1990). Kinetic and equilibrium approaches to estuarine chemistry. *Sci. Total Environ.* **97/98**, 253–266.

Muller-Karger, F.E., McClain C.R. and Richardson, P.L. (1988). The dispersal of the Amazon's water. *Nature* **333**, 56–59.

Newman W.S. and Fairbridge, R.W. (1986). The management of sea-level rise. *Nature* **320**, 319–320.

Nichols, M.M. (1986). Consequences of sediment flux: escape or entrapment? *Rapp. P.-v. Cons. int. Explor. Mer.* **186**, 343–351.

Nixon, S.W. (1988). Physical energy inputs and the comparative ecology of lake and marine ecosystems. *Limnol. Oceanogr.* **33**, 1005–1025.

Nowicki, B.L. and Oviatt, C.A. (1990). Are estuaries traps for anthropogenic nutrients? Evidence from estuarine mesocosms. *Mar. Ecol. Prog. Ser.* **66**, 131–146.

O'Connor, B.D.S., Costelloe, J., Keegan B.F. and Rhoads, D.C. (1989). The use of "REMOTS" technology in monitoring coastal enrichment resulting from mariculture. *Mar. Poll. Bull.* **20**, 384–390.

Officer, C.B. and Ryther, J.H. (1980). The possible importance of silicon in marine eutrophication. *Mar. Ecol. Prog. Ser.* **3**, 83–91.

Oppenheimer, M. (1989). Climate change and environmental pollution: Physical and biological interactions. *Clim. Change* **15**, 225–270.

Paasche, E. (1988). Pelagic primary production in nearshore waters. In: *Nitrogen Cycling in Coastal Marine Environments* (Ed. by T.H. Blackburn and J. Sorensen). *SCOPE* **33**, 33–57. Wiley, Chichester.

Paerl, H.W. (1988). Nuisance phytoplankton blooms in coastal, estuarine, and inland waters. *Limnol. Oceanogr.* **33**, 823–847.

Palanques, A., Plana, F. and Maldonado, A. (1990). Recent influence of man on the Ebro margin sedimentation system, northwestern Mediterranean Sea. *Mar. Geol.* **95**, 247–263.

Pauly, D. and Thia-Eng, C. (1988). The overfishing of marine resources: Socioeconomic background in southern Asia. *Ambio* **17**, 200–206.

Payne, A.I.L., Gulland J.A. and Brink, K.H. (1987). *The Benguela and Comparable Ecosystems. S. Afr. J. mar. Sci.* **5**.

Pegler, K. and Kempe, S. (1988). The carbonate system of the North Sea: Determination of alkalinity and TCO_2 and calculation of PCO_2 and SI_{cal} (spring 1986). In: *Biogeochemistry and Distribution of Suspended Matter in the North Sea and Implications to Fisheries Biology* (Ed. by S. Kempe, V. Dethlefsen, G. Liebeziet and V. Harms). *Mitt. Geol.-Palaont. Inst. Univ. Hamburg, SCOPE/UNEP, Sonderbd.* **65**, 35–87.

Peltier W.R. and Tushingham A.M. (1989). Global sea level rise and the greenhouse effect: Might they be connected? *Science* **244**, 806–810.

Pentreath, R.J. (1987). The interaction with suspended and settled sedimentary materials of long-lived radionuclides discharges into coastal waters. *Contin. Shelf Res.* **7**, 1457–1469.

Peterson, D.H., Hager, S.W. and Schemel, L.E. (1988). Riverine C, N, Si and P transport to the coastal ocean: An overview. In: *Coastal–Offshore Ecosystem Interactions. Lecture Notes on Coastal and Estuarine Studies. Vol. 22* (Ed. by B.-O. Jansson), pp. 227–253. Springer-Verlag, Berlin.

Petts, G.E. (1984). *Impounded Rivers. Perspectives for Ecological Management.* Wiley, Chichester.

Phillips, D.J.H. and Tanabe, S. (1989). Aquatic pollution in the Far East. *Mar. Poll. Bull.* **20**, 297–303.

Pope, J. (1989). Fisheries research and management for the North Sea; The next hundred years. *Dana* **8**, 33–43.

Postma, H. (1988). Physical and chemical oceanographic aspects of continental shelves. In *Ecosystems of the World*. **27**. *Continental Shelves*. (Ed. by H. Postma and J.J. Zijlstra), pp. 5–37. Elsevier, Amsterdam.

Postma, H and Zijlstra, J.J. (Eds) (1988). *Ecosystems of the World*. **27**. *Continental Shelves*. Elsevier, Amsterdam.

Pye, K. (1987). *Aeolian Dust and Dust Deposits*. Academic Press, Orlando, Florida.

Queen, W.H. (1977). Human uses of salt marshes. In: *Ecosystems of the World*. *1. Wet Coastal Ecosystems* (Ed. by V.J. Chapman), pp. 363–368 Elsevier, Amsterdam.

Radach, G., Berg J. and Hagmeier E. (1990). Long-term changes of the annual cycles of meteorological, hydrographic, nutrient and phytoplankton time series at Heligoland and at LV ELBE 1 in the German Bight. *Contin. Shelf Res.* **10**, 305–328.

Ray, G.C. (1989). Sustainable use of the coastal ocean. In: *Changing the Global Environment*. *Perspectives on Human Involvement* (Ed. by D.B. Botkin, M.F. Caswell, J.E. Estes and A.A. Orio), pp. 71–87. Academic Press, London.

Richardson, K. (1989). Algal blooms in the North Sea: *the Good, the Bad and the Ugly*. *Dana* **8**, 83–93.

Rogers, C.S. (1990). Responses of coral reefs and reef organisms to sedimentation. *Mar. Ecol. Prog. Ser.* **62**, 185–202.

Ronner, U. (1985). Nitrogen transformations in the Baltic proper: Denitrification counteracts eutrophication. *Ambio* **14**, 134–138.

Rosenberg, R. (1990). Negative oxygen trends in Swedish coastal bottom waters. *Mar. Poll. Bull.* **21**, 335–339.

Rowe, G.T. (1987). Seasonal growth and senescence in continental shelf ecosystems: a test of the SEEP hypothesis. *S. Afr. J. Mar. Sci.* **5**, 147–161.

Ruddle, K. (1987). The impact of wetland reclamation. In: *Land Transformation in Agriculture* (Ed. by M.G. Wolman and F.G.A. Fournier), *SCOPE* **32**, 171–201. Wiley, Chichester.

Rueness, J. (1989). *Sargassum muticum* and other introduced Japanese macroalgae: Biological pollution of European coasts. *Mar. Poll. Bull.* **20**, 173–176.

Salge, U. and Wong, H.K. (1988). The Skagerrak: A depo-environment for recent sediments in the North Sea. In: *Biogeochemistry and Distribution of Suspended Matter in the North Sea and Implications to Fisheries Biology* (Ed. by S. Kempe, V. Dethlefson, G. Liebeziet and U. Harms). *Mitt. Geol.-Palaont. Inst. Univ. Hamburg, SCOPE/UNEP, Sonderbd.* **65**, 367–380.

Salomons, W., Bayne., B.L. Duursma, E.K. and Forstner, U. (Eds) (1988). *Pollution of the North Sea. An Assessment*. Springer-Verlag, Berlin.

Seitzinger, S.P. (1988). Denitrification in freshwater and coastal marine ecosystems: ecological and geochemical significance. *Limnol. Oceanogr.* **33**, 702–724.

Skreslet, S. (1986). Freshwater outflow in relation to space and time dimensions of complex ecological interactions in coastal waters. In: *The Role of Freshwater Outflow in Coastal Marine Ecosystems* (Ed. by S. Skreslet), NATO ASI Ser. Ecol. Series Vol 7, pp. 1–13. Springer-Verlag, Berlin.

Smetacek, V., Bathmann, U., Nothig E.-M. and Scharek, R. (1991). Coastal eutrophication—causes and consequences. In: *Ocean Margin Processes in Global Change*. (Ed. by R.F.C. Mantoura, J.M.-Martin and R. Wollast). Wiley, Chichester.

Smith, D. and Dawson, A. (1990). Tsunami waves in the North Sea. *New Scientist* **127**, (1728), 46–49.

Smith, S.E. and Abdel-Kader, A. (1988). Coastal erosion along the Egyptian delta. *J. Coast. Res.* **4**, 245–255.

Smith, S.V. (1984). Phosphorus versus nitrogen limitation in the marine environment. *Limnol. Oceanogr.* **29**, 1149–1160.

Smith, S.V. and Mackenzie, F.C. (1987). The ocean as a net heterotrophic system: Implications for the carbon biogeochemical cycle. *Global Biogeochem. Cycles* 1, 187–198.

Smith, S.V., Hollibaugh, J.T., Dollar S.J. and Vink, S. (1989). Tomales Bay, California: A case for carbon-controlled nitrogen cycling. *Limnol. Oceanogr.* 34, 37–52.

Soutar, A. and Isaacs, J.D. (1974). Abundance of pelagic fish during the 19th and 20th centuries as recorded in anaerobic sediment off the Californias. *Fish. Bull.* 72, 257–273.

Southward, A.J. (1980). The western English Channel—an inconstant ecosystem? *Nature* 285, 361–366.

Southward, A.J., Boalch, G.T. and Maddock, L. (1988). Fluctuations in the herring and pilchard fisheries of Devon and Cornwall linked to change in climate since the 16th century. *J. Mar. Biol. Ass. UK* 68, 423–445.

Spencer, T. (1988). Coastal biogeomorphology. In: *Biogeomorphology* (Ed. by H.A. Viles), pp. 255–318. Blackwell, London.

Spitzy, A. and Ittekkot, V. (1991). Dissolved and particulate organic matter in rivers. In: *Ocean Margin Processes in Global Change*. (Ed. by R.F.C. Mantoura, J.-M. Martin and R. Wollast), pp. 5–18. Wiley, Chichester.

Stanley, D.J. (1990). Recent subsidence and northeast tilting of the Nile Delta, Egypt. *Mar. Geol.* 94, 147–154.

Starkel, L. (1989). Global paleohydrology. *Quatern. Internat.* 2, 25–33.

Stebbing, A.R.D. and Harris, J.R.W. (1988). The role of biological monitoring. In: *Pollution of the North-Sea. An Assessment* (Ed. by W. Salomons *et al.*), pp. 655–665. Springer Verlag, Berlin.

Stevenson, J.C. (1988). Comparative ecology of submersed grass beds in freshwater, estuarine, and marine environments. *Limnol. Oceanogr.* 33, 867–893.

Stevenson, J.C., Ward, L.G. and Kearney, M.S. (1988). Sediment transport and trapping in marsh systems: Implications of tidal flux studies. *Mar. Geol.* 80, 37–59.

Stumm, W. (Ed.) (1987). *Aquatic Surface Chemistry. Chemical Processes at the Particle–Water Interface*. Wiley, Chichester.

Tans, P.P., Fung, I.Y. and Takahashi, T. (1990). Observational constraints on the global atmospheric CO_2 budget. *Science* 247, 1431–1438.

Ters, M. (1987). Variations in Holocene sea level on the French Atlantic coast and their climatic significance. In: *Climate. History, Periodicity and Predictability* (Ed. by M. Rampino, J.E. Sanders, W.S. Newman and L.K. Konigsson), pp. 204–237. Van Nostrand Reinhold, New York.

Townsend, D.W. and Cammen, L.M. (1988). Potential importance of the timing of spring plankton blooms to benthic-pelagic coupling and recruitment of juvenile demersal fishes. *Biol. Oceanogr.* 5, 215–229.

Turner, R.E., Rabelais N.N. and Nan, Z.Z. (1990). Phytoplankton biomass, production and growth limitations on the Huanghe (Yellow River) continental shelf. *Contin. Shelf Res.* 10, 545–571.

Turner, S.M., Malin, G., Liss, P.S., Harbour, D.S. and Holligan, P.M. (1988). The seasonal variation of dimethyl sulfide and dimethylsulfoniopropionate concentrations in nearshore waters. *Limnol. Oceanogr.* 22, 264–375.

Twilley, R.R. (1988). Coupling of mangroves to the productivity of estuarine and coastal waters. In: *Coastal–Offshore Ecosystem Interactions. Lecture Notes on Coastal and Estuarine Studies. Vol. 22* (Ed. by B.-O. Jansson), pp. 155–180. Springer-Verlag, Berlin.

Underdal, B., Skulberg, O.M., Dahl, E. and Aune, T. (1989). Disastrous bloom of *Chrysochromulina polylepis* (Prymnesiophyceae) in Norwegian coastal waters 1988—mortality in marine biota. *Ambio* 18, 265–288.

UNEP (1990). The State of the Marine Environment. *Reports and Studies GESAMP No. 39.*

UNESCO (1989). The Ocean as a Source and Sink for Atmospheric Trace Constituents. Final report of SCOR W.G. 72, UNESCO *Tech. Paper Mar. Sci.*

UNESCO (1990). Relative Sea-level Change: a Critical Evaluation. UNESCO *Rep. Mar. Sci.* No. 54,

Vellinga, P and Leatherman, S.P. (1989). Sea level rise, consequences and policies. *Clim. Change* **15**, 175–189.

Walsh, G.E. (1977). Exploitation of mangal. In: *Ecosystems of the World. 1. Wet Coastal Ecosystems* (Ed. by V.J. Chapman), pp. 347–362. Elsevier, Amsterdam.

Walsh, J.J. (1988). *On the Nature of Continental Shelves.* Academic Press, New York.

Walsh, J.J. (1989). How much shelf production reaches the deep sea? In: *Productivity of the Ocean: Present and Past* (Ed. by W.H. Berger *et al.*), pp. 175–191. Wiley, Chichester.

Walter, H. (1983). *Vegetation of the Earth and Ecological Systems of the Geobiosphere* (2nd Edn). Springer-Verlag, Berlin.

Warrick, R. and Farmer, G. (1990). The greenhouse effect, climatic change and rising sea level: Implications for development. *Trans. Inst. Br. Geogr. N.S.* **15**, 5–20.

Warwick R.M., Clarke, K.R. and Suharsono, (1990). A statistical analysis of coral community responses to the 1982–83 El Nino in the Thousand Islands, Indonesia. *Coral Reefs* **8**, 171–179.

Wassmann, P. (1990). Calculating the load of organic carbon to the aphotic zone in eutrophicated coastal waters. *Mar. Poll. Bull.* **21**, 183–187.

Wells, J.T. and Coleman, J.M. (1987). Wetland loss and the subdelta lifecycle. *Estuar. Cstl. Shelf Sci.* **25**, 111–125.

Wells, S. and Edwards, A. (1989). Gone with the waves. *New Scientist* **124**, (1690), 47–51.

Wilson, J.A., Townsend, R., Kelban, P., McKay, S. and French, J. (1990). Managing unpredictable resources: Traditional policies applied to chaotic populations. *Ocean & Shoreline Mgmt* **13**, 179–197.

Wolfe, D.A. and Kjerfve, B. (1986). Estuarine variability: An overview. In: *Estuarine Variability* (Ed. by D.A. Wolfe), pp. 3–17. Academic Press, New York.

Worrest, R.C. and Hader, D.-P. (1989). Effects of stratospheric ozone depletion on marine organisms. *Environ. Conserv.* **16**, 261–263.

Wright, D.A. and Phillips, D.J.H. (1988). Chesapeake and San Francisco Bays. A study in contrasts and parallels. *Mar. Poll. Bull.* **19**, 405–413.

Wright, L.D., Wiseman, W.J., Yang, Z-S., Bornhold, B.D., Keller, G.H., Prior, D.B. and Suhayda, J.N. (1990). Processes of marine dispersal and deposition of suspended silts off the mouth of the Huanghe (Yellow River). *Contin. Shelf Res.* **10**, 1–40.

Wulff, F., Stigebrandt A. and Rahm, L. (1990). Nutrient dynamics of the Baltic Sea. *Ambio* **19**, 126–133.

Zektzer, I.S., Ivanov V.A. and Meskheteli, A.V. (1973). The problem of direct groundwater discharge to the seas. *J. Hydrology* **20**, 1–36.

Zektzer, I.S., Dzhamalov, R.G. and Safronova, T.I. 1983. The role of submarine groundwater discharge in the water balance of Australia. In: *Groundwater and Water Resource Planning. Proceedings of the International Symposium. Vol 1*, pp. 209–219. UNESCO-IAH-IAHS. UNESCO, Paris.

The Past as a Key to the Future: The Use of Palaeoenvironmental Understanding to Predict the Effects of Man on the Biosphere

J.M. ADAMS and F.I. WOODWARD

I. SUMMARY

If science can understand how the world has changed in the past, it may be possible to use some of this knowledge to predict the future effects of human activities on the biosphere. In this chapter we discuss some of the relationships between atmospheric composition, climate, and the distribution and composition of biological communities, which are apparent from the historical, geological and palaeontological record, and consider the extent to which these are relevant to issues of future global change.

There is a need to be cautious about drawing analogies. Many of the best understood changes in the past occurred at speeds or in environmental settings very different from those associated with man's activities, and in many cases the resolution and accuracy of data are still insufficient to allow firm conclusions to be drawn. However, provided that these limitations are taken into account, the information which has been gleaned already has implications for our view of the biosphere's response to changes in radiative forcing caused by human activities. It appears that

ADVANCES IN ECOLOGICAL RESEARCH VOL. 22
ISBN 0–12–013922–7

there are many intruiging analogies from the past which might be regarded as clues or warnings of what to expect. The information that is most relevant to predicting future changes comes from the detailed study of climatic and biotic changes of the Quaternary (the last 2·4 million years), in which the geographical and biotic setting is most closely similar to the present, and in which there is a good enough environmental record to obtain a detailed understanding of the nature and mechanisms of changes.

At the basic level of predicting the responses of climate to changes in radiative forcing, a knowledge of past climatic and oceanic conditions offers a testing ground for general circulation models (GCMs), which have been designed on the present-day world in order to predict future conditions. Some GCMs seem to be fairly accurate in simulating the distribution of temperature conditions during the past glacial cycle, but they are less accurate in simulating water-balance changes. This level of error indicates possible limitations in the accuracy of current predictions of vegetation and agricultural changes resulting from greenhouse warming.

Other potentially important clues to the future come from studying the responses of sources and sinks of greenhouse gases to climatic change, and their potential role in feedback mechanisms. Changes in such reservoirs are known to have occurred in parallel with climatic changes during the Quaternary period, and they may provide warning of the possibility of rapid positive feedback effects on greenhouse gas levels.

Some general indications of the response of biotas to rapid climate change can also be obtained from past evidence of postglacial migrations, and from biological introductions in more recent times. The slow response times of tree species to climatic fluctuations during the Quaternary indicate that the ecology of a greenhouse world will take many centuries or millennia to adjust to changed climatic conditions, although some groups (such as insects and arable weeds) might be able to keep track with the rapidity of change. We can expect that as climate changes, the familiar communities of animals and plants will be broken up, and new combinations of species will occur, as a result of both ecological disequilibrium and changing combinations of climatic parameters.

Overall, it seems that the perspective of the past has made significant contributions to our ability to predict future effects of man on the biosphere. It has provided a testing ground for hypotheses about the future which have been based on present-day knowledge of the world, and the study of past events has also led to the formulation of new hypotheses. During the next few years, further understanding of the response of the biosphere to anthropogenic perturbation is likely to result from a rapidly advancing knowledge of the climatology and biogeography of the Quaternary.

II. INTRODUCTION

The other chapters in this volume have dealt with some of many ways in which human activity may ultimately affect the broad-scale nature of the biosphere. Extrapolations into the future are mainly based on reasoning from our best understanding of patterns and processes in the present-day world. However, there is a sense in which many of the types of changes which are being forecast may already have happened in the past (Chaloner, 1990). From the geological record it is becoming increasingly apparent that the earth's history has been characterized by continual change. Perhaps it is possible to draw on this past record of change to improve our understanding of what to expect, just as the bitter lessons of human history are often used to predict the consequences of a particular course of action?

In this chapter we will explore some of the ways in which data from various times in the past—on a time-scale ranging from centuries to hundreds of millions of years—can be used to help improve understanding of what to expect if greenhouse gas emissions continue on their present upward trend. We will also consider the ways in which these data *cannot* be used, and the extent to which questionable analogies have been drawn between the past and the future.

Our aim here is two-fold. Firstly, we hope to offer some specifically useful items to those more accustomed to thinking from the present-day perspective, by providing additional information, ideas and comments against which to consider the issues of global change. Secondly, at a broader philosophical level we hope to demonstrate how issues of future change can be approached by referring back to past events, and to point out the general sorts of potential pitfalls which are encountered.

III. THE RECORD OF THE PAST

There are many ways of gathering information about past changes in the biosphere, and imaginative new methods are continually being discovered. The general level of confidence in our understanding of the earth's history for all times in the past has increased greatly over recent years as new analytical techniques have been developed, and as more and more basic data have been gathered on past environments and life forms.

Despite the progress, there are some unalterable principles which must always be borne in mind when comparing a past world with the modern world. Firstly, the level of resolution in the information generally becomes less the further back in time one looks, as the chemical composition of sediments and fossils becomes more altered, and as a greater proportion of the rocks which were laid down have been eroded away or deeply buried under other deposits. This loss of resolution may give an ambiguous or misleading picture of past events. Secondly, as one goes back further in time the biotic and geographical setting resembles our modern world less and less. Trying to draw parallels with a future world may thus be misleading because the basic parameters of the system were so different in the past. Because of these two sets of problems, comparisons with our present-day situation must become more and more vague the further back one goes, to the extent that they are not really trustworthy as analogies at all. It is essentially only during the Tertiary (which began 65 myr ago and ended 2·4 myr ago), with modernization of floras and faunas, and present continental and circulation patterns coming into play, that the analogues start to seem useful.

The most useful data on the potential effects of human activities will surely come from the relatively recent past of the Quaternary (the past 2·4 myr). Importantly, the geographical setting of the Quaternary is closely similar to the present, with the major continental arrangements and mountain ranges essentially the same as they are today. It is also during the Quaternary that we start to see regional floras and faunas that are very similar to the present, and the quality of fossil preservation in sediments allows us to build up a much more detailed picture of the changes in species distribution patterns which have occurred. Details of climatic conditions and atmospheric and ocean chemistry are obtainable from isotopes in ocean and ice cores, from gas bubbles trapped in ice (e.g. Barnola *et al.*, 1987), and from the distribution of plankton and other organisms (e.g. COHMAP, 1988). The most striking feature of the Quaternary era is that it has been dominated by dramatic fluctuations in climate (there have been several "ice ages" interspersed with warmer periods), atmospheric composition and in the distributions of animals and plants. It is thus a time of environmental change rather than stasis. Overall, this period of time appears to present some promising analogues against which to predict the knock-on effects of anthropogenic perturbations to the biosphere.

Despite its attractions, the existing record of the Quaternary that is available in the scientific literature offers a rather biased view of the world. For example, a detailed knowledge of post glacial migrations of species is so far largely restricted to particular tree species in certain limited areas of the northern temperate zone. For most of the time span of the Quaternary, even the very general nature of the vegetation over large areas of the tropics remains unknown. It is necessary to bear in mind such limitations in the record when drawing conclusions which relate to the future.

For the most recent past, extending back over the past few centuries or millenia, the standard palaeoecological indicators are supplemented by direct historical records from old manuscripts, and (for the last 200 years or so) scientific data. Although these might seem a most promising source of information, coming as they often do from direct observation, there are many problems in using them. Verbal descriptions are often difficult to put into quantitative terms, and there is still always the risk that the authors might not have given an accurate account. Early scientific data were often gathered using relatively crude instrumentation, with poorly standardized techniques. Nevertheless, direct observation from the past is often a welcome source of information which can be considered in combination with the information from other sources.

A. Greenhouse Gases and Climate Warming

The current surge of concern about global change is primarily due to the observation that atmospheric levels of CO_2, CH_4, N_2O and CFCs are rapidly increasing (Watson et al., 1990, and Chapter 1, this volume). It is difficult to extrapolate into the future to predict how fast these gases will build up if no action is taken to curtail their production, because many political and economic factors could affect this (in addition to some of the possible biospheric feedbacks discussed in this chapter). However, it seems that if present trends continue, CO_2 will reach 560 p.p.m. by around the mid-to-late twenty-first century, which is double the pre-industrial level CO_2 level of 280 p.p.m. (Neftel et al., 1985). Other gases (e.g. CH_4) seem likely to double much sooner than this (Watson et al., 1990). This chapter concentrates on the effects which CO_2 and the other so-called greenhouse gases may have on climate and biosphere, as their levels increase during the next 100 years or so.

The greenhouse gases are known to have radiative properties which a priori would lead one to suppose that they will trap heat and so warm the earth's surface. There is virtually no question that for any given set of conditions of solar input, vegetation cover, etc., an increase in the concentrations of these gases will produce an initial warming effect at the surface (Chapter 2, this volume). What is much less certain is the actual quantitative effect which a given increase in concentration of each gas will tend to have on temperatures, and also the global distribution of the warming effect. The scenario on which most attention is currently focused is the radiative equivalent of a doubling of the pre-industrial concentration of CO_2, which may well occur by the middle of the next century or so (Chapter 1, this volume).

Various mathematical models (known as general circulation models, or GCMs) have been used to attempt to simulate conditions in a world with greenhouse gas concentrations higher than they are today, as an indication of what we might expect in the future (Chapter 1, this volume). However, it is difficult to know how much trust to put in them. The models disagree with one another to a very significant extent in both the amount and the distribution of warming which they predict for any given greenhouse gas increase (Mitchell et al., 1990, and Chapter 2, this volume). There is the possibility that they could all be wrong because some important climatic feedback effect has been overlooked or poorly accounted for, and that this could radically alter the way in which the climate will respond to future greenhouse gas increases (Raval and Ramanathan, 1989).

The growing record of the past includes information on changes in climate, on time-scales varying from centuries to millions of years, and also of changes in the concentrations of greenhouse gases in the atmosphere. Studying the relationships (or lack of relationships) between past climatic changes and past variations in gas concentrations offers some possibility of improving our understanding of how strongly and how rapidly these gases can operate as a control on climate. This enables the testing and improvement of the GCM models which have been designed to predict the effects of future greenhouse gas increases on temperature. What sources of information from the past can be brought to bear on the problem of knowing what to expect, in terms of warming, from the addition of a given amount of a gas to the atmosphere?

1. Historical Temperature Records

i. Sources of information. Historical information, from the past century-and-a-half, is one of the pillars of present concern about global warming. Meteorological data from stations around the world reveal a mean annual temperature increase of around 0·6°C since 1850, with a particularly rapid temperature rise during the 1980s (Fig.1). It is well established that during this time-span, the mean CO_2 concentration in the atmosphere has increased substantially, from around 280 p.p.m. to 350 p.p.m. (Watson *et al.*, 1990). Likewise there has been an overall increase in measured global sea surface temperatures during the same period (Fig.2), and other possible temperature indicators—such as the raising of altitudinal limits of montane glaciers around the world (Fig.3)—also seem to indicate an overall rise in temperature.

Overall, the apparent rise in temperature in the last century-and-a-half is of the order predicted by GCMs for the CO_2 rise which has already occurred, although it is towards the lower end of the predicted range of temperature responses.

ii. Are the data reliable? The historical data might thus be regarded as an indication that greenhouse warming is already occurring, to the approximate degree that GCMs predict. This would provide much-needed confidence in our ability to predict the future with an acceptable degree of accuracy. However, the data cannot be simply accepted as they stand. For example, take the records of surface air temperature over the last 150 years. Firstly there is the problem that the records themselves are not necessarily a reliable

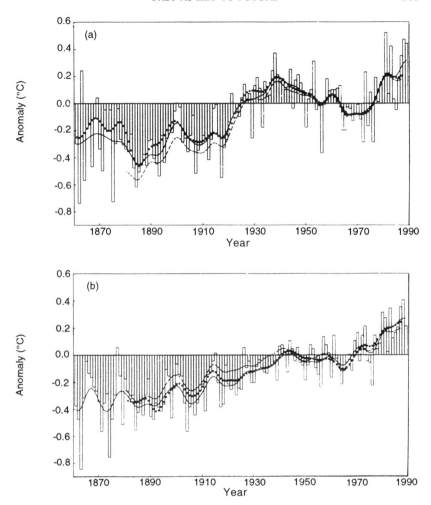

Fig. 1. Mean annual air temperature records for the world since 1850. (a) Northern hemisphere, (b) Southern hemisphere. (Source: Folland *et al.*, 1990.)

reflection of what was happening to global mean temperatures. The number of stations is fairly small, and they tend to be clustered in particular parts of the world, leaving huge areas virtually unrepresented. Instrumentation has changed to some extent during this period, and there have been changes in the precise ways in which readings were taken (e.g. the siting of thermometers) which cannot easily be taken into account (Folland *et al.*, 1990). Another factor tending to reduce one's confidence in results for air temperature readings is the fact that the meterological stations tend to be clustered around centres of population. As urbanization and industrialization have

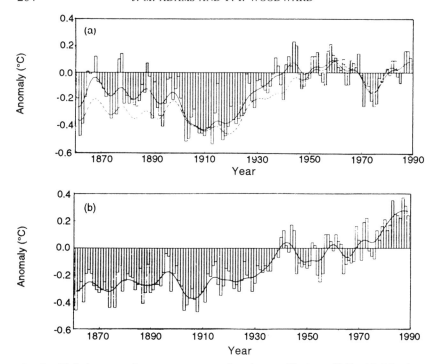

Fig. 2. Global sea surface temperatures for the world since 1850. (a) Northern hemisphere, (b) Southern hemisphere. (Source: Folland *et al.*, 1990.)

increased in many of these areas during the last 150 years, there is the distinct possibility that a local "heat island" effect (which is certainly detectable in large conurbations as compared with their surrounding countryside) may have been a factor tending to increase temperature readings. A correction factor has been introduced by Hansen and Lebedoff (1987), who suggest that the warming trend be reduced by 0·1°C to account for such errors, to leave a temperature rise of around 0·5°C since 1850.

Similar sorts of sampling problems are encountered for the sea surface temperature readings; for example the effects of heat from ships' engines on thermometer readings taken over board. Bottomley *et al.* (1990) have attempted to correct for such errors, and they find a smaller but still significant amount of warming in the sea surface temperature, of around 0·4°C during the time series from 1860 to 1990. The record of glacier retreat around the world is difficult to quantify in terms of a required temperature rise, but Lamb (1977) concludes that it would seem to support a substantial (~0·5°C) rise in mean temperatures since the early 1800s.

Overall, from analysis of these data and from the fact that the different sources of data show a similar trend, it seems reasonable to conclude that

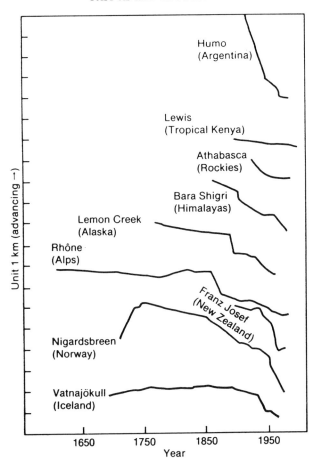

Fig. 3. Worldwide fluctuations in glacier termini. (Source: Folland *et al.*, 1990.)

there has been an overall warming of the earth's surface by around 0·5°C (perhaps slightly more or slightly less) since 1850, which is still within the range predicted by GCMs (Folland *et al.*, 1990).

iii. A background fluctuation? Accepting that the global climate has in fact warmed in the past century-and-a-half, how are we to know if the greenhouse gas rise has indeed been a major component in causing this warming? The difficulties become apparent when we examine the evidence of "background noise" in temperatures over the last 1000 years or so, going back to before the present CO_2 rise. Lamb (1977) has used historical indicators, such as agricultural records and records of freezing-over of rivers and bays, to indicate that at

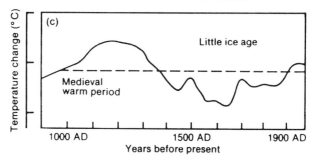

Fig. 4. Schematic diagram of North Atlantic region temperature variation during the last thousand years, from various historical indicators. (Source: Folland *et al.*, 1990.)

least in Europe and the North Atlantic region (and, with less certainty, for other parts of the world), the mean annual temperature in the mid-1600s to mid-1700s was another 0·5°C lower than it was in 1850 (Hooper, 1990). The accuracy of such estimates is open to question, but it would seem that several different sources of historical evidence point to the same general pattern. The precise factors which could have caused these climatic fluctuations are not clear, but there do *not* appear to have been detectable fluctuations in atmospheric CO_2 associated with these pre-industrial temperature changes (Folland *et al.*, 1990) The general picture obtained from these historical records (Fig.4) is that the earth's mean annual temperature can fluctuate by about a degree or so on the time-scale of centuries. In this sense, the much-vaunted climatic warming which has occurred since 1850 could be viewed as just another natural oscillation about the longer-term mean. On the other hand, it could be that background fluctuations have masked a stronger warming effect, which would otherwise have become apparent by now.

Another possibility is that the true amount of ocean–atmosphere warming in the last 150 years has been temporarily disguised by the export of heat to the deeper layers of the oceans. It is not yet possible to account confidently for this factor using GCMs as they either completely ignore the potential effects of lags in warming caused by the oceans, or disregard the rate of heat transfer down from the surface layers of the oceans (Folland *et al.*, 1990).

There is also uncertainty as to how strongly human activity has changed key atmospheric parameters other than greenhouse gas levels, such as atmospheric haze, during the last 150 years. Because these values are not known, it has not been possible to account for them with any confidence in GCM simulations of the post-industrialization CO_2 increase. For example, it is possible that increased levels of atmospheric haze resulting from burning of fossil fuels and forests have acted to mask the full underlying increase in the climate forcing potential of greenhouse gases since the early 1800s. Without these factors, perhaps the climatic warming potential of CO_2 and other gases might turn out to be closer to the mean for GCM predictions, or even well above the mean? We might find out in the future, if for whatever reasons (such as desulphurization of fuels) haze production is decreased, allowing the greenhouse gases that are already in the atmosphere to exert their full warming effect (Wigley and Barnett, 1990).

iv. The spatial distribution of warming since 1850.   In addition to predicting an overall global warming, GCMs make definite predictions about the spatial distribution of warming which will occur as the CO_2 concentration in the atmosphere is increased. How well does the spatial pattern of post-1850 warming correspond to that predicted by GCMs for a CO_2-induced warming?

One interesting observation is that although there seems to have been an overall global warming since 1850, certain parts of the world have shown a net cooling during this period. An observation of local cooling does not necessarily conflict with the models of CO_2 warming; other influences may dominate climates locally, and some of the models themselves indicate that certain areas are likely to cool initially as circulation patterns are changed by global warming.

A potentially more serious criticism is that the principal GCM models, on which most work on the prediction of future effects of a CO_2 rise has been based, all predict a much stronger warming effect at the highest latitudes than at lower latitudes as atmospheric CO_2 is increased. The warming should be strongest in the winter half of the year, due to the effects on reflectance by ice and snow (Folland *et al.*, 1990). The post-1850 annual mean temperature data do indeed suggest that there has been a slightly greater warming of the higher latitudes than the equatorial latitudes (Folland *et al.*, 1990), but the increase in polar temperatures is nowhere near as great as the models predict. This seems to indicate that *either* the CO_2 rise is not the principal cause of the temperature increase, in which case the models are incorrect in their predictions of the way in which climate will warm in response to greenhouse gases or that there is something basically wrong with the way that models treat the higher latitudes.

In conclusion, the overall picture for the post-1850 temperature record is one of considerable ambiguity, on which it seems too risky to place much emphasis in interpreting the accuracy of climate models. However, the rapidly accelerating temperature rise through the 1980s and early 1990s seems more strongly suggestive of a genuine greenhouse warming which is becoming strong enough to be noticeable against the background of fluctuation and oceanic damping. A more confident analysis of the responsiveness of global temperatures to the present phase of CO_2 increase, and accuracy of GCMs in predicting this, will only be possible when the network of standard measuring stations has been operating for a greater length of time under standard conditions. The level and speed of temperature rise will need to exceed the rate of background fluctuation in global temperature consistently if it is to be concluded that there is indeed a relationship between greenhouse gases and temperature as strong as the models predict (Wigley and Barnett, 1990). Of course, by the time this scientific confidence has been achieved, it may be too late to do anything about moderating the effects.

2. Longer-term Relationships Between Greenhouse Gases and Climate

i. Mesozoic–Tertiary warm climates: Due to greenhouse gases? During the Cretaceous and early Tertiary (about 135–40 myr ago), global climates appear to have been very much warmer than they are at present. This is indicated by evidence from the distribution of fossils, which shows for example that broad-leaved deciduous forests grew in areas which were very near to the earth's poles at that time (Wolfe, 1985), and also from oxygen isotope ratios in carbonates laid down at the time (Barron, 1983). Various authors have speculated that the warmer climates at that time might have been due to higher greenhouse gas concentrations in the atmosphere. Budyko *et al.* (1987) have used a series of geochemical models to suggest that CO_2 levels in the late Cretaceous/early Tertiary atmosphere were tens or perhaps even hundreds of times higher than today.

Whilst it is very interesting to speculate about the role of a high atmospheric CO_2 in such distant times, this has very little relevance to predicting the effects of the present greenhouse gas rise. Firstly, there is too much uncertainty as to the actual CO_2 concentration in the Cretaceous and Tertiary for the evidence to be quantitatively useful. If indeed CO_2 *was* orders of magnitude higher than it is today, it is unlikely that the present anthropogenic phase of greenhouse gas addition will ever reach such extremes, even without regulatory controls on production. Secondly, the broad geographical setting in terms of topography and continental arrangement was very clearly different (its details are still in much dispute). Since such factors are crucial in determining global circulation patterns and heat balance, it is not yet possible to extricate the role of atmospheric composition in producing these warmer climates.

ii. The Quaternary. The 2·4 myr of the Quaternary period offers more hope of a relevant perspective on the role of greenhouse gases in driving climatic change. Its familiar geographical and biotic setting, and the considerable detail with which its climates, oceanography and atmospheric composition are known, seem to make it promising ground for understanding the effects of these gases.

An interesting observation comes from studies on the composition of gas bubbles trapped in ancient polar ice; that the levels of CO_2 and other atmospheric components have varied closely in parallel with temperature changes (Fig. 5) (Neftel et al., 1988; Chapellaz et al., 1990). The warm interglacial phases are associated with relatively high levels of CO_2, CH_4, and N_2O and low levels of dust and airborne non-sea salt sulphate. The much colder glacial periods have low levels of the gases and high levels of the particulate matter. During the relatively rapid temperature increases at the start of interglacial phases, the gases CH_4 and CO_2 show a similarly rapid increase in concentration (Ruddiman et al., 1990). During the previous interglacial (125 000–115 000 years ago i.e. 125–115 kyr) when global temperatures were slightly warmer (perhaps as much as 2–3 °C higher) than they are today, CO_2 and CH_4 were also at fairly high concentrations in the atmosphere (Fig.5), although not as high as they have reached in the past century. All these atmospheric components measured in ice cores may be expected from their basic physics to have had an effect on temperatures; positive in the cases of CO_2, CH_4 and N_2O, and negative in the case of sulphate particles and dust. It thus seems reasonable to suggest that they may have had some role in causing or amplifying the climatic changes which occurred during the Quaternary.

Taken at face value, this observation certainly seems to have important implications for the future. The gases in question are the same ones which are rapidly accumulating in our atmosphere; and if an 80 p.p.m. rise in CO_2 could cause the ending of an ice age, what could it do to future climates? A global temperature rise of 4–5°C (the amount which occurred at the end of the last glacial) would certainly be greater than the response predicted by GCMs for an 80 p.p.m. CO_2 addition to our present-day world (they predict only an 0·5–1·5°C rise in temperature). Perhaps this indicates that these models underestimate the sensitivity of our present climate to CO_2?

Such a view would be far too simplistic. Firstly, the correlation between greenhouse gases and global temperature is not absolute. For example the early-to-mid Holocene (around 9–6 kyr ago), which may have been slightly warmer on a global scale (Folland et al., 1990; Mitchell, 1990) appears to have had CO_2 concentrations which were somewhat lower than during the most recent two millennia (Neftel et al., 1988). In any case, the change in CO_2 and CH_4 which occurred at the end of the last glacial was less than the increase which has already occurred since 1750, and no such drastic changes have occurred yet. Lorius et al. (1990) have used multivariate analysis to study the relationship between greenhouse gas concentrations and temperature in the ice-core record, and conclude that only part of the

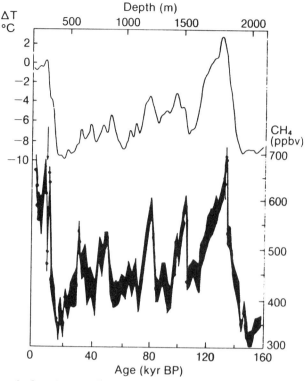

Fig. 5. Record of methane and temperature for the late Quaternary, from ice cores. (Source: Barnola *et al.*, 1987, Chappellaz *et al.*, 1990; IPCC p. 18, p. 11).

lowered temperature during glacial conditions was actually due to these gases. From this, they suggest that the climate-forcing potential of a future CO_2-doubling would be very much as GCMs predict, with a 3–4°C temperature rise.

Additional confidence in our general understanding of greenhouse gas–climate relationships comes from the testing of GCMs based on the present-day world against the earth of the relatively recent past. Attention has concentrated on the temperature conditions at the last glacial maximum (18 kyr ago), which represents the coldest phase in the last 100 kyr (Lorius *et al.*, 1990; Street-Perrott, 1991). When the climate models are adapted to the changed geographical conditions of a different seasonal radiation pattern, lower sea level and more extensive ice-sheets—together with the lower glacial CO_2—they show that the global distribution of temperatures at 18 kyr can be fairly accurately simulated in the context of present understanding of atmospheric physical processes. In a simulation by Broccoli and Manabe

(1987) using the Geophysical Fluid Dynamics Laboratory (GFDL) model, the reduced CO_2 seems to lower temperature by less than 1°C, with other factors such as ice sheet and sea ice albedo providing the main 3–4°C drop in temperature. The extra cooling due to the ice albedo seems to have been due to long-term feedbacks of both CO_2 and of the seasonal radiation input on ice-sheet growth.

Two other leading GCMs, the GISS and CCM models, also seem capable of accurately predicting the global temperature situation during the last glacial stage (COHMAP, 1988). Obviously, what is regarded as *sufficiently* accurate for a model to be a good simulation of climate is a rather subjective issue, but the simulated sea-surface temperatures at 18 kyr ago are within 2°C or less of the observed temperature (as deduced from palaeoenvironmental data) in most parts of the world.

The conclusion to be drawn from the apparent success of these modelling attempts, is that rather than invalidating the GCMs as predictors of greenhouse gas effects on climate, observations from the past provide evidence of their accuracy in simulating climatic responses to radiative forcing. It appears that the climate system at the end of the last glacial was particularly sensitive to changes in radiative forcing brought about by greenhouse gases, with an additional and more important forcing being provided by changes in the distribution of seasonal radiation. This explains how such a large temperature change could occur when the CO_2 and CH_4 changes were relatively small. Both the ice-core record, and modelling studies support the predictions of a number of GCMs that a future CO_2 doubling to around 600 p.p.m. will give a global temperature rise of about 3–4°C at equilibrium. Certain mis-matches *are* still apparent from examination of the past (e.g. the indications from palaeo-tree-line data that there may have been a higher lapse rate of temperature with altitude than the models predict), and these may motivate re-examination and improvement of the GCMs. The stage is now set for further palaeoclimatic experiments with versions of the models that include additional components of the climate system, such as aerosol content and ocean circulation (COHMAP, 1988).

In addition to looking back into the colder phases of the past, there may be insights to be had from some of the more recent warmer phases. The apparently warmer-than-present phases of the mid-Holocene, the last inter-glacial and the late Pliocene, have been used for *qualitative* comparison between theory and observation in climatic modelling (Folland *et al.*, 1990). The biological and isotopic data for all three warm phases indicate that the temperature difference relative to now was much more pronounced at higher latitudes (Fig.6). This agrees with the general predictions of GCM models which show that a change in radiative forcing from greenhouse gases will tend to have its greatest effect at high latitudes. It also seems to bolster the view that the post-1850 temperature rise seen in the climatic records has the

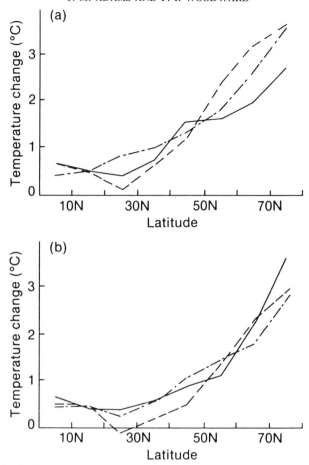

Fig. 6. Latitudinal warmth relative to the present, for three "warm" phases in recent earth history. (a) Winter, (b) summer. (—) Holocene; (- - -) last interglacial; (—·—·—) Pliocene. (Source: Cubasch and Cess, 1990.)

wrong geographical distribution to be the outcome of radiative forcing by CO_2, and that it could instead be a relatively short-term outcome of "random" changes in circulation and albedo.

However, there are problems with the general comparison of model predictions for the future against the past. Warmer climates during the mid-Holocene and the last interglacial (and possibly also the Pliocene) seem to have been related to changes in the seasonal distribution of solar radiation, and perhaps different ocean circulation modes (Rind and Chandler, 1991; Covey, 1991), so it could be that their outcome would not show the same latitudinal pattern as a greenhouse gas effect. There is even some uncertainty as to whether the global mean temperature of the mid-Holocene really was warmer than it is now (Mitchell, 1990). Thus, GCM simulations of the future could still be incorrect in their prediction of a large latitudinal difference in warming, because the evidence from the past which is seen as supporting them is derived from situations which may not really be analogous. The criticism of the post-1850 record on the basis of these latitudinal temperature differences in the late Quaternary might not be justified either: the global palaeotemperature distributions are likely to have represented an overall equilibrium situation, whilst the post-1850 pattern of warming may be different because it represents an initial phase in the warming process, being complicated by the processes of heat uptake by the oceans.

In general, it seems that a detailed knowledge of Quaternary palaeoclimatology and oceanography may provide a useful testing ground for the GCMs that are intended to predict future responses of temperature to greenhouse gases. The comparisons which have been carried out offer some encouraging evidence of the accuracy of the GCM predictions. However, it is necessary to proceed carefully in using other more general observational comparisons to predict the future on the basis of the past. The boundary conditions for climate are not identical, so if we trust these analogies too faithfully we may be misled.

Whilst caution in drawing analogies is necessary, it may be good to keep an eye out for evidence of novel, previously unconsidered aspects of the system which could change as the earth warms. Various possible greenhouse gas feedbacks will be considered in Section C of this chapter, and an observation from palaeoclimate in the late Quaternary hints at the sort of suprises which may be in store as the climate system begins to change. About 11 kyr ago, after a millennium of relatively warm conditions and rapid retreat of ice sheets, temperatures in the North Atlantic region quickly plummeted to near-glacial levels. A cold tundra flora returned to Britain, and mountain glaciers began to form in the uplands. The Younger Dryas phase (so-called because of the return of the arctic plant *Dryas octopetala* to lowland Britain) lasted for about 1000 years, ending very abruptly some time around 10 kyr ago (Ruddiman and McIntyre, 1981). This temporary cold phase is thought to have been induced by a flooding of the North Atlantic with low-salinity meltwater from the retreating ice sheets, which may have disrupted the ocean circulation pattern and thereby cut off the export of heat to high latitudes (Broecker *et al.*, 1990; Jones, 1991). Recent preliminary GCM runs which incorporate a coupled ocean and atmosphere circulation seem to support the idea that a similar event could also be triggered by the melting of ice sheets following an anthropogenic warming phase. It would indeed be ironic if our greenhouse world turned out in the shorter term to be more like an ice age.

From the relatively familiar historical perspective of the Quaternary, man is venturing into little-known territory by continuing to raise the greenhouse gas concentrations of the atmosphere. If the sort of global temperature rise predicted by the GCMs *is* correct, by the end of the twenty-first century temperatures will start to exceed anything that has occurred during the Quaternary. The world may enter what Imbrie and Imbrie (1979) have called a "super-interglacial", with warmth at a level which has not been known since the Late Tertiary. Thinking in an historical context, Imbrie and Imbrie also point out that by raising greenhouse gas levels we may be inadvertently delaying the global temperature decline towards another ice age. Taking the Quaternary as their model, they point out that interglacials have tended to last about 10 kyr. Our present interglacial has already lasted about 10 kyr, so by Imbrie and Imbrie's reckoning, the earth should now be due for the temperature decline towards the next glacial phase; indeed, it has been suggested that the Little Ice Age might even have been just the start of this cooling process, but that the anthropogenic rise in greenhouse gases has already brought us out of it. It is interesting to consider that we may just have saved ourselves from problems which could have been even worse than those which we now face from global warming. This is all very pleasant specula- tion, but there is little more to it than that. Judging by the isotopic and environmental records of the past few interglacials, the temperature decline into full-glacial conditions is a rather unpredictable process involving a great deal of fluctuation, and variation in timing between one interglacial and the next. The smoothed "average" of a "typical" interglacial may not really be representative of any of them in this sense. It is essentially impossible to tell whether or not the "little ice age" itself really was the onset of a longer-term temperature dip that would have continued to the present day without the intervention of greenhouse gases. Similarly, judging by the considerable differences between the isotope signatures of interglacials during the Qua- ternary it seems unwise to predict that a substantial temperature decline would necessarily have occurred over the next few millenia. One cannot rule the possibility out, but one cannot really rule it in, either. If an overall pattern can be discerned from the history of previous interglacials, it is that the decline in the long-term average temperature from full-interglacial warmth to full-glacial cold takes several tens of thousands of years, so the process would probably not in any case have proceeded very far along its course during the next few millenia.

B. Precipitation and Water-balance Changes Associated with Climate Warming

If the earth warms, there are likely to be significant changes in other climatic parameters. Particularly important amongst these are changes in the amount and distribution of precipitation, or rather the balance between precipitation and evaporation. The water balance greatly influences agricultural yields over much of the world, especially in tropical and subtropical countries. It is also a major factor in determining the distribution of natural and semi-natural vegetation types, and of the species which live in these environments (Woodward, 1987a).

GCMs have been used to attempt to predict how precipitation and water balance will change as the greenhouse gas concentration in the atmosphere is increased. As a comparison between the results of two major models in Fig.7 shows, there is still considerable disagreement between the predictions of such models. Many of the areas which the GFHI model predicts will get moister under a CO_2-doubling, the UKHI model predicts will get drier, and vice versa. This is an indication of the level of uncertainty which exists in attempting to forecast how precipitation and water balance will change as a result of greenhouse warming (Mitchell et al., 1990). Since the whole future of a country's economy could well depend on whether its climate gets wetter or drier, and by how much, it is obviously important to try to improve the accuracy of the predictions as much as possible. It is against such a background of general uncertainty that information from the past may be able to make useful contributions, by providing general analogues for comparing with future warm conditions, and by allowing us to find out which—if any—of the current GCM predictions are sufficiently accurate for our needs to predict the future.

1. Past Analogues for Moisture Conditions in a Warmer Climate

During the very warm era of the early Tertiary, there appear to have been remarkably moist climates over nearly all of the earth's surface (Wolfe, 1985). Most of the low latitudes seem to have been covered in tropical rainforest vegetation, whilst the high latitudes were dominated by temperate rainforest of a general type which now survives in only a few favoured localities such as the west coast of North America (Tallis, 1990). As a global cooling set in during the mid-to-late Tertiary, arid-zone vegetation types started to appear over increasing areas (Tallis, 1990). Cores from ocean sediments show that the amount of dust blowing in off the continents increased greatly, providing another indication of the spread of deserts (Rea et al., 1985).

Such observations bolster the view that when the global climate is warmer, it tends to be moister, due to increasingly effective export of water vapour from the oceans to the land. However, there is a need to be cautious here. The geographical setting has changed greatly since the early Tertiary. The world's major mountain ranges have all formed during the Tertiary, and it is likely that their rain-shadow effect has been an important influence in causing the

276   J. M. ADAMS AND F. I. WOODWARD

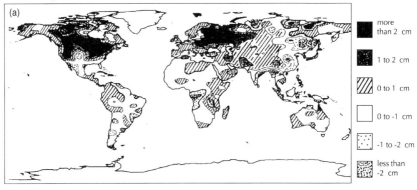

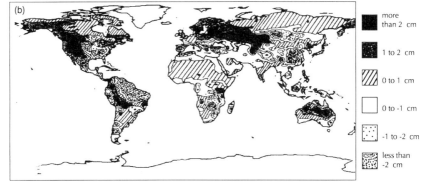

Fig. 7. Comparison of the change in soil moisture due to a CO_2 doubling, for the months December, January and Feburary, predicted by two different GCM models: (A) GFHI model, (B) UKHI model. (Source: Mitchell *et al.*, 1990.)

observed drying (Tallis, 1990). It is also worth bearing in mind that the preceding Cretaceous period, which was also much warmer than today, seems to have had strong seasonal aridity in many parts of the world (J. Chapman, pers. comm.), so it is not clear that climatic warmth has always been associated with increased moisture.

The Pliocene period (5–2·4 myr ago) is likely to be the closest analogue to a greenhouse world which is available from the Tertiary. During the warmest part of the Pliocene, between about 4 and 3 myr ago, summer temperatures in the mid-latitudes seem to have been about 4°C higher than they are today (Folland *et al.*, 1990). This is about the same amount of warming that GCMs predict for the mid-latitudes of the northern hemisphere by the end of the next century. From what is known of the distribution of Pliocene vegetation, there is tentative evidence that climates were moister than at present in most parts of the northern hemisphere (although there is uncertainty over whether the "moist" climates recorded for the Pliocene in different places did actually occur simultaneously with one another) (Folland *et al.*, 1990). How closely this situation resembles a future greenhouse world is debatable; for instance several large mountainous areas had not yet reached their present altitude, so their blocking influence on rainfall may have been less at that time. There is also a general lack of information on whether climates in the southern hemisphere were also moister during this time of warmth.

For the more familiar territory of the Quaternary era, more detailed reconstruction of palaeoenvironments allows a better picture to be built up for studying how water balance has changed with temperature. There were various phases of the Quaternary when mean global temperatures appear to have been higher than they are today. For the last (Eemian) interglacial, about 125–115 kyr ago, various sources of evidence suggest that mean annual temperatures in the mid-latitudes were 2–3°C higher than at present. As for the Pliocene, information is so far mainly limited to the northern hemisphere, but Folland *et al.* (1990) report that all the indicators are of significantly moister climates with around 30–50% higher rainfall . Again, this appears to support the idea that a greenhouse world will be a moister place.

The early and mid-Holocene "warm" phase, between 9 and 6 kyr ago, has been suggested as a potential indicator of moisture conditions in a future greenhouse world (Folland *et al.*, 1990), although the extent to which it actually represented global warmth (rather than just northern hemisphere summer warmth) is uncertain (Mitchell, 1990). As the palaeoenvironmental evidence summarized by Folland *et al.* (1990) and Street-Perrott *et al.* (1990) emphasizes, this phase seems to have been associated with moister-than-present conditions over much of the northern hemisphere. Most strikingly, the Sahara seems to have been almost entirely covered by grasslands and scrublands at that time (Lezine, 1989). However, certain areas in the southern hemisphere such as parts of southern Africa, seem to have been drier than at present (Street-Perrott *et al.*, 1990).

For contrast, it is also possible to look in the opposite direction towards times when global temperatures were cooler than at present. Palaeoenvironmental data from the last glacial maximum (LGM 18 kyr ago) also indicate there has been a general relationship between temperature and moisture availability during the Quaternary. Deserts were much more widespread than they are today, and moist climate vegetation types such as rainforests seem to have been reduced to a fraction of their present extent (Grove, 1986; Adams *et al.*, 1990). Evidence from desert dust in ice cores and ocean sediments seems to confirm that the overall aridity of the earth has been much greater during Quaternary cold periods than during warm periods (Petit *et al.*, 1981).

However, there are exceptions to this trend. Some areas, such as southwestern USA seem to have been wetter than they are today (COHMAP, 1988), although this does not negate the overall picture of increased glacial aridity.

The overall impression derived from the supposed relationship between temperature and moisture balance during the Quaternary, is that warmer conditions will tend to be moister on a global scale. However, as for earlier periods there is the problem that the basic physical conditions of the system were not the same as they will be in a greenhouse world. Perhaps crucially, the seasonal distribution of sunlight was different during the warmer-than-present phases of the last interglacial and early Holocene, due to differences in the position of the earth in its orbit (Folland et al., 1990; Mitchell, 1990). As mentioned previously, the potential power of these "Milankovitch" variations in driving climate is indicated by the evidence that they have been a critical factor behind the temperature variations of glacial–interglacial cycles. On an a priori level, it could be argued that the seasonal radiation differences were more important in causing changes in moisture balance than global temperature as such. So, a future greenhouse world might *not* show the same sorts of relationships between temperature and moisture (Folland et al., 1990; Mitchell, 1990). It is obviously important to proceed with caution when making such general comparisons between the past and the future.

2. Testing the Moisture-balance Component of GCMs by Applying Them to the Past

Rather than simply making vague general analogies between past and future, it may be possible to put the palaeodata to good use as a specific testing ground for the GCMs. A rigorous test of the assumptions inherent in GCMs may be made possible by specifically altering the boundary conditions of the models so that they are projected backwards in time. This "experimental" approach has already been discussed in the context of temperature changes, and several authors have used the same basic approach to test the moisture-balance component of models. It offers the possibility of examining more thoroughly whether any discrepancies which may arise from the general comparative approach discussed above are actually due to problems in the modelling assumptions, or to changed boundary conditions.

In order to compare GCM predictions of moisture balance against palaeoevidence, it is necessary to bring both the model predictions, which are in terms of precipitation and water-balance figures, and the palaeoevidence of lake levels, fossil dune distributions, vegetation cover and composition, into a common currency. This is often a difficult task, but one which is becoming increasingly refined as more information is gathered.

In the case of vegetation indicators, the matching of palynological evidence to specific moisture balance (and also temperature) conditions is possible. This approach uses pollen "response surfaces", based on studying how the relative and absolute abundance of pollen types appearing in sediments varies in relation to *present-day* climatic variables. This flexible quantitative method has been used by the COHMAP (1988) group to validate the accuracy of climate modelling attempts for eastern North America in the late-glacial and Holocene. The use of response surfaces offers considerable promise for future testing of climate models, although to be of use in global testing of climatic models it will have to be dependent on a very detailed pollen record and subsequent analysis over large areas.

Alternatively, a rough comparison can be made using bioclimatic schemes (Woodward, 1987a, and Chapter 3, this volume), which are based on the observed correlations between vegetation structure and climatic zones in the present world. Using such a scheme, the data from a GCM can be put into the form of predicted vegetation maps, for whatever time in the past is being simulated. These maps can then be compared in a *general* way with the vegetation which seems from palaeoevidence to have actually existed. However, no bioclimatic scheme can accurately predict the distribution of all vegetation types everywhere in the present-day world (because vegetation is so much an outcome of subtle combinations of many climatic, soil and site characters). This problem may be especially great in terms of past conditions, because it seems that very unfamiliar climatic patterns must have been the cause of certain vegetation types that do not exist in our present world, such as the rather strange "steppe–tundra" or "mammoth steppe" (Guthrie, 1990) which existed over much of the northern hemisphere during glacial phases (Adams *et al.*, 1990). In such cases, we may have no real way of knowing whether the GCM prediction of climate is right as we have no diagnostic vegetation types to match up against it. What constitutes a sufficiently close match between a bioclimatically predicted vegetation type and an actual vegetation type for it to be treated as "correct" is a rather subjective decision in itself, and may give a misleading result. Despite these caveats it does seem reasonable to regard vegetation zones and other general qualitative evidence from the Quaternary as a reliable *general* indicator of the moisture con-ditions of past climates, with sufficient accuracy to allow their use as a test of climate model results.

An example of the application of palaeodata to the past as a testing ground for models comes from the work of Adams *et al.* (1990) and Prentice (1990). Prentice (1990) has applied one of the standard GCM models (the GISS model), to the global conditions 18 kyr ago at the LGM, under altered boundary conditions of seasonal radiation, ice cover, sea level and atmospheric CO_2 concentration. When Prentice's vegetation maps from the GCM model applied (Fig. 8a) are compared with results of a compilation of palaeovegetation evidence by Adams *et al.* (1990) (Fig. 8b), the contrast is striking. The GISS model used by Prentice predicts moister conditions over large areas of the land surface for 18 kyr, with a 90% reduction in the area of deserts and a 20% increase in the area of rainforests. The more direct palaeoevidence shows that in fact the opposite occurred; there was a major expansion of deserts and a substantial decrease in rainforest area. It appears that there is something seriously wrong with the model in its treatment of water balance under glacial conditions, probably due to it overestimating precipitation under the cooler glacial climate. It is possible that the lower atmospheric CO_2 level at that time may have had a significant effect in reducing the water-use efficiency of the vegetation (Woodward, 1987b), but it is unlikely that it could account for all of the difference between the GCM predictions and palaeoevidence. Such apparent errors in the prediction capabilities of a GCM do not add to our confidence that it will be capable of accurately forecasting how water-balance patterns will change in response to climate warming. When applied to a future greenhouse scenario, with a doubled CO_2 level, the GISS model predicts moister conditions over much of the world. Arguing just on the basis of its apparent errors when applied to the 18-kyr situation, it may be that the GCM generally underestimates how strongly moisture balance improves with climatic warming, so in fact the world might actually become a still moister place than it predicts.

The example of the GISS prediction is perhaps unrepresentative of current GCMs, as it even tends to simulate present-day climate conditions rather poorly in comparison with other models. Guetter and Kutzbach (1990) have used the NCAR CCM model (which is more typical of current GCMs in general) to produce global vegetation zone "predictions" under the Koppen bioclimatic scheme, for 126, 18, 9 and 6 kyr. The broad climate/vegetation zone categories used in the Köppen scheme may tend to confuse the comparison, but it at least allows a quick preliminary overview. Palaeomapping for vegetation at 126 kyr is too sketchy for any meaningful comparison between the CCM predictions and palaeoevidence, but a few qualitative comparisons are possible with the later dates. As for the GISS model in Prentice's study, when applied to 18 kyr the CCM seems to overestimate moisture availability for much of the earth's surface. For example, it predicts an expanded area of rainforest in Africa for that time, when the rainforest is actually known to have been of much smaller extent. Extensive areas of moist boreal forest are predicted for Europe and Asia, whereas palaeoevidence shows that this was not the case. For the more recent Holocene times, the CCM seems to be better able to predict the conditions which actually occurred. For example, at 9 and 6 kyr the prediction of moister conditions in the Sahara is fulfilled. There is also a qualitative agreement in that the GCM correctly predicts that the Asian desert areas seem to have been smaller at that time (Grove, 1986).

Rather vague comparisons of this sort are only using part of the potential power of palaeoevidence as a test of the predictive abilities of climate models. More thorough analysis of the palaeoclimate implications of past vegetation distributions offers considerable scope. The use of pollen "response surfaces" has already been mentioned: this technique can be applied to both moisture balance conditions and temperature conditions (e.g. COHMAP, 1988). Even if response surfaces are not used, thorough mapping and description of palaeovegetation has definite benefits. A recent result of mapping of the palaeoecology of the Sahara and Sahel (although not using pollen response surfaces) has been Street-Perrott *et al.* (1990)'s test of moisture conditions simulated by the CCM model (the same model used by Guetter and Kutzbach, above) for 9 kyr. The CCM GCM has been shown to be capable of reasonably accurately predicting the wetter palaeoclimate in the North African region on the basis of the changed conditions of seasonal radiation influx. Its accuracy is further increased when the model is also supplied with prescribed data on surface albedo resulting from the changed distribution of

Fig. 8. Comparison of predictions of vegetation zones from a GCM with palaeo-evidence for 18 kyr. (a) Model results, (b) reconstruction from palaeoevidence. (Source: (a) is from Adams *et al.* (1990), (b) is from Prentice (1990). To aid comparison, ice-sheet areas in both maps are the same, and are taken from Adams *et al.*, 1990.) Key to vegetation types; 1. Tropical forest, 2. Woodland, 3. Grassland, 4. Temperate forest, 5. Conifer forest, 6. Tundra, 7. Semi-desert, 8. Desert, 9. Polar desert, 10. Ice, 11. and 12. Steppe-tundra (northern and southern types).

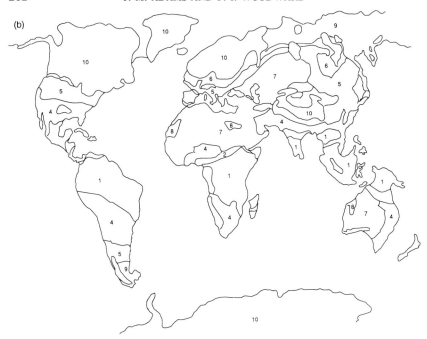

Fig. 8. Continued

vegetation zones (note that although this method uses vegetation as both an indicator of moisture conditions and a boundary condition of the model, this does not introduce circularity. If the workings of the model had been incorrect the simulated palaeoclimate conditions could still have worked out to be *wrong* for the vegetation whose albedo conditions were specified in the model) (Street-Perrott *et al.*, 1990). It appears that an enhanced monsoonal effect at these times can explain the palaeoclimate conditions on the basis of existing GCM modelling assumptions and techniques. This is certainly a promising indicator of the potential accuracy of these modelling methods in predicting moisture balance in a future greenhouse climate, and an indication of the increased accuracy which may be achievable if albedo feedbacks are brought into these models (Street-Perrott, 1991). It may also be partly the lack of this albedo factor which leads to the apparent inaccuracy of Guetter and Kutzbach's projections into the last glacial maximum. The additional accuracy obtained from including albedo effects is however slightly unsettling from the point of view of trusting existing greenhouse model projections which do not include this factor.

The success of the modelling attempt by Street-Perrot *et al.* also in itself confirms that a simple general analogy between past and future moisture conditions would be incorrect because of the importance of the changed boundary conditions of seasonal sunlight distribution. Using GCMs we know that this particular analogy between past and future would be a poor one, and we also know *why* it is a poor analogy.

C. Feedback Effects Involving Sinks and Sources

Over the coming thousands of years, much of the extra CO_2 which man is adding to the atmosphere will be taken up by the oceans, to form organic matter in sediments and bicarbonates dissolved in sea water. The other greenhouse gases will decline more rapidly from their raised levels if anthropogenic production ceases, as they are broken down by ultra-violet light or by reacting with such oxidizing molecules as the hydroxyl radical. Many would disappear back down to natural levels within decades or centuries of their anthropogenic sources being removed (Watson *et al.*, 1990).

The long-term perspective seems of minor relevance to a situation in which greenhouse gases are rising rapidly, and look set to rise for some time to come. The natural mechanisms which tend to remove these gases are obviously being overwhelmed in the short term. Yet, they *do* seem to be having at least some effect on the rate of rise. For instance, in the case of the carbon cycle, around 50% of the extra carbon which is added to the atmosphere each year goes "missing", in the sense that it does not contribute to the rising CO_2 level (Watson *et al.*, 1990). There are various possible ways in which this carbon might be getting removed from the atmosphere, but no-one is quite sure how important each route is, and how it is likely to respond to a continuing input of CO_2. Perhaps in time, these sinks could start to saturate and so become relatively less effective in taking up CO_2. Perhaps they might even respond to greenhouse warming by going into reverse, and acting as net *sources* of CO_2 and other greenhouse gases? Obviously, if such changes are going to occur it is important to allow for them in planning quotas or cutbacks in greenhouse gas emissions. In order to do this, we must understand how the biosphere's sources and sinks operate in response to perturbations. There is much to be gained from studying biogeochemical cycles in the present-day world, but the more sophisticated understanding of past changes in the these cycles which has come about in recent years may also be able to give some important perspectives.

1. Vegetation, Soils and Feedbacks on CO_2

It is estimated that there are presently around 550 Gt (1 Gt $= 10^9$ t) of carbon in live vegetation, and 1500 Gt in soils and plant litter (Melillo *et al.*, 1990). This represents a major reservoir of carbon in exchange with atmospheric CO_2; for example it seems that around 50 Gt of carbon is exchanged between vegetation and the atmosphere each year in photosynthesis and respiration (Melillo *et al.*, 1990). If the climate changes due to greenhouse gas emissions, we can expect from the observed relationships between climate and the amount of carbon in vegetation (Olson *et al.*, 1982) and soils (Post *et al.*, 1982), that there will be some change in overall carbon storage and CO_2 exchange. This represents the potential for either negative or positive feedbacks on the atmospheric concentrations of CO_2 and on the climate, as the soil–vegetation system alters its carbon balance in response to climate change. Another potentially important factor interacting with the indirect climate effect, is the direct biological effect of CO_2 on metabolism and water balance (Woodward, 1987b; Melillo *et al.*, 1990).

i. Estimating carbon storage under past conditions of warmer climates. As was discussed earlier in this chapter (Section B), the warmer phases in the earth's recent history seem to have been associated with a different moisture balance over the continents. From present-day observations of natural vegetation, it is known that the warmer and/or moister the climate, the greater the amount of carbon which tends to be stored in vegetation and soils (Olson *et al.*, 1982; Post *et al.*, 1982). Those reconstructions of vegetation zones which are available for the warmer phases of the previous interglacial and the Pliocene indicate that global carbon storage in vegetation and soils would have been favoured by the moister climates that seem to have prevailed. In contrast, the cold phase of the LGM (18 kyr) seems to have been associated with much greater aridity over much of the earth's surface. Some calculations by the authors and their co-workers, based on the global distribution of vegetation zones at that time, indicate that there may have been less than half as much carbon in vegetation and soils at the LGM, than during most of the present interglacial (Adams *et al.*, 1990).

Reconstructions of vegetation zones for the apparently "warm" period of the early-to-mid Holocene also indicate a much greater spread of forests and other dense vegetation in areas of the northern hemisphere which are now naturally barren desert (Folland *et al.*, 1990). Calculating on the basis of observed present-day carbon storage in particular vegetation zones, H. Faure and colleagues (unpublished data) extrapolate back to the early Holocene distribution of vegetation (Lezine, 1989) to suggest that perhaps as much as 100 Gt of carbon (about 0·05% of the present terrestrial total) has been lost from Saharan and sub-Saharan Africa alone since that time. (However, evidence from certain other parts of the world indicates drier conditions than at present, and presumably a decreased carbon storage.)

It seems that as the world's climate in the past has become warmer, the moister conditions and the resulting spread of vegetation that are associated with warming may have tended to favour increased carbon storage. This suggests that vegetation and soils would tend to act as a negative feedback on atmospheric CO_2 and on climate. However, as was discussed in Section B of this chapter, past phases of warm climate are not necessarily identical with a future phase induced by the greenhouse effect. The general prediction of increased carbon storage, based on vegetation zone changes in the past, is in general agreement with the prediction of Prentice and Fung (1990) using a GCM simulation for a CO_2 doubling. However, it is difficult to do anything more than regard such general analogies from the past as a rough qualitative indicator of how carbon storage will tend to change as the earth warms.

ii. Constraints on response times of carbon storage. Whilst a future climatic warming seems likely to favour increased carbon storage, the rapid pace of the climate change which may occur means that the response times of vegetation and soils will be crucial in determining how much CO_2 is actually taken up. Warming phases in the past, such as occurred at the end of the last glacial phase, were probably associated with major increases in carbon storage over large areas of the land surface. This offers a superficially tempting analogy against which to forecast the speed of future carbon storage changes.

Schlesinger (1990b) has suggested that the estimates of Adams *et al.* (1990), which show a contrast between full-glacial and full-interglacial carbon storage, are consistent with very slow potential response times for increases in the amount of carbon storage on land (about 0·13 Gt of carbon per year, which is a small fraction of the rate at which carbon is presently being added to the atmosphere). However, this is not really a relevant approach to the problem. Comparing points in time where the distribution of vegetation seems to have been in overall equilibrium with climate does not properly address the issue of how fast carbon storage could potentially respond to a very rapid change in climate. Several such rapid climate changes are known to have occurred in the North Atlantic region during the last 100 kyr, the most recent being the post-Dryas warming around 10 kyr. Phases of rapid climate change (or the periods of relative stasis which followed them) seem to be the best analogy for the future, as they are likely to reveal the processes of ecosystem change "unleashed" and able to operate at their maximum potential rate.

Although phases of rapid vegetation change might potentially be useful analogues for the future, it is impossible to quantify the rates of carbon uptake response without very detailed direct information on the actual rates of soil and vegetation change at those times. A few qualitative indications of the response times are available, however. The palaeoevidence for lags in vegetation change following rapid postglacial warming events (Section C) suggests that in many areas the vegetation may not reach its potential stature and biomass for centuries or millenia after a greenhouse warming gets under way. Of course, this will depend on the extent to which increase in vegetation stature can or cannot be achieved by species already present *in situ*. If the pre-

286 J. M. ADAMS AND F. I. WOODWARD

existing vegetation is able to respond, the process might still be a slow one. Quantitative indications of how long vegetation carbon might take to build up come from the more recent past, by studies of temperate and tropical forests which have been logged, or cleared and allowed to regenerate, in the last few hundred years. These studies show that whilst carbon storage in forest vegetation often takes centuries to reach its maximum potential, the initial build-up of biomass is much faster (e.g. figures summarized by Cannell, 1982). Thus, a major part of the feedback potential of developing vegetation on CO_2 might occur on a time-scale of decades.

In addition to migration lags, other factors could be important in limiting future build-up of vegetation biomass in certain regions. Soil factors may be significant in limiting the colonization of bare areas by vegetation. Magri (1989) suggests that slow rates of soil maturation limited the build-up of forest biomass for thousands of years at sites in central Italy, following episodes of rapid climatic amelioration during the Quaternary. A similar set of soil factors may have been limiting to the build-up of forest biomass for over 1000 years in Britain following rapid warming events during the late Quaternary (Pennington, 1986). In a future greenhouse world, such lags due to soil development time might occur beyond the present arctic and alpine treelines. They might perhaps also be important in semi-arid areas, especially those which have been denuded of their soils over the past millenia because of heavy grazing pressure on the existing vegetation (although there has been no detailed study of Quaternary vegetation lags in such areas). However, studies on the more recent history of colonization of mineral soils in many parts of the world suggest that these conclusions of very long lags due to soil maturation may not be justified. For example, at Glacier Bay in Alaska, glacial morianes are colonized by dense, tall conifer forest within 150 years of their initial formation (Burrows, 1990).

Some clues to the maximum rates at which carbon can build up in the soils themselves come from dating studies of soils which have been formed anew at various times during the Holocene (under conditions of relatively stable climate), in such places as sand dunes and glacial outwash deposits. Schlesinger (1990a) suggests, on the basis of these studies, that in a maturing soil the rate of carbon accumulation is often only about $0 \cdot 1$–$0 \cdot 01$ g m^{-2} year^{-1}, so that many soils in the present world must have taken thousands of years to build up their organic content of kilograms of carbon per square metre. This gives some idea of how slowly soils will be able to respond fully to the moister and warmer conditions of a greenhouse world. Even if we imagine new soils forming on areas of arid desert or polar desert as the climate becomes moister and warmer, and an increase in the carbon storage potential of existing soils under forests or grasslands, it seems unlikely that a major fraction of the carbon being added to the atmosphere each year by man could be taken up by soils (Schlesinger, 1990a,b). However, Schlesinger's figures seem suprisingly slow in comparison with estimates of soil carbon build-up presented in many studies on soil development during primary colonization of areas by vegetation (Burrows, 1990). As Schlesinger points out, the initial rate of carbon build-up over the first few centuries may be much higher than its overall rate in the long term, so perhaps the potential feedback from newly vegetated areas will be stronger than for areas with a pre-existing vegetation cover.

Whilst a climatic warming can be expected to increase carbon storage in many soils, a decrease in carbon storage can be expected in certain areas. For instance, in the major boreal peatlands the lowering of water tables is likely to result in a net oxidation of peat. Episodes of net peat oxidation are recorded from many peat bogs (Clymo, 1984) during past phases of relatively warm climatic conditions such as the mid-Holocene (in these areas the trend with warming seems to have been a switch towards decomposition over production, partly related to falling water tables). Carbon dating indicates that most peat bogs have spread and built up only gradually over many thousands of years (Clymo, 1984). Whilst the initiation of peat build-up further north than at present might be favoured by climate warming, it is possible that the rate of oxidation in existing peat bogs could greatly exceed this, resulting in a major net efflux of carbon to the atmosphere in the shorter term.

All this evidence seems to indicate that if, in the future, climate switches towards a state favouring *increased* carbon storage, the maximum potential storage of carbon by vegetation and soils is not likely to be reached for centuries or millennia. However, in the short term, a relatively rapid initial build-up of carbon might provide a very important carbon sink. On the other hand, decomposition of peat may result in a net release of carbon to the atmosphere, thereby acting as a positive feedback on climate. If in fact the greenhouse climate shifts to favour *decreased* carbon storage in many areas, we may see the sort of rapid decomposition and release of soil carbon which has been observed in decade-timescale studies on soils under changed land-use conditions.

iv. The CO_2 fertilization effect. It is known from growth experiments in plant physiology that an increase in CO_2 levels above the present ambient, up to about 1000 p.p.m., tends to increase photosynthesis in plants, and results in more efficient water use (Mooney *et al.*, 1990). If a similar effect occurs in natural habitats as CO_2 levels increase, this may tend to result in increased carbon storage in vegetation and soils. However, there is great uncertainty as to how strong the effect will actually be. If indeed there *is* a strong CO_2-fertilization effect, it could be of considerable importance in providing an additional negative feedback on CO_2 levels in the atmosphere.

Whilst changes in atmospheric CO_2 during the later Quaternary are well documented, it is virtually impossible to separate any vegetation changes due to CO_2 fluctuations from the effects of the climatic changes which accompanied them. An interesting confirmation that the CO_2 rise over the past two centuries has had *some* effect on vegetation physiology is the finding of Woodward (1987b) and Penuelas and Matamda (1990) that the stomatal numbers of temperate tree leaves have decreased during that time, presumably as a result of decreased water stress/increased carbon availability. Studies are currently under way to determine whether stomatal indices of leaves were any different under the low CO_2 levels of glacial phases. Whether or not such changes in stomata have occurred, the correlative effects on standing biomass are impossible to judge simply from this indicator alone. Another qualitative indication consistent with a direct CO_2 effect on woody plant growth, and possibly biomass, comes from the study of tree rings that have formed over the past two centuries. Graumlich (1991) has studied trends in tree-ring width in conifers growing under semi-arid conditions at high altitudes. On physiological grounds, one would expect that plants growing in this type of environment would show a relatively strong CO_2-enhancement of growth, compared with plants growing in less water-limited conditions under higher total CO_2 pressures. There appears to have been a slight significant increase in mean ring width over this time series at some, but not all, sites. However, such changes could partly have come about through the climatic changes which have occurred during that time (Section A).

Generally, there are some worrying uncertainties as to what role the terrestrial system plays, and will continue to play, in taking up anthropogenic CO_2. Around 4 Gt of carbon from fossil fuels and tropical forest clearance seems to go "missing" each year, with only 3 Gt accumulating in the atmosphere (Tans *et al.*, 1990). Oceanographers have studied the rate at which carbon is being taken up by the oceans, and on the basis of these studies Tans *et al.* (1990) have suggested that only 1 Gt actually goes into the oceans, with around 3 Gt being absorbed by terrestrial ecosystems as a result of CO_2-fertilization. An historical perspective may be relevant to considering the potential future role of the terrestrial system in taking up carbon.

It does seem that looking back over the past century-and-a-half, a great deal of the anthropogenically produced CO_2 has gone "missing", with vegetation and soils possibly being one of the major sinks. Calculations indicate that around 190 Gt (59%) of the anthropogenically released carbon has been removed rather than having accumulated in the atmosphere (Watson *et al.*, 1990). It would be much easier to try to account retrospectively for this large "lump" of carbon than for a much smaller increment of carbon going missing each year. If we can solve the mystery of where all this missing carbon has gone, we may then be able to tell where future additions of carbon will go. Could the missing 190 Gt be the result of CO_2-fertilization, which Tans *et al.* suggest is a major sink? To take an extreme situation, consider that all 190 Gt of missing carbon had been taken up by vegetation and soils during the last 150 years. If indeed all of this carbon had gone into the terrestrial system, it would represent almost a 10% increase in total carbon mass, with about 5% being during the last 50 years. This sort of figure, or even a lower figure such as 140 Gt (allowing for the Tans *et al.* estimate of a quarter of the carbon being taken up by the oceans), seems rather unlikely when one considers the evidence of strong nutrient limitation on biomass accumulation in CO_2-fertilized ecosystems (Melillo *et al.*, 1990). Likewise the historical soil data of Schlesinger (1990a), showing very slow accumulation of soil carbon over thousands of years, seem to make it very unlikely that the soil carbon reservoir could have responded so rapidly as to have acted as a major sink. Although Schlesinger's figures may underestimate the accumulation rates of carbon in certain soils, evidence generally indicates that mature soils would tend to respond over many decades or centuries (Burrows, 1990). Tans *et al.* (1990) suggest that the mid-latitude and boreal forest zone may be responding to CO_2 to act as the major sink for "missing" carbon. If it is, the proportional increase in carbon storage in this area would have to have been much greater than the average for the world as a whole. This is made particularly unlikely by the evidence for strong nutrient limitation of temperate and boreal forest growth. This scenario illustrates that there are reasons to doubt that CO_2-fertilization alone could have resulted in most of the "missing" carbon having been taken up by existing areas of vegetation on the land. If in fact much of the carbon *has* been going into vegetation and soils as a result of a CO_2-fertilization effect, it seems unlikely that they will be able to continue responding so strongly for much longer before nutrient limitation begins to operate on productivity. However, here as always the historical perspective is fraught with complexities and uncertainties. For instance, acidic pollutants from fossil fuel burning might have been important in providing additional nutrients to forest growth, thereby facilitating a rapid build-up of carbon storage. No-one is quite sure just how land use has changed over the last 200 years, and just how much carbon has actually been released from cleared soils and vegetation, and taken up by regenerating vegetation in other areas. There is the possibility that changes in agricultural and forestry activities over the last 50 years may have allowed a huge amount of carbon uptake by regenerating forests in the mid-latitudes of the northern hemisphere (R. Sedjo, reported in *New Scientist* article 4.5.91). Thus, it may be *land use* rather than CO_2-fertilization which has been taking up the missing carbon. Many of the same problems apply to tracing the fate of the missing CO_2 on a year-by-year basis in the present day, and overall the longer-term historical perspective may be better equipped to deal with them.

290   J. M. ADAMS AND F. I. WOODWARD

Work is now proceeding apace to try to narrow the uncertainties on the question of the missing carbon.

2. CH_4 Production by the Biosphere

As a greenhouse gas being produced by the biosphere, CH_4 constitutes another potential feedback in the greenhouse warming process. Various biotic and non-biotic sources have been documented for the present-day world, and how these would change in response to climate change is uncertain (Watson et al., 1990).

The ice-core record of the Quaternary shows that CH_4 levels have closely paralleled past temperature changes (Chappellaz et al., 1990) (Fig. 7), varying between about 0.35 p.p.m. during glacials and 0.65 p.p.m. during the early stages of interglacials. Chappellaz et al. (1990) suggest that these fluctuations are mainly due to variations in the global area and productivity of wetlands, in response to changes in temperature and moisture balance. If the climate generally becomes moister as a result of future greenhouse warming, it could be that CH_4 production from natural sources will be favoured. This will tend to produce a positive feedback, amplifying the climate warming. Taking the glacial–interglacial fluctuation (~ 0.30 p.p.m.) as an indicator of the responsiveness of natural CH_4 production to climate change, it seems unlikely that the amplification effect would be of major importance against an anthropogenically produced background level (presently ~ 1.6 p.p.m. and rising rapidly). However, the present high level of CH_4 is at least partly due to rice production and cattle farming (Watson et al., 1990), which are themselves constrained by climate. If climate changes so as to favour the spread of these activities, a substantial amplification of the warming effect might ultimately occur.

In the present world, a huge amount of CH_4 lies trapped in permafrost and beneath the seas in the form of hydrates and clathrates, apparently formed from "fossil" methane migrating up out of older organic-rich rocks (Kvenvolden, 1988). On the basis of the present-day distribution and stability characteristics of these reservoirs, Kvenvolden warns that an initial greenhouse warming might be amplified by the release of this CH_4 over the coming centuries. However, the actual sizes of the hydrate and clathrate reservoirs, and their potential to respond to a rapid climate warning, remains an area of considerable uncertainty. There may be an interesting perspective from the past which can be brought to bear on this subject. For instance, Nisbet (1989, 1990) notes that the link between orbital forcing changes and deglaciation seems to require a high-latitude amplifying agent acting on climate. He suggests that this link may partly have been supplied by the release of high-latitude reservoirs of CH_4. To back up this hypothesis Nisbet assembles various sources of evidence from the geological record. He argues (Nisbet, 1989) that the rapid rises and peaks of CH_4 levels in ice cores (around 13 kyr and 10·5 kyr) at the start of the present interglacial seem to have occurred too early for them to be the product of either high-latitude or low-latitude wetland activity. Instead, he suggests that part of the initial CH_4 rise observed in the ice core record was due to rapid destabilization of CH_4 hydrates and clathrates, under permafrost, ice sheets and sea floors, during summer radiation maxima in the high latitudes. Because CH_4 is such a labile gas in the atmosphere, the CH_4 "peaks" in the ice-core evidence may under-represent the true extent of transient bursts of high CH_4 levels. Nisbet (1990a) points to widespread late glacial/early Holocene "pock marks" on land and on the continental shelves as the product of massive CH_4 bursts from these sources, occurring as the climate warmed and as the sea-level rose. Once an initial high latitude summer warming occurred due to Milankovitch forcing, the release of CH_4 may thus have reinforced the warming of climate and played a critical role in the glacial–interglacial transition. Newly released CO_2 from the oceans may then have taken over from CH_4 as the major forcing agent. The validity of Nisbet's scenario seems open to dispute, as the historical evidence is rather tenuous. In any case, as Nisbet points out, the amount of CH_4 required to set off the deglaciation process may have been relatively small in terms of actual radiative potential, as the global system may have been especially sensitive to small changes in forcing at that time. This may thus be unrepresentative of the future climatic effect of the release of a similar amount of the gas. However, we should perhaps bear in mind this evidence of "pock marks" and rapid CH_4 rises in the ice cores as a possible indicator of the future. The CH_4 reservoirs involved will have been kept from building up during glacials by the repeated warming events which have occurred during the Quaternary. If these CH_4 reservoirs have indeed been capable of playing a significant role in amplifying changes on a glacial–interglacial scale, the response of as-yet undisturbed reservoirs to a further rapid warming—and a global temperature level which has not occurred within the last 4 million years—might have unfortunate implications for mankind (Huntley, pers. comm.).

3. The Oceans

This chapter has unavoidably been biased towards the terrestrial system, because of the backgrounds of the authors. However, the oceans will clearly be a major component in determining the future of CO_2 levels in the atmosphere, because of their vast potential as a source or sink for carbon. Their potential importance is emphasized by the history of the carbon cycle during the Quaternary, as revealed by ice-core and ocean-sediment data. It appears that changes in ocean circulation and plankton activity during glacial–interglacial cycles may have been largely responsible for the large changes in the CO_2 content of the atmosphere which occurred during those times (although the details are still very uncertain; see Berger, 1991). On the basis of this general relationship from the past between ocean carbon storage and temperatures, it seems possible that the changes in oceanic temperature and circulation which occur during greenhouse warming could result in a decreased ability of the "oceanic pump" to take up CO_2. Combined with the effects of warming on solubility of CO_2 in the surface layers, such changes might even result in the oceans becoming a net *source* of CO_2. Some results from ice-core studies in Greenland (Stauffer *et al.*, 1984) suggest that during the Quaternary, abrupt CO_2 concentration shifts of the order of 50 p.p.m. may have occurred within less than 100 years, associated with sudden climatic changes (with temperature shifts of around 5°C). The cause and effect relationship here is not clear, but it may indicate that a greenhouse warming could set off a very rapid positive feedback process between temperature and ocean CO_2 output.

Other feedbacks due to oceanic changes could also occur. The possibility of CH_4 clathrate and hydrate destabilization has already been mentioned. Another interesting process suggested from observations on the past, is that ice-core evidence from Antarctica suggests that the gas dimethyl sulphide (DMS) is produced by ocean plankton at a greater rate under changed glacial oceanic conditions (Legrand *et al.*, 1991). From its possible effects on cloud droplet formation and albedo, Charleson *et al.* (1987) have argued that DMS production could be an important factor in reinforcing the colder conditions during glacial phases. If all the links in their argument are correct, it may be that changes in ocean ecology during a future greenhouse warming will further decrease DMS production and this will further reinforce the warming process.\

4. The Gaia Hypothesis

One subject which is unavoidable in the context of biospheric feedbacks is the Gaia hypothesis of Margulis and Lovelock (1974) and Lovelock (1989). This is a general interpretation of the geological record and of present-day biogeochemical cycles, which emphasizes the role of the biosphere as a regulator of climate and environment in the past and present world. Lovelock and Margulis suggest that the complexity and efficacy of these feedbacks are so great that the whole biosphere (together with the inorganic components of the oceans, atmostphere and even the earth's crust) might reasonably be compared to the metabolism of a single living organism, forming an intergrated whole (known as *Gaia*) which is capable of responding homeostatically to future challenges. Putting aside the question of whether the analogy between the biosphere and a living organism runs very deep, we have to ask whether the generalized picture of powerful negative feedbacks really does offer useful insights to the immediate future. As Lovelock has pointed out, a simplistic interpretation of the Gaia hypothesis might lead to the view that the system will be able to cope with whatever we can throw at it. There is clearly a need for caution here!

The true role of biotic feedbacks in regulating the earth's climate and environment over the last several billion years remains uncertain and controversial within the scientific community, although an acceptance of their importance does seem to be gaining ground amongst younger scientists. If indeed one accepts that the earth's history has been dominated by the moderating homeostatic influences of the biosphere, many of the feedbacks involved will have operated broadly and gradually in response to long-term changes over many millions of years. This may have little relevance to understanding their quantitative response to relatively rapid anthropogenic perturbations. The oscillating system of positive feedbacks by which glacial–interglacial cycles occur does not exactly look like the operation of efficient homeostasis on a shorter time-scale. It appears that there are some intrinsically destabilizing factors within the system (perhaps, for example, DMS production), which tend to be "swamped out" to varying extends by other feedbacks. Lovelock (1989) himself qualifies his view of a homeostasic world with the suggestion that if anthropogenic effects push the components of negative feedback loops too far, these could also switch to driving positive feedbacks. We could thus throw parts of the global system outside the realm of homeostasis and into a runaway positive feedback if we are incautious enough.

If there is a guiding principle for the future to be gained from the Gaian perspective, it is that we should watch where we are treading as we modify the world around us. Things often turn out to interact in subtle and complex ways, in a most unexpected manner. Having said this, one has to remember that Gaia is just a generalization. It would be incautious to predict something simply on the basis of this generalization without also examining the specifics of each situation, just as the principle of cellular homeostasis is not in itself a very satisfactory guide to predicting the effects of a drug on metabolism.

D. Effects on Floras and Faunas

For many reasons, the effects of climate change on floras and faunas should be a major part of any assessment of the costs or benefits of allowing a greenhouse gas build-up to continue. In order to make an informed decision on this, it will be necessary to use as many different sources of information as possible. There is clearly much to be gained from the study of present-day biogeography and ecology (Mellilo et al., 1990). However, in addition to this there is a potentially vast record of past changes in the distribution of animals and plants in response to climatic change and other disturbances.

1. Expansion of Ranges and Populations

In the present-day world, there are many obvious relationships between climatic factors and the distribution ranges of species. Within the areas in which a species occurs, the actual abundance of individuals also often varies in relation to climate. From these relationships it appears that if the climate zones shift, many species will be capable of living in places in which they do not now occur naturally, and will also become more common within certain parts of their ranges.

If a global climate warming occurs, it is likely that many mid-latitude and high-latitude species will be capable of surviving north of their present-day ranges. For instance, a typical model prediction is that the mean annual temperature of Britain will rise by about 2·5°C between 1990 and 2050. In geographical terms, this is equivalent to a northward shift of climatic isotherms by 300–400 km, and to keep up with this shift a species' northern range boundary will have to show an overall rate of movement of about 5–6 km year^{-1} (Cannell, 1990; Tallis, 1990). In the tropics, range boundaries are more often related to variation in rainfall or water balance. If climates become moister than they are now, we can likewise expect that many species will tend to expand their ranges.

However, the new biogeography of a greenhouse world will not instantly snap into place as fast as the climate changes. In order for a species to expand its range, individuals must physically spread out and establish themselves over the new territory. For them to reach their potential level of abundance, scattered individuals behind the new range boundary must multiply up into populations. It is of great interest to know how quickly or how slowly such range and population adjustments will occur. For any particular set of species, we must ask whether they will be able to keep up with the climate as it changes, or whether they will lag behind and take centuries or millennia to catch up.

i. Postglacial migrations: Temperate-zone trees.   There is an obvious appeal in the analogy between the impending greenhouse scenario and the great migrations of species which occurred as climate warmed at the ends of glacial phases. However, although we know from the fossil record that migrations and range extensions did occur, the dynamics are still very sketchy for most groups of organisms. The greatest detail is known for the migrational histories of woody plant species in the mid-latitudes of Europe and eastern North America, during the warming phase of the last postglacial. These trees and shrubs represent a tiny fraction of the floras in which they occur, but they are also of value to us for the character they give to landscapes and plant communities. Apart from their intrinsic importance, there is also the hope that they might give clues to how other groups of species will behave as climatic conditions change.

Migration rates since 14 kyr have been estimated for 500-year intervals from the pollen record of central and northern Europe (Huntley and Birks, 1983) and 1000 year intervals for eastern North America (Davis, 1976, 1983, 1986; Delcourt and Delcourt, 1987). For Europe the maximum recorded rates for species vary between about 0·02 and 2 km year^{-1}, with a mean of around 0·4 km year^{-1}. For North America the rates tend to be somewhat lower, ranging from 0·08 to 0·4 km year^{-1}, with a mean of around 0·2 km year^{-1}. The fact that there is such variation in migration rates seems to suggest that not all of these species were migrating fast enough to keep up with climate, and that there is tremendous variation in the potential of tree species to migrate. Going simply by these published figures for migration rates, it seems the geographical ranges of a few species might almost be able to keep track in a greenhouse warming scenario in which climate moved northward by about 5 km year^{-1}, and would perhaps show a migration lag of the order of a few decades. The future rate of warming may be expected to vary with latitude, according to climate model predictions, so the rate at which tree species' ranges will need to shift to keep step with climate warming will be greatest at the more northerly latitudes. If the reported migration rates are representative of the boreal and temperate zones of the world, it appears that at all latitudes most tree species would be left far behind, but might catch up on a time-scale of centuries or millennia if the warming stabilized.

However, just relying on the reported migration rates of trees in the pollen record is a very simplistic way of trying to forecast their future responses to climate change. There are various reasons for thinking that it might not be a very good predictor. Firstly, there is the problem of actual interpretation of what was happening. The major causal and limiting factors in tree migrations are very difficult to disentangle. There is a need to distinguish between migration patterns limited by the slow amelioration of climate, and those limited by the dispersal capabilities of species under climatic conditions favourable to range expansion (Prentice, 1986). This is made especially difficult by the uncertainties over the actual time course of climate change during the deglaciation; no-one is quite sure to what extent the climatic amelioration occurred, by sudden jumps or slow century-by-century change, and whether different areas warmed at different times. The pollen record itself may be regarded as imperfect, in the sense that it may have failed to pick up the initial sparse "wavefront" of tree migration (Bennett, 1986). This may have given a misleading impression of the speeds at which range boundaries were shifting, and the times when the main range-boundary movements occurred. Secondly, there are basic differences between the ecological and climatic context of postglacial migrations and future greenhouse migrations. The major limiting factors on migration rates might have been very different from those which will become important in a future rapid greenhouse warming. As mentioned above, part of the tree species migration in the postglacial may have kept step with climatic changes occurring at a much slower pace than are likely to occur over the coming centuries. Also, the influence of humans, as agriculturalists and foresters, is a pervasive influence in the modern world.

If we are determined to persist with an analogy between tree migrations in the postglacial world and the greenhouse world, it seems most appropriate to concentrate attention upon times and places in the postglacial where there is evidence that the climate did change extremely rapidly, in the hope that this will reveal something of the "true" migration potential of temperate trees. There appear to have been two very rapid warming phases in the North Atlantic region at some time around 13 and 10 kyr, which show up in the marine and ice-core record of the North Atlantic as sudden oxygen isotope and plankton shifts (Dansgaard et al., 1989), and in the terrestrial record as abrupt changes in both the sedimentology and in the composition of invertebrate faunas (Atkinson et al., 1987). The rate of climate warming which has been suggested from these sources is quite remarkable; for example from the terrestrial record there is the suggestion of a change of perhaps 7°C in July temperature and 20°C in January temperature in under a century (Coope and Brophy, 1972). At these times, there is considerable evidence of ecological disequilibrium on the western edge of Europe (Coope and Brophy, 1972; Coope, 1987; Ponel and Coope, 1991) and in parts of the north-eastern USA (Davis, 1984). Insects, molluscs and water plants characteristic of temperate conditions were present, and yet many temperate forest trees were completely absent (or present in very low numbers) from these areas for more than a thousand years. One interpretation is that the trees were simply unable to expand their ranges or populations fast enough to exploit climatic conditions which would have been suitable for them (Coope and Brophy, 1972; Coope, 1987).

It is thought that during such phases of free migration, the seed-dispersal capabilities of trees would have been a major constraining factor on range-boundary movement. The sequences of arrival of species in the pollen record, and their individual patterns of northward spread observed during the earlier phases of postglacial migration may relate to differences in their intrinsic seed dispersal capacities, and the differences in positions of refugia and topographical barriers (Davis, 1986). It is interesting to observe that contrary to what one might expect from present-day ecological considerations, there was no simple relationship between seed weight, wind dispersal and observed rates of spread in the early Holocene (Sauer, 1988).

Other disequilibrium factors may also have been important. Although landscapes in Britain remained predominantly treeless following the rapid 13 kyr warming event, tree birches (*Betula*) were present but confined to scattered patches in valley bottoms for almost a thousand years until the climate deteriorated again (Pennington, 1986). Pennington suggests that the slow development of soils and mycorrhizal fungi on the postglacial landscapes was preventing population expansion of these birches up on to the valley slopes and hilltops, and also possibly slowing the pace of migration of other tree species into the region, by offering few places in which a dispersed propagule would be able to establish and grow.

It may in fact be that sometime during the period of their "absence" from the pollen record, trees other than birch had also reached these latitudes as very sparse populations, not detectable in the pollen record, and that the lag in their "arrival" was more in population expansion rather than migration. That this could be an issue perhaps shows how artificial the borderline between the two processes of range expansion and population expansion actually is, as the effect is more-or-less the same in terms of the character or the general ecology of the area. In a broad biogeographic sense, relating to the ultimate survival of species in response to climate change, this is a more important distinction, however.

If the general interpretation put forward by Coope, Pennington and their colleagues is correct, the long migration lags in the spread of trees through the open postglacial landscapes of Europe may well be repeated north of the Arctic treeline, as greenhouse warming takes effect. It is at these high latitudes that the warming is expected to be greatest, perhaps matching the magnitude of mid-latitude warming at the beginning of the Holocene.

The scenario of striking vegetation disequilibrium lasting for many centuries or millennia is not universally accepted, however. Huntley (pers. comm.) suggests that the open landscapes studied by Coope and by Pennington could have been the product of relatively continental climatic conditions which allowed the survival of some temperate elements (the insects and waterplants) and not others (trees), although this does not seem to tally with the evidence of oceanic beetle species (Atkinson et al., 1987) and other general indicators of mild winters and fairy moist summers. Webb (1986) doubts the importance of the pace of soil development as a limiting factor, pointing to evidence of much more rapid colonization of glacial debris by trees in the more recent past, although Pennington (1986) argues that this comparison is not a valid one. Climatic modelling of response surfaces has been used to suggest that in North America at least, where the climatic warming appears to have been somewhat more steady and gradual than in western Europe, some of the early Holocene migration patterns of trees were in equilibrium with climate, rather than lagging behind (Webb, 1986; Webb et al., 1987; COHMAP, 1988). As with so many issues in biology, the choice between equilibrium or disequilibrium scenarios in the early

Holocene is not a simple one. Both may be correct for the past to some extent, and both may likewise be correct for the future.

The migration of tree species through existing forests is likely to be more representative of the situation which will occur during greenhouse warming in most parts of the temperate zone or elsewhere. There may be some indications from the past of the effect which pre-established forests will have on migration rates. As forest developed over the postglacial landscapes, the slower-migrating species would have had to compete more and more with other species which had already established themselves. It is not surprising to find various indications that this had a further slowing-down effect on migration rates, delaying the full range expansion of some species by further thousands of years. For example, hazel (*Corylus avellana*) seems to have migrated much more slowly through eastern Europe than through western Europe, where there was much less forest (Deacon, 1974). In North America, phases of increased forest fire frequency during the Holocene seem to have opened up the forests of Nova Scotia to phases of immigration of temperate tree species, which remained in the region after the fire frequency decreased again (Delcourt and Delcourt, 1987a, b). Part of the reason that European trees generally seem to have migrated faster than their North American counterparts in the postglacial, may be that there was less pre-existing forest in Europe at the times of most rapid warming (Delcourt and Delcourt, 1987a, b). Even so, the picture is not a simple one; as mentioned previously, the detailed study of migration patterns of trees in eastern North America has led to the view that many of the other idiosyncratic migration patterns observed for different tree species between the early and mid-Holocene could actually have been the product of equilibrium responses to shifting combinations of climatic variables (Webb, 1986; Webb et al., 1987). However, there does appear to be a fairly general acceptance amongst Palaeoecologists that the competition factor did indeed introduce migration lags of millennia in at least some tree species which underwent relatively late postglacial range expansions. The palaeoevidence thus seems to confirm the general expectation that natural range expansion of tree species through existing forest will be a particularly slow process, with response times to be measured in millennia if a several degree rise in temperature occurs. However, this will also depend crucially on the influence of the climate change on such factors as fire frequency and death rates in the existing forests. There may not be a true analogy to this in the postglacial situation, in which most forest development apparently occurred either during slow climatic warming, or *after* the major rapid warming events.

As mentioned previously, there are likely to be some important differences between the Holocene tree species migrations and those under a future greenhouse climate change. In the temperate zones, the forests are now generally so fragmented that trees will only be able to expand their ranges fully if their propagules can cross distances of several kilometres or more of cultivated land. There are in fact some signs of this relatively "long distance" dispersal having occurred in postglacial migrations. From the recorded rates of migration in the pollen record it seems that tree species must have advanced stepwise across the postglacial landscape, through occasional long-distance dispersal events of perhaps tens of kilometres, followed by expansion out from these population nuclei (Davis, 1984; Sauer, 1988; Tallis, 1990). This might be regarded as a sign that tree populations could indeed manage to make the "leap" between forest fragments. However, in a man-made environment there is also the difference that a localized woodland fragment will offer a much less effective "target" for propagules than an entire landscape, so the analogy may not be a good one. It is interesting that some of the species which spread fastest in the early Holocene had relatively heavy seeds which are bird-dispersed (Sauer, 1988). It may be that the spread of these species will be favoured as their seeds are carried between forest fragments by birds.

We should not neglect another very important difference between the present anthropogenic situation and the postglacial; the deliberate planting of trees. Many trees are already extensively planted well outside their natural climatic ranges, where they may exist as poorly reproducing and poorly competitive individuals. As climate changes, certain of these species may start to compete on more even terms with the natural flora, and begin to expand their populations. In this sense, many of the elements of the tree flora of the future could already be *in situ*, ready for when the change occurs.

Range expansion is only the initial phase in the adjustment of tree floras to climate change. Population expansion following colonization is what makes a species ecologically significant rather than confined to a few scattered outposts. This process of population expansion is likely to occur over a much longer time-span than the initial migration. Pennington (1986) and Magri (1990) discuss the influence of soil factors in delaying population expansion of trees in deglaciated landscapes, a factor which may become important in some areas as ice sheets and glaciers retreat, and as polar desert landscapes become available for colonization. Bennett (1986) has calculated the doubling times of tree populations in the postglacial environments of the East Anglia region of Britain. He finds doubling times which range from 31 to 460 years, depending on the species and individual locality, with a mean of around 158 years. Interestingly, a rare piece of information on tropical tree communities, for the Queensland rainforests (Chen, 1988) indicates very similar doubling times for tree populations following the moistening of climate in the early Holocene. It generally takes several doubling times before a tree species reaches its ecological equilibrium density after reaching a new area, so this process of adjustment in tree floras is likely to take millennia to be completed. Even this seems likely to be an underestimate of the time necessary for adjustment, as these figures may represent relatively open landscapes in which competition with established species was less important. In a future warming event, there may be the influence of pre-existing forest communities in slowing population expansion, although the disturbance effects of logging activities and perhaps the rapidity of the climate change itself may help to open up these forests to the newly established colonists.

The overall conclusion from what is known of postglacial tree migrations, is that most temperate tree floras are likely to be left far behind by greenhouse climate change, and may take millennia to adjust. However, this adjustment process may well be greatly facilitated by the incidental and deliberate intervention of humans.

ii. Other postglacial migrations. What is the evidence for postglacial migrations in all the species of organisms which are *not* temperate trees? There is not much evidence at all, unfortunately, and that which is available is biased towards the same regions of Europe and North America as for the trees.

For the apparently very rapid warming events in northwestern Europe at around 13 and 10 kyr ago, the available fossil evidence indicates that several groups of organisms responded much faster than trees (Coope and Brophy, 1972; Coope, 1987). For instance in the 10 kyr warming event beetles, waterplants, snails, mayflies and diatoms characteristic of temperate conditions all appear suddenly and simultaneously in the fossil record (Coope, 1987; Ponel and Coope, 1991; Coope, pers.comm.), within the context of the limited resolution available in this record. Coope (1987) regards the simultaneous arrival of so many species, and the correlation of their arrival with changes in the sedimentology, as indicating very short response times in these groups. If Coope and his colleagues are indeed correct in this interpretation, it seems that the time required for both geographical range extension and population expansion in these organisms may have been of the order of decades, rather than the centuries and millennia taken by trees. As future greenhouse warming occurs we might expect to observe these species, and the members of the groups most closely similar to them ecologically, migrating virtually in concert with the rate of climate change. The result could be the sorts of bizarre ecological situations which seem to have occurred in Europe following sudden late Quaternary warming events, with areas having some of their characteristic elements but lacking other important forms (such as trees) which have failed to catch up. As mentioned previously, Coope's interpretation of the palaeoevidence remains contentious amongst some botanists, who suggest that the observed lack of trees had more to do with a peculiarly continental climate at that time than any real disequilibrium in migration (Huntley, pers.comm.).

In any case, the response times of insect or water-plant migrations over open postglacial landscapes may not be representative of the future situation.

These postglacial migrations might be regarded as moving into very sparse or non-existent assemblages of plants and animals. It is difficult to gauge what influence competition with the existing floras and faunas will have on the process in the future greenhouse world (but see Section C2).

Outside the northern temperate zones, very little information is available. It would of course be very interesting and relevant to know how tropical species migrated in response to the climatic fluctuations of the Quaternary. Chen's (1988) work on population expansion times in Australian rainforest trees has already been mentioned. There are some indications of long delays in geographical range extension of rainforest trees in the early postglacial, possibly related to some sort of disperal lag (Grove, 1986). Talbot *et al.* (1984) give some tentative evidence of a delay of millennia in the spread of rainforest trees onto grasslands in Central Africa in the early Holocene, after the climate *may* have become suitable for them. Even the vertical migration of a thousand metres when tropical vegetation zones gradually moved back up mountains may have been accompanied by migration lags detectable in the fossil record (Grove, 1986). Much has been written about the supposed existence of centres of diversity in the tropics as the locations of Quaternary forest refugia (e.g. Prance, 1982). If indeed such refugial centres of diversity do exist, thousands of years after the climate again became favourable for the spread of forest, they provide an indication of just how slowly many tropical species of animals and plants are able to migrate in response to change. However, the existence of certain of these supposed centres of diversity has been called into question (Nelson *et al.*, 1990), and whether these areas actually correspond to the positions of Quaternary forest refugia is in further doubt (Colinvaux, 1987). There is certainly a need for further palaeoecological studies of tropical forests, to indicate the potential of their component species to undergo migration and population expansion as conditions change to favour these processes.

iii. Recent introductions and other invasions. Various range extensions have been observed during the last few centuries, and documented in varying degrees of detail. Some of these expansions appear to have occurred as a result of climatic fluctuations, but many have occurred after species have been introduced (accidentally or deliberately) by humans into new regions of the world. Even if a species has been newly introduced, the situation might perhaps be regarded as analogous to the sudden appearance of a zone of favourable climate adjacent to the existing natural range of a species.

The types of organisms which have been introduced around the world are many and varied, and the patterns of invasion which they have shown are correspondingly varied. Data on biological invasions are biased towards the mid-latitudes of Europe, North America, Australia, New Zealand and South Africa because of the high concentration of botanists and zoologists in these regions. As in cases of range extension in the more distant past, it is often difficult to define where a geographical range boundary lies at any point in time. Even if one knows everything about the distribution of a species, the decision of where to draw a line around the scatter of individuals and populations is a subjective one. Whilst it is a necessary step, it is important to bear in mind this element of ambiguity in any published account of a range migration. The rates of advance reported in some of the observed range extensions have been very rapid, paticularly for birds and mammals. Range extensions of tens of kilometres per year have been reported for mammals which have invaded Britain (Holdgate, 1986), and up to 49 km year^{-1} in the case of the collared dove (*Stretopelia decaocto*) in Europe (Hengeveld, 1989). In the case of migratory birds, which in any case travel hundreds or thousands of kilometres each year, the potential to respond to a progressive climate zone shift seems almost limitless, as is illustrated in the many past records of southerly migrants turning up in northern Europe in unusually warm summers. Plants seem to have invaded somewhat more slowly than animals, despite the potential of many for wind dispersal; for example about 3 km year^{-1} in the Oxford ragwort (*Senecio squalidus*) in Britain, even with the aid of railway transport (Holdgate, 1986; Hengeveld, 1989). As with the ragwort, the further spread of introduced animals and plants *within* a new region in the past may have been unintentionally aided by man's activities in transporting goods. It is possible that under climatic conditions favouring range extension in the future, many native species will also benefit from this mode of travel.

Species able to extend their ranges at this sort of rate could potentially keep up with or exceed the projected rates of greenhouse warming, at the speed with which it is projected to occur in the mid-latitudes. Following the initial wavefront of invasion, there is a somewhat slower process of population expansion and "filling in" of gaps in the distribution. Cases of rapid population increase in the mid-latitudes are rhododendron (*Rhododendron ponticum*) and sycamore (*Acer pseudoplatanus*) in Britain, which have spread from widespread but scattered planting to become woodland undershrub and incipient canopy dominants respectively in many areas, within the space of several centuries (Rackham, 1986). An adjustment of woody floras on this time-scale seems remarkably rapid in comparison with estimates based on the postglacial spread of trees, and perhaps illustrates how effectively humans can facilitate the establishment of a species.

These examples might represent what will happen with native species in *certain cases* following greenhouse climate change, but they can in no way be seen as a representative sample of native floras and faunas. Those species which have spread rapidly and widely after artificial introduction are in many ways at the extreme end of a spectrum of invasiveness, representing only a fraction of the total number which have been introduced around the world.

In the British flora, most of the introduced species which have successfully escaped to establish populations in semi-natural habitats seem to have remained very localized (Clapham *et al.*, 1987). It is possible that highly invasive introduced species are particularly unrepresentative of floras and faunas in general, in that they have left behind a burden of parasites and diseases which would slow down the expansion of native populations, or in other characteristics of their biology.

There may be a general lesson for the future to be learnt in that the introduced species which have managed to spread rapidly often tend to be invading habitats disturbed by man, perhaps indicating the resistance of many unaltered natural communities to invasion by either natives or non-natives (Holdgate, 1986). Possibly, the disruptive effects of rapid climate change on existing plant communities could help to open them up for invasion by natives and non-natives alike.

In contrast, a look back at the history of the semi-natural woodland flora of Britain illustrates just how slowly many native woodland flowers may be capable of spreading between woodland fragments in response to new opportunities for range extension. Many patches of secondary woodland which have existed for centuries remain depauperate in these "ancient woodland indicators", even where the nearest populations are only a few hundred metres away (Rackham, 1986). However, in other parts of their climatic ranges in Europe (and even within particular parts of Britain) many of these same species seem to have been capable of much more effective spread into secondary woodland (Grubb, pers. comm.), suggesting that their behaviour might change if the climatic background were to alter.

2. Range Contraction and Extinction

As climatic patterns shift in response to rising concentrations of greenhouse gases, many plants and animals are likely to be left under conditions which exceed their physiological and ecological tolerance ranges. If populations die out over a broad area, this will become apparent as a contraction of the geographic range of that species. The eventual outcome depends partly on the dispersal abilities, and partly on the rate at which population decline occurs as conditions change. If the rate of contraction of its range in one direction exceeds the rate at which the range can expand to take advantage of new zones of favourable climate, the species may ultimately go extinct.

The possibility that many species could be driven totally or virtually extinct by greenhouse warming is a worrying scenario. Doubtless, the way in which any particular species responds to change will depend on its own particular combination of characteristics and circumstances, but it is neces-sary to try to draw conclusions which are as broad as possible. In addition to studying the population biology and dispersal of organisms in the present day, it is important to realize that the past may offer the opportunity for studying the circumstances associated with the local or total extinction of organisms.

i. Extinctions associated with postglacial climatic warming. In addition to the immigration of warmer-climate species, the warming phase in the mid-latitudes was associated with the regional elimination of many forms which had previously dominated the periglacial landscapes. Likewise, in the low-latitudes, desert and grassland species were replaced by forms suited to wetter climates.

In western Europe, the rapid warming events of 13 kyr and 10 kyr resulted in the loss of many cold-climate species of plants and insects (Coope, 1987; Ponel and Coope, 1991). The limited time-resolution available in the fossil record prevents a very detailed understanding of just how rapidly these forms were lost, and whether it was strongly associated with the immigration of more southerly species which could have out-competed them. Overall, the loss of the arctic species and their replacement appears to have occurred in less than a century. The fate of the plant species which were the cold-stage floras of western Europe is known in much less detail. Grasses and sedges were important elements of the pollen flora both before and after the warming events, but because it is not possible to distinguish particular species in the pollen record we do not know to what extent there was a rapid replacement of northerly species by southerly ones (Bennett, pers. comm.). Certain species of arctic plants—such as *Dryas*—did die out in the lowlands at around this time, but the timing of their extinction is not known in any detail. Not all of the species in the flora and fauna were eliminated, however. Some beetle species which have wide climatic distributions in the present-day world seem to have been present both before and after the warming events (Coope, 1987). A number of relatively cosmopolitan plants such as docks (*Rumex*) and wormwood (*Artemisia*) become considerably more abundant in the pollen record after the warming event. However, as is the case for the grasses and sedges, the pollen record does not allow identification of these plants to the species level, so we cannot be sure to what extent more southerly species or races had replaced the pre-existing ones. Despite the uncertainty which surrounds them, the ecological events associated with these rapid climatic changes may be the closest analogue which we have to compare against the predicted greenhouse scenarios. As one would expect from present-day ecological principles, species with short generation times (insects) were rapidly eliminated. Species which have wide ecological and geographical ranges (certain insects, and possibly *Rumex* and *Artemisia*) were best able to survive the change *in situ*. These may be the sorts of patterns which will emerge again as greenhouse warming takes effect, with many of the familiar "weed" species of animals and plants remaining with us, but those which do die out may go very quickly.

An interesting observation is that of the insect and plant species eliminated from lowland Europe during the rapid warming phases, all have survived at either higher altitudes or higher latitudes (Coope, 1987). This is perhaps a reassuring sign of the future resilience of these organisms against complete extinction, although how widely such an observation could be applied to other taxonomic groups and other parts of the world is questionable.

Certainly, the mammals of the late Quaternary do not appear to have escaped so lightly. The postglacial warming was associated with complete extinction of many species of large mammal in various parts of the world. In Europe, North and South America and in Africa, the last fossil records of many species are concentrated around 11–12 kyr, with indications that many species had been becoming scarcer during the preceeding millennium or so (Marshall, 1986). Various authors have blamed this widespread extinction on Stone-Age hunters, expanding in range and population density, and using more effective weapons and trapping techniques to kill their prey. Amongst other evidence, advocates of this view point out that these faunas had survived many previous large climatic fluctuations without a similar concentration of extinctions having occurred. However, the coincidence of the extinctions with a time of such widespread vegetational change seems to imply that climatic relationships had at least some part to play. As Marshall has pointed out, it may well be that extinction occurred as the outcome of a *combination* of hunting pressure from Stone-Age man, and the effect of climatic change in reducing and fragmenting the populations of these mammals. If correct, this example supports the widely held view that man's influence on natural populations may make them less able to survive future climatic changes. It is also interesting to observe that smaller species of mammals seem to have suffered much less extinction than larger ones at the beginning of the Holocene (A. Lister, pers. comm.), as one might have expected from their larger total population sizes and reproductive rate.

Taking the Quaternary as a whole, it is perhaps suprising that so many dramatic fluctuations in climate did not result in more extinctions of species. On a regional scale, Europe seems to have lost a number of southerly moist-climate genera of trees during the Quaternary (Sauer, 1988), continuing a process of depletion which began during the Pliocene. In this sense, the southern European moist-climate tree flora has been left relatively depauperate compared with certain other parts of the temperate zone (Adams and Woodward, 1989). However, despite having been confined to very sparse full-glacial refugia (Bennett *et al.*, 1991), the western and central European tree flora is no poorer in numbers of species than climatically similar areas of eastern Asia and North America in which the refugial areas were probably much more extensive (Adams and Woodward, 1989). To take a very different group of organisms, of more than 2000 species of insects (mainly beetles) known from the fossil record of the Quaternary in Britain, all survive today either in Britain or elsewhere in Eurasia (Coope, 1987). Similarly, the excellent fossil record of marine and terrestrial molluscs indicates that nearly all forms which existed in the late Pleistocene survive today.

The above examples are from the biota of the temperate zone, which many would regard as more resilient than the tropics. Recently, a clearer picture of the level of fluctuation which tropical communities were exposed to during the Pleistocene has begun to emerge. Tropical sea-surface temperatures seem to have fluctuated by about 2°C between glacials and interglacials, (COHMAP, 1988) a change similar to that which is predicted for a greenhouse gas doubling by the end of next century, but this change is not really comparable because it apparently occurred over thousands of years rather than a century or so, (in contrast to the much more rapid changes in the North Atlantic). Despite the rapid and sustained sea-level changes which occurred during the Pleistocene (a rise of about 1·3–2 m per century occurred between 10 and 11 kyr, which is about the same as rates of projected greenhouse sea-level rise), there are no concentrations of reef coral extinctions associated with these phases (Rosen and Veron, pers. comm.).

Those other reservoirs of diversity, the rainforests, do not appear to have been existing in static environments as they were once thought to have done. In Africa and South East Asia, there is considerable evidence that arid conditions greatly reduced the forest areas (Grove, 1986; Adams *et al.*, 1990). Colinvaux (1987) estimates that the lowering of equatorial mean temperatures during glacial phases may have further squeezed many tropical species by confining them to the very lowest altitudes near to what was then sea level (however, this is based on high-altitude vegetation zone lowering, which might not be a good indicator of temperature decrease in the lowlands. Such estimates tend to disagree with those from other palaeoenvironmental sources). Despite this, the rainforests throughout the tropics contain remarkable levels of diversity, although there is no detailed pollen record available for most of the tropics to show how much extinction was actually associated with these Pleistocene fluctuations. The pollen record for the lowland rainforest in northern Queensland (Dodson, 1989), indicates that there was very little extinction in this area despite the forest being confined to very small refugia during arid glacial phases. It is necessary to bear in mind that a fossil pollen record may well be very unrepresentative of the compositional history of a flora in which most tree species are insect pollinated. However, it is interesting to note only one taxon present in this pollen record seems to have gone extinct during the upper Pleistocene, a species of the gymnosperm tree *Dacrydium* (Dodson, 1989).

These observations from the past seem to provide a broadly reassuring indication of the resilience of many groups of species to climatic change. It may well be that the species we see in our present-day world are those which have been selected during the Quaternary for resilience to climatic change (Coope, 1987). However, we should not forget that the conditions of change which these species have survived are not identical to those which will result under greenhouse warming. The speed and magnitude of warming may eventually exceed what is survivable, there will be the added influence of very high CO_2 levels on physiology and competition, and also there is the direct and capricious influence of man in destroying these communities.

3. Community Types

The broad community assemblages of species in the world are a familiar and important aspect of the environment. Thus, from a conservation point of view, the way in which *communities* will respond to climate change is important. Implicit in the preceeding sections of this chapter is the concept that species will respond to climate change as independent units, not as communities. This is certainly the impression one gets from the Quaternary record of the mid-latitudes. Looking back at the changing plant distributions of Europe and North America, we see that tree species which nowadays grow together, migrated independently and formed unfamiliar combinations (Davis, 1984). As discussed above, these changing combinations may partly have been the outcome of competitive disequilibrium, allowing species to grow on soils and under climate conditions in which they would eventually be outcompeted as the forests matured. But they also appear to have been the outcome of the differing and individualistic climatic requirements of tree species, at times when the combinations of important climatic parameters were different from today (Delcourt and Delcourt, 1987a, b; Huntley, 1990a, b). Another example of a "no present analogue" community from the past is the herbaceous steppe–tundra which existed over much of the northern hemisphere at around the LGM (Guthrie, 1990). As the name suggests, it combined species of grassy steppes with others which occur in present-day tundra, and nothing like it exists today. The steppe–tundra must have been the outcome of an unfamiliar set of climatic conditions bringing together a set of species whose requirements happened to overlap. "No present analogue" communities have also been described for Pleistocene beetles, snails and burrowing mammals respectively. It seems that as climate changes in the future, our familiar communities may be lost, and the new combinations of species which form may be very unfamiliar.

It is to be expected that species which are linked through a predator–prey or parasite–host relationship will show a much stonger tendency to remain together than species on the same trophic level. Likewise, species of invertebrates and herbaceous plants which presently live in habitats dominated by particular species of trees might show a specialized requirement for the conditions which they provide. However, even here the past provides an interesting perspective which indicates that our expectations may be too simplistic. Although it feeds on other insects, and not directly on oak, the carabid beetle *Calosoma inquisator* is nowadays found only in oak woodlands (Coope, 1987). However, it was present in the post-Dryas landscapes in which birch was the only common tree. Apparently, this supposed oak-habitat specialist was able to make use of birch woodland as a habitat under the peculiar conditions which existed at that time (Coope, 1987). It would be fascinating to know more about the ways in which such interactions can alter in response to rapid climate changes, and detailed examination of the fossil record may give further clues to this.

IV. CONCLUSIONS

This volume on global change has concentrated on predicting causal relationships between anthropogenic activities and changes in the earth's biosphere. From what has been summarized in this chapter, it seems that a knowledge of the past can help to improve these predictions in two general ways.

Firstly, the past can be used to validate, or invalidate, various sorts of causal links which have already been suggested on the basis of observations in the present-day world. For example, it is useful to know how much warming will occur from the addition of a given amount of a greenhouse gas to the atmosphere, or how fast a species will migrate in response to a given amount of climatic warming; we can see how climatic or biogeographic models perform against comparable changes which occurred in the "testing grounds" of the past. Thus, if we have good enough data to dissect out examples of such links from the past, it may be possible to use them to test our preconceived ideas on what should happen if a given anthropogenic perturbation occurs.

Secondly, observations from the past can be used to suggest new causal links which might potentially operate in the future, but which had not been previously considered. This could at least provide some advance warning of processes which might otherwise have provided unpleasant suprises in the future. Examples of possible causal links which have been suggested at least partly as a result of observations from the past are the feedbacks between climate and plankton activity, and the occurrence of rapid ocean–climate switches.

The analogies which may be drawn have varying degrees of usefulness, according to the precision with which their details are known and the differences in their general setting. It is very important to bear in mind that these past events are not true "experiments" which exactly reproduce conditions 50 or 100 years hence. There are many problems in drawing analogies from the past, and these problems become more severe the further one looks back in time. Many factors have varied in parallel, and in many cases the scale or speed of past events were very different from those likely to occur in the future. Generally, there has been a strong element of caution in the academic scientific literature in the interpretation of past events as an indicator of the future. Predictably, this has not always been true of popular science writing. There are various books and articles which have simplified and misrepresented the evidence from the past, in the context of predicting future change. Whilst many scientists would agree with the general message which these popular writings put forward, it is important to point out possible flaws in their arguments. Only in this way will the public perception of the global change issue be based on sound reasoning rather than wild speculation.

It is hoped that this chapter has helped to emphasize the usefulness of one particular scientific approach for predicting future global change, and highlighted the importance of gathering more and better information on the past in order to improve our understanding of the future. There is also a general lesson to be drawn from all of this, which is the the way is which today's pure research can become tomorrow's applied research. A few decades ago, much of the basic work which provided the palaeoecological and palaeoenvironmental information could have been regarded as of little or no relevance to the concerns of most of humanity. Suddenly, the questions it is helping to answer may have a bearing on the well-being of billions of people, and of a great many of the other plants and animals in our world.

ACKNOWLEDGEMENTS

We thank Keith Bennett, Jennifer Chapman, Russell Coope, Quentin Cronk, Hugues Faure, Janice Fuller, Peter Grubb, David Hall, Brian Huntley, Adrian Lister, James Lovelock, Peter Moore, Alayne Perrott, Line Rochefort, Charles Turner and Chronis Tzedakis for discussions and comments on the manuscript. Janice Fuller's help with checking the proofs is also much appreciated.

REFERENCES

Adams, J.M. and Woodward, F.I. (1989). Patterns in tree species richness as a test of the glacial extinction hypothesis. *Nature* **339**, 699–701.
Adams, J.M., Faure, H., Faure-Denard, L., McGlade, J.M. and Woodward, F.I. (1990). Increases in terrestrial carbon storage from the Last Glacial Maximum to the present. *Nature* **348**, 711–714.
Atkinson, T.C., Briffa, K.R. and Coope, G.R. (1987). Seasonal temperatures in Britain during the past 22 000 years, reconstructed using beetle remains. *Nature* **325**, 587–592.

Barnola, J.M., Raynaud, D., Koroketvich, Y.S. and Lorius, C. (1987). Vostok ice core provides 160 000-year record of atmospheric CO_2. *Nature* **329**, 408–411.

Barron, E.J. (1983). A warm, equable Cretaceous: The nature of the problem. *Earth Sci. Rev.* **19**, 305–338.

Bennett, K.D. (1986). The rate of spread and population increase of forest trees during the postglacial. *Phil. Trans. R. Soc. B.* **314**, 523–531.

Bennett, K.D., Tzedakis, P.C. and Willis, K.J. (1991). Quaternary refugia of north European trees. *J. Biogeogr.* **18**, 103–116.

Berger, W.H. (1991). No change down under. *Nature* **351**, 186–187.

Bottomley, M., Folland, C.K., Hsiung, J., Newell, R.E. and Parker D.E. (1990). *Global Ocean Surface Temperature Atlas.* Joint Meteorological Office/Massachusetts Institute of Technology project. HMSO, London.

Broccoli, A. and Manabe, S. (1987). The influence of continental ice, atmospheric CO_2 and land albedo on the climate of the last glacial maximum. *Climate Dynamics* **1**, 87–89.

Broecker, W.S., Bond, G., Klas, M., Bonani, G. and Wolfli, W. (1990). A salt oscillator in the glacial Atlantic? *Palaeoceanography* **5**, 469–477.

Budyko, M.I., Ronov, A.B. and Yanshin, A.L. (1987). *History of the Earth's Atmosphere* (English translation). Springer-Verlag.

Burrows, C.J. (1990). *Processes of Vegetation Change.* Unwin Hyman, London.

Cannell, M.G.R. (1982). *World Forest Biomass and Primary Production Data.* Academic Press.

Cannell, M.G.R. (1990). Carbon dioxide and the global carbon cycle. In: *The Greenhouse Effect and Terrestrial Ecosystems of the UK* (Ed. by M.G.R. Cannell and M.D. Hooper), pp. 6–9.

Chaloner, W.G. (1990). Global change: the past as a key to the future? *Science in Public Affairs*, **5**, 3–10.

Chappellaz, J., Barnola, J.M., Raynaud, D., Koroketvich Y.S. and Lorius, C. (1990). Ice-core record of atmospheric methane, over the past 160 000 years. *Nature* **345**, 127–131.

Charleston, R.J. Lovelock, J.E., Andreae, M.O. and Warren, S.G. (1987). Oceanic phytoplankton, atmospheric sulphur, cloud albedo and climate. *Nature* **326**, 655–661.

Chen, Y. (1988) Early Holocene population expansion of some rainforest trees at Lake Barrie basin, Queensland. *Aus. J. Ecol.* **13**, 225–233.

Clapham, A.R., Tutin, T.G. and Moore, D.M. (1987). *The Flora of the British Isles*, 3rd Edn. Cambridge University Press, Cambridge.

Clymo, R. (1984). The limits to peat bog growth. *Trans. Roy. Soc. Lond. B* **303**, 605–654.

COHMAP members (1988). Climatic changes of the last 18 000 years, observations and model simulations. *Science* **241**, 1043–1052.

Colinvaux, P.A. (1987). Amazon diversity in the light of the palaeoecological record. *Quat. Sci. Rev.* **6**, 63–114.

Coope, G.R. (1987). The response of Late Quaternary insect communities to sudden climatic changes. In: *The Organization of Communities. British Ecological Society 27th Symposium* (Ed. by J.H.R. Gee and P.S. Giller), pp. 421–438. Blackwell Scientific, Oxford.

Coope, G.R. and Brophy, J.A. (1972). Late Glacial environmental changes indicated by a coleopteran succession in North Wales. *Boreas* **1**, 97–142.

Covey, C. (1991). Credit the oceans? *Nature* **352**, 196–197.

Cusbasch and Cess (1990). Processes and modelling. In: *Climate Change, the IPCC Scientific Assessment* (Ed by J.T. Houghton, G.J. Jenkins and J.J. Ephraums), pp. 69–92. Cambridge University Press, Cambridge.

Dansgaard, W., White, J.W.C. and Johnsen, S.J. (1989) The abrupt termination of the Younger Dryas climate event. *Nature* **339**, 532.

Davis, M.B. (1976). Pleistocene biogeography of temperate deciduous forests. *Geoscience and Man* **13**, 13–26.

Davis, M.B. (1983). Quaternary history of deciduous forests of eastern North America and Europe. *Ann. Miss. Bot. Gard.*, **70**, 550–563.

Davis, M.B. (1986). Holocene vegetational history of the eastern United States. In: *Late-Quaternary Environments of the United States. Vol. 2. The Holocene* (Ed. by H.E. Wright, Jr), pp. 166–181. Longman, London.

Deacon, J. (1974). The location of refugia of *Corylus avellana* L., during the Weichselian glaciation. *New Phytol.* **73** 1055–1063.

Delcourt, P.A. and Delcourt, H.R. (1987a). Late Quaternary dynamics of temperate forests: applications of palaeoecology to issues of global environmental change. *Quat. Sci. Rev.* **8**, 129–145.

Delcourt, P.A. and Delcourt, H.R. (1987b). *Long-term Forest Dynamics of the Temperate Zone.* Springer-Verlag, New York.

Dodson, J.S. (1989). Late Pleistocene vegetation and environmental shifts in Australia, and their bearing on faunal extinctions. *J. Arch. Sci.* **16**, 207–217.

Folland, C.K., Karl, T.R. and Vinnikov, K.Y.A. (1990). Observed climate variation and change. In: *Climate Change, the IPCC Scientific Assessment* (Ed. by J.T. Houghton, G.J. Jenkins and J.J. Ephraums), pp. 201–238. Cambridge University Press, Cambridge.

Graumlich, L.J. (1991). Subalpine tree growth, climate, and increasing CO_2: an assessment of recent growth trends. *Ecology* **72**, 1–11.

Grove, A.T. (1986). Climate change from 30 000 to 10 000 BP: the environmental circumstances underlying human cultural developments in tropical latitudes. In: *Human Cultural Adaptations in Late Pleistocene and Early Holocene Times. Final Papers.* Vol. 1. (Ed. by Mimes) World Archaeological Congress.

Guetter & Kutzbach, J.E. (1990). A modified Koppen classification applied to model simulations of glacial and interglacial climates. *Climate Change* **16**, 193–215.

Guthrie, R.D. (1990). *The Frozen Fauna of the Mammoth Steppe: the Story of Blue Babe.* University of Chicago Press, Chicago.

Hansen, J. and Lebedeff, S. (1987). Global trends of measured surface air temperature. *J. Geophy. Res.* **92**, 13345–13372.

Hengeveld, R. (1989). *Dynamics of Biological Invasions.* Chapman & Hall, London.

Holdgate, W.M. (1986). Summary and conclusions; characteristics and consequences of biological invasions. *Phil. Trans. Roy. Soc. Lond. B.* **314**, 733–742.

Hooper, M.D. (1990). Historical evidence of climatic change effects. In: *The Greenhouse Effect and Terrestrial Ecosystems of the UK* (Ed. by M.G.R. Cannell and M.D. Hooper), pp. 54–62.

Huntley, B. (1990a). European vegetation history; palaeovegetation maps from pollen data—13 000 BP to present. *J. Quat. Sci.* **5**, 103–122.

Huntley, B. (1990b). European postglacial forests: compositional changes in response to climatic change. *J. Veg. Sci.* **1**, 507–518.

Huntley, B. and Birks, H.J.B. (1983). *An Atlas of Past and Present Pollen Maps for Europe 0–13 000 Years Ago.* Cambridge University Press, Cambridge.

Imbrie, J.I. and Imbrie, K.P. (1979). *Ice Ages: Solving the Mystery.* Harvard University Press, Harvard.

Jones, G.A. (1991). A stop–start ocean conveyer. *Nature* **349**, 364–365.

Kvenvolden, K.A. (1988). Methane hydrates and global climate. *Global Biogeochemical Cycles* **2**, 221–229.

Lamb, H.H. (1977). *Climate History and the Future. Climate Present, Past and Future, Vol. 2.* Methuen.

Legrand, M., Feniet-Saigne, C., Saltzman, E.S., Germain C., Barkov, N.I. and Petrov, V.N. (1991). Ice-core record of oceanic emissions of dimethylsulphide during the last climate cycle. *Nature* **350**, 144–146.

Lezine, A.M. (1989). Late Quaternary vegetation and climate of the Sahel. *Quat. Res.* **32**, 317–334.

Lorius, C., Jouzel, J., Raynaud, D. and Le Treut, H. (1990). The ice-core record: climate sensitivity and future greenhouse warming. *Nature*, **347**, 139–145.

Lovelock, J.E. (1989). *The Ages of Gaia*. Oxford University Press, Oxford.

Magri, D. (1988). Interpreting long-term exponential growth of plant populations in a 250 000 year pollen record from Valle di Castiglione (Roma). *New Phytol.* **112**, 123–128.

Margulis, L. and Lovelock, J.E. (1974). Biological modulation of the Earth's atmosphere. *Icarus* **21**, 471–489.

Marshall, L.G. (1986). Who killed Cock Robin? An investigation of the extinction controversy. In: *Quaternary Extinctions* (Ed. by Martin and Kleine), pp. 785–806. University of Arizona Press, Tucson.

Melillo, J.M., Callaghan, T.V., Woodward, F.I., Salati, E. and Sinha, S.K. (1990) Effects on Ecosystems. In: *Climate Change, the IPCC Scientific Assessment* (Ed. by J.T. Houghton, G.J. Jenkins and J.J. Ephraums), pp. 283–310. Cambridge University Press, Cambridge.

Mitchell, J.F.B. (1990). Greenhouse warming: Is the mid-Holocene a good analogue? *J. Clim.* **3**, 1177–1192.

Mitchell, J.F.B., Manabe, S., Meleshko, V. and Tokioka T. (1990). Equilibrium climate change—and its implications for the future. In: *Climate Change, the IPCC Scientific Assessment* (Ed. by J.T. Houghton, G.J. Jenkins and J.J. Ephraums), pp. 131–172. Cambridge University Press, Cambridge.

Mooney, H.A., Drake, R.J., Luxmore, B.G., Oeschel, W.C. and Pitelka, L.F. (1990). How will terrestrial ecosystems interact with the changing CO_2 concentration of the atmosphere and anticipated climate change? *Bioscience*, in press.

Neftel, A., Moore, E., Oeschger, H. and Stauffer, B. (1985). Evidence from polar ice cores for an increase in atmospheric CO_2 in the past two centuries. *Nature* **315**, 45–47.

Neftel, A., Oeschger, H., Stauffelbach, T. and Stauffer, B. (1988). CO_2 record in the Byrd ice core 50 000–5000 years BP. *Nature* **331**, 609–611.

Nelson, B.W., Ferreira, C.A.C., Silva, M.F. and Kawasaki M.L. (1990). Endemism centres, refugia and botanical collection density in Brazilian Amazonia. *Nature* **345**, 714–716.

Nisbit, E.G. (1989). The end of the last ice age. *Can. J. Earth Sciences* **27**, 148–157.

Nisbit, E.G. (1990). Climate change and methane. *Nature* **347**, 23.

Olson, J.S., Watts, J.A. and Allinson, L.J. (1983) Carbon in Live Vegetation in Major World Ecosystems. Environmental Sciences Division, Publication No. 1997 (Oak Ridge National Laboratory).

Penuelas, J. and Matamda, R. (1990). Changes in N and S leaf content, stomatal density and specific leaf area of 14 plant species during the last three centuries of CO_2 increase. *J. Exp. Bot.* **41**, 1119–1124.

Pennington (1986). Lags in adjustment of vegetation to climate caused by the pace of soil development: evidence from Britain. *Vegetatio* **67**, 105–118.

Petit, J.R., Briat, M. and Royer A. (1981). Ice age aerosol content from East Antarctic ice core samples and past wind strength. *Nature* **293**, 391–394.

Ponel, P. and Coope, G.R. (1991). Lateglacial and Early Flandrian coleoptera from La Taphanel, Massif Central, France: climatic and ecological implications. *J. Quat. Sci.* **5**, 235–249.

Post, W.M., Emanuel, W.R., Zinke, P.J. and Stangenburger, A.G. (1982). Soil carbon pools and world life zones. *Nature* **298**, 156–159.

Prance, G.T. (1987). Biogeography of neotropical plants. p. 182–220. In: *Biogeography and Quaternary History in Tropical Latin America* (Ed. by T.C. Whitmore and G.T. Prance), pp. 182–220. Oxford University Press, New York.

Prentice, I.C. (1986). Vegetation responses to past climatic variation. *Vegetatio* **67**, 131–141.

Prentice, K.C. (1990). Bioclimatic distribution of vegetation for general circulation model studies. *J. Geophys. Res.* **95**, 11811–11830.

Prentice, K.C. and Fung, I.Z. (1990). The sensitivity of terrestrial carbon storage to climate change. *Nature* **346**, 48–51.

Rackham, O. (1986). *The History of the Countryside*. Dent, London.

Ravel, A. and Ramanathan, V. (1989). Observational determination of the greenhouse effect. *Nature* **342**, 758–761.

Rea, D.K., Leinen, M. and Janek, T.R. (1985). Geologic approach to long-term history of atmospheric circulation. *Science* **227**, 721–725.

Rind, D. and Chandler, M. (1991). Increased ocean heat transport and warmer climates. *J. Geophys. Res.* **96**, 7437–7461.

Ruddiman, W.F. and MacIntyre, A. (1981). The North Atlantic ocean during the Last Deglaciation. *Palaeogeog. Palaeoclimat. Palaeoecol.* **35**, 145–214.

Sauer, J.D. (1988). *Plant Migration: the Dynamics of Geographic Patterning of Plant Species*. University of California Press, California.

Schlesinger, W.H. (1990a). Evidence from chronosequence studies for low carbon storage potential of soils. *Nature* **348**, 232–234.

Schlesinger W.H. (1990b) Vegetation an unlikely answer. *Nature* **348**, 690.

Street-Perrott, F.A. (1991). General circulation (GCM) modelling of palaeoclimates: a critique. *The Holocene* **1**, 74–80.

Street-Perrott, F.A., Mitchell, J.F.B., Marchland, D.S. and Brunner J.S. (1990). Milankovitch and albedo forcing of the tropical monsoons: a comparison of geological evidence and numerical simulations for 9000 BP. *Trans. Roy. Soc. Edinburgh: Earth Sciences* **81**, 407–427.

Talbot, M.R., Livingstone, D.A., Palmer, D.A., Maley, J., Melack, J.M., Delibrias, G., and Gulliksen, S. (1984). Preliminary results from sediment cores from Lake Bosumti, Ghana. *Palaeoecol. Africa*, **16**, 173–192.

Tallis, J.H. (1990). *Plant Community History*. Chapman & Hall, London.

Tans, P.P., Fung, I. and Takahashi, T. (1990). Observational constraints on the global atmospheric CO_2 budget. *Science* **247**, 1431–1438.

Watson, R.T., Rhode, H., Oeschger, H. and Siegenthaler, U. (1990). Greenhouse gases and aerosols. In: *Climate Change, the IPCC Scientific Assessment* (Ed. by J.T. Houghton, G.J. Jenkins and J.J. Ephraums), pp. 1–40. Cambridge University Press, Cambridge.

Webb, T. (1986). Is vegetation in equilibrium with climate? How to interpret Late-Quaternary pollen data, *Vegetatio* **67**, 75–91.

Webb, T., Bartlein, P.J., and Kutzbach, J.E. (1987). Climatic change in eastern North America during the past 18 000 years: comparisons of pollen data with model results. In: *North America and Adjacent Oceans During the Last Deglaciation* (Ed. by W.F. Ruddiman and H.E. Wright, Jr), pp. 447–462. Geological Society of America, Boulder.

Wigley, T.M.L. and Barnett, T.P. (1990). Detection of the greenhouse effect in observations. In: *Climate Change, the IPCC Scientific Assessment* (Ed. by J.T. Houghton, G.J. Jenkins and J.J. Ephraums), pp. 245–255. Cambridge University Press, Cambridge.

Wolfe, J.A. (1985). Distribution of major vegetational types during the Tertiary. In: *The Carbon Cycle and Atmospheric CO₂: Natural Variations Archaean to Present* (Ed. by E.T. Sundquist and W.S. Broecker), pp. 357–375. Geophysical Monographs 32. American Geophysical Union, Washington, DC.

Woodward, F.I. (1987a). *Climate and Plant Distribution.* Cambridge University Press, Cambridge.

Woodward, F.I. (1987b). Stomatal numbers are sensitive to increases in CO₂ from preindustrial levels. *Nature* **327**, 617–618.

Woodward, F.I. (1990). Global change: translating plant ecophysiological responses to ecosystems. *Trends Ecol. Evol.* **5**, 308–311.

Index

Lagoons, 222, 245
LAI *see* leaf area index
Land
  margin, 241
  surface parameterizations, 5, 54
  use, 82, 288
    forest, 85–86
    see also agriculture
Landfill, 167
Landscape dynamics, 242
Large system studies, integrated, 244
Large-scale changes, 42
Lasioderma serricorne (cigarette beetle), 132
Late glacial maximum *see* LGM
Latitudinal warmth, 272
Laurentide ice sheets, 16
Leaching, 180, 183
Leaf area index (LAI), 100–101, 111
Leaf litter, 177
Leafhoppers, 129
Lepidopterous pests, 131
Leptinotarsa decemlineata (Colorado beetle), 129, 133
Leptoglossus clypealis, 139
LGM
  carbon storage, 284
  community types, 307
  moisture balance, 279
Lignin, 180, 181, 195
  climate resource quality interactions, 182, 183, 185
  decomposition/humification, 190
  plant-soil interactions, 194
  soil organic matter formation, 188
LINKAGES model, 194, 195, 196
Litter, 176
  decomposition, 180–186
    climate effects isolation, 181–182
    climate resource quality interactions, 182–186
  production, 174
Little Ice Age, 274
Locusta sp. (locusts), 123, 129
Lodgepole pine, 181
Logging activities, 300
Low Countries, 85
Low status pests, 152
Lymantria sp., 126

Macroalgae, 226

Macrosiphum euphorbiae (potato aphid), 147
Macrosiphum persicae (aphid), 130
Madagascar, 228
Magnesium, 192
Maine, Gulf of, 233
Maize, 66, 137
  plants and water use, 68
  sea-level rise, 81
  silage, 72
  stem borer (*Chilo* sp.), 138
  yields, 77, 84
  see also grain maize
Malacosoma disstria (forest tent caterpillar), 133
Maldive Islands, 80
Mammals, burrowing, 307
Mammoth steppe *see* steppe-tundra
Man, effects of on biosphere, 259–309
Mangroves, 213, 223, 231, 237, 245
  ecological responses, 240
  ecosystems, 237
  temperature, 233
Mannose, 190
Mariculture, 241
MATS, 181
Matter
  exchange, 218–220
  suspended, 220, 227–228
Mayflies, 300
Mediterranean, 38, 86, 191, 228
  fruit fly (*Ceratitis capitata*), 120, 132, 133
Mesic forest, 97, 98
  areal coverage, 99
  plant energy balance model, 101
Mesic grasslands, 177
Mesozoic-Tertiary warm climates, 268
Metals, heavy, 230
Meteorological data, 243, 262
Methane (CH_4), 6, 10, 23, 200, 201, 222
  biogeochemical processes, 224, 240
  carbon pools, 170, 172, 176
  clathrate, 214, 241
  climate and ecological interactions, 53
  emisaion and removal, 28
  "fossil", 290
  greenhouse effect, 34, 261
  oceans, 292
  plant-soil interactions, 193, 199

Advances in Ecological Research
Volumes 1–22

Cumulative List of Titles

A century of evolution in *Spartina anglica*, **21**, 1

Aerial heavy metal pollution and terrestrial ecosystems, **11**, 218

Analysis of processes involved in the natural control of insects, **2**, 1

Ant-plant-homopteran interactions, **16**, 53

Biological strategies of nutrient cycling in soil systems, **13**, 1

Bray-Curtis ordination: an effective strategy for analysis of multivariate ecological data, **14**, 1

Can a general hypothesis explain population cycles of forest lepidoptera? **18**, 179

The climatic response to greenhouse gases, **22**, 1

Communities of parasitoids associated with leafhoppers and planthoppers in Europe, **17**, 282

Community structure and interaction webs in shallow marine hard-bottom communities: Tests of an environmental stress model, **19**, 189

The decomposition of emergent macrophytes in fresh water, **14**, 115

Dendroecology: A tool for evaluating variations in past and present forest environments, **19**, 111

The development of regional climate scenarios and the ecological impact of greenhouse gas warming, **22**, 33

Developments in ecophysiological research on soil invertebrates, **16**, 175

The direct effects of increase in the global atmospheric CO_2 concentration on natural and commercial temperate trees and forests, **19**, 2

The distribution and abundance of lake-dwelling Triclads – towards a hypothesis, **3**, 1

The dynamics of aquatic ecosystems, **6**, 1

The dynamics of field population of the pine looper, *Bupalus piniarius* L. (Lep., Geom.), **3**, 207

Earthworm biotechnology, and global biogeochemistry, **15**, 379

Ecological aspects of fishery research, **7**, 114

Ecological conditions affecting the production of wild herbivorous mammals on grasslands, **6**, 137

Ecological implications of dividing plants into groups with distinct photosynthetic production capabilities, **7**, 87

Ecological implications of specificity between plants and rhizosphere micro-organisms, **21**, 122

Ecological studies at Lough Ine, **4**, 198

Ecological studies at Lough Hyne, **17**, 115

The ecology of the Cinnabar moth, **12**, 1

Ecology of the coarse woody debris in temperate ecosystems, **15**, 133

Ecology, evolution and energetics: a study in metabolic adaptation, **10**, 1

Ecology of fire in grasslands, **5**, 209

Ecology of mushroom-feeding Drosophilidae, **20**, 225